A Quantum Approach to Alloy Design

Related titles

Metallurgy and Design of Alloys With Hierarchical Microstructures (ISBN 978 0128120682)

Materials Informatics: Novel Analytics for Materials Data (ISBN 978 0124103948)

High Entropy Alloys (ISBN 9780128002513)

Materials Today

A Quantum Approach to Alloy Design

An Exploration of Material Design and Development Based Upon Alloy Design Theory and Atomization Energy Method

Masahiko Morinaga
Emeritus Professor, Nagoya University, Nagoya, Japan

ELSEVIER

Elsevier
Radarweg 29, PO Box 211, 1000 AE Amsterdam, Netherlands
The Boulevard, Langford Lane, Kidlington, Oxford OX5 1GB, United Kingdom
50 Hampshire Street, 5th Floor, Cambridge, MA 02139, United States

Notices
Knowledge and best practice in this field are constantly changing. As new research and experience broaden our understanding, changes in research methods, professional practices, or medical treatment may become necessary.

Practitioners and researchers must always rely on their own experience and knowledge in evaluating and using any information, methods, compounds, or experiments described herein. In using such information or methods they should be mindful of their own safety and the safety of others, including parties for whom they have a professional responsibility.

To the fullest extent of the law, neither the Publisher nor the authors, contributors, or editors, assume any liability for any injury and/or damage to persons or property as a matter of products liability, negligence or otherwise, or from any use or operation of any methods, products, instructions, or ideas contained in the material herein.

British Library Cataloguing-in-Publication Data
A catalogue record for this book is available from the British Library

Library of Congress Cataloging-in-Publication Data
A catalog record for this book is available from the Library of Congress

ISBN: 978-0-12-814706-1

Publisher: Matthew Deans
Acquisition Editor: Christina Gifford
Editorial Project Manager: Gabriela D. Capille
Production Project Manager: Debasish Ghosh
Cover Designer: Christian J. Bilbow

Typeset by MPS Limited, Chennai, India

Contents

Preface

More than 35 years have passed since I started the research of alloy design at the beginning of the 1980s. During this period much progress has been made in the quantum approach to alloy design. For a long time I have been planning to write a book on this subject, because I have felt that it is my primary duty to present this quantum approach in as straightforward a way as possible.

Complex alloy problems are treatable with the aid of the molecular orbital calculations of electronic structures of alloys. In view of the electron bonding, it is no longer needed to distinguish between iron and steels and nonferrous alloys, and even the current classification into metals, semiconductors, ceramics, and polymers is not necessarily needed. Therefore, this book covers various topics of not only metals and alloys, but also metal oxides, hydrides, and even hydrocarbons. Besides the phase stability of alloys, the physical and chemical properties such as strength, corrosion resistance, hydrogen storage, and catalysis are treated with the aid of molecular orbital theory. Recently, an atomization energy method has been developed to express a role of each constituent atom in the chemical bond formation of the material in an energy scale. Thus, the theory for alloy design is still in progress. New challenges and trials are important for the further advancement of the theory.

The opportunity and incentive to prepare this book have been provided by Professor Ma Qian of Royal Melbourne Institute of Technology in Australia and Professor Hirohiko Adachi of Kyoto University in Japan. Professor H. Adachi also helped me in various aspects, so that the present work could not be completed without his collaboration. I express my gratitude to the late Professor N. Yukawa of Toyohashi University of Technology, as he introduced me to the field of alloy design. Professor H. Nakai of Waseda University guided me in the energy density analysis that he developed. I am grateful to Professor Y. Murata, Professor H. Yukawa, Professor M. Yoshino, all of Nagoya University; Professor S. Kamado of Nagaoka University of Technology; Professor H. Ezaki of National Institute of Technology, Suzuka College; and Professor M. Morishita of University of Hyogo for their collaboration with me on the study for alloy design. Finally my wife joins me in expressing our appreciation to the graduate students and young men who worked with me at both Toyohashi University of Technology and Nagoya University.

Masahiko Morinaga

NAGOYA, JAPAN

September 2018

1

Introduction

1.1 Alloy Properties

Superior alloy has a beautiful microstructure. For example, the micrograph shown in Fig. 1-1 is taken from a heat resistant single-crystal Ni-based superalloy for use of the blades in jet engines and gas turbines. The cuboidal γ' (Ni_3Al) precipitates with the submicrometer size are arranged three-dimensionally and periodically in the γ (fcc Ni) matrix. The volume fraction of the γ' phase is as high as 60%–65%. This beautiful microstructure is, however, no longer sustained, if any harmful phase (i.e., the σ phase) precipitates in the γ matrix, which deteriorates mechanical properties of alloys. Therefore, it is strongly needed to control the alloy composition to be free from the σ-phase precipitation. For this purpose, the phase computation (PHACOMP) method was developed in 1964. Since then, we have used the term "alloy design".

Needless to say, mechanical strength is one of the most important properties of the structural alloys. Plastic deformation takes place readily in most alloys, and both the size and the shape change easily by either cold or hot working. This is a great advantage of metals and alloys over other materials. Even new functions could emerge in some alloy by cold working. For example, gum metals (Ti alloys, e.g., Ti-9%Nb-3%V-6%Zr-1.5%O and Ti-23%Nb-0.7%Ta-2%Zr-1.2%O (mol%)) have been developed recently [1]. A typical stress–strain curve is shown in Fig. 1-2A. By applying the 90% cold working to this alloy, nonlinear elastic deformation takes place as in gum material. Also, as shown in Fig. 1-2B, in case gum metal is 90% cold worked, elastic constant and linear thermal expansion coefficient keep constant in the temperature range of $-200°C$ (73 K) to 400°C (673 K). Such unique properties are called the elinvar and the invar properties, respectively.

Recently, there have been several trials to get superior mechanical properties by introducing severe plastic deformation to metals and alloys by using special processes such as equal channel angular processing (ECAP), high pressure torsion (HPT), and accumulative roll bonding (ARB) [2–4]. No size or shape changes are produced by these processes, but instead ultrafine grained (UFG) materials with 200–20 nm in grain sizes are produced. For example, very high strength Al and Fe metals are developed by the ARB process [5,6].

Besides the mechanical properties, for example, oxidation resistance is required for high-temperature alloys such as Ni-based, Co-based, and Fe-based alloys. If these alloys are exposed to air at high temperatures, metal oxide of Cr_2O_3, Al_2O_3, or SiO_2 is formed depending on the temperature. Such oxides formed on the surface protect the alloy from further oxidation.

To understand the oxidation process, it is important to know the defect structure in the oxide. For example, as shown in Fig. 1-3A, the formation energies of lattice defects are

A Quantum Approach to Alloy Design. DOI: https://doi.org/10.1016/B978-0-12-814706-1.00001-7

FIGURE 1-1 The γ' phase precipitated in a single crystal Ni-based superalloy. The γ phase exists in the interface region between adjacent cubes of the γ' phase. *By courtesy of* Dr. T. Ohno (Hitachi Metals Ltd.). *The micrograph used in the cover of this book is the same as this picture, that is taken from Hitachi Metals Technical Review, vol. 13, 1997.*

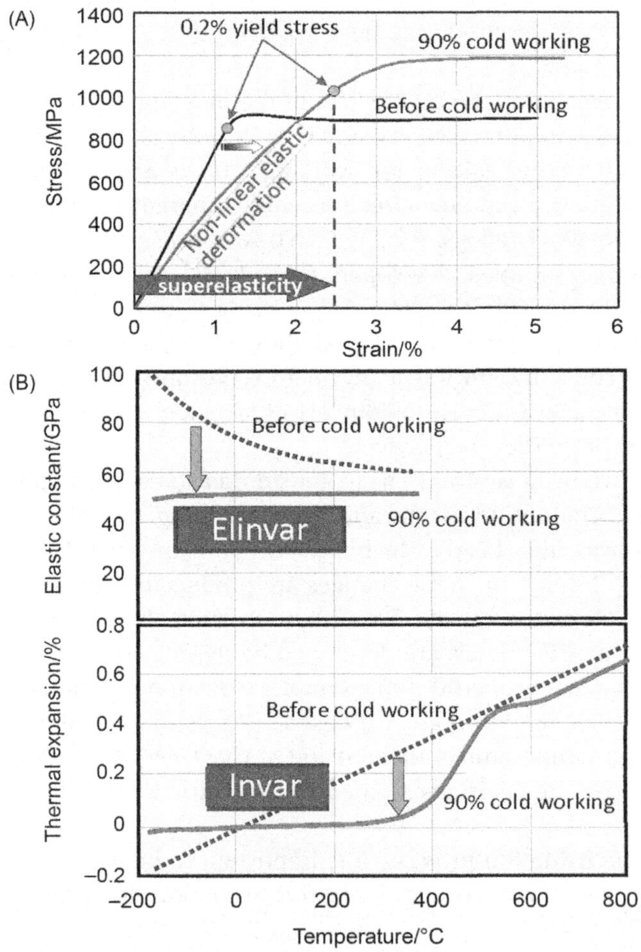

FIGURE 1-2 (A) Stress vs. strain curves for gum metals (Ti alloys) and (B) elinvar and invar properties emerge in them by cold working. *By courtesy of Dr. T. Saito (Toyota Physical and Chemical Reserch Institute).*

(A) Calculated formation energies of lattice defects

	Schottky defect (eV)	Cation Frenkel defect (eV)	Anion Frenkel defect (eV)
SiO_2	7.01	9.10	4.85
Al_2O_3	2.14	1.81	5.87

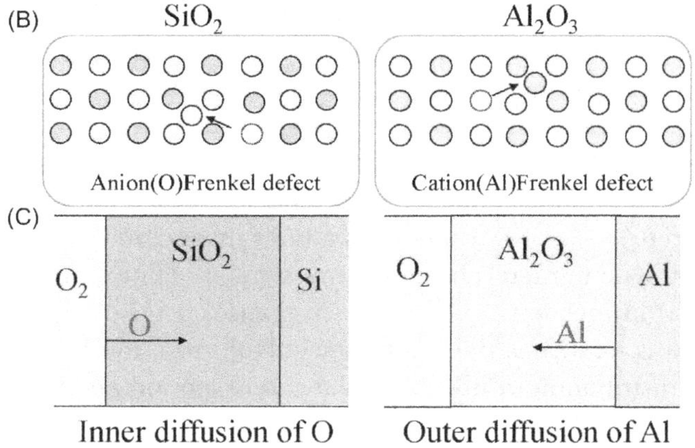

FIGURE 1-3 Formation of oxide layer formed on metal surface. (A) Calculated formation energies of lattice defects, (B) lattice defects in SiO_2 and Al_2O_3, and (C) ionic diffusion mechanism during oxidation at high temperature.

calculated in SiO_2 and Al_2O_3 [7]. The lattice defect in the oxide is assumed here to be either the Schottky defect, or the cation or anion Frenkel defect. The Schottky defect is a pair of anion and cation vacancies, and the cation (or anion) Frenkel defect consists of a vacancy and an interstitial of cation (or anion) [8]. Judging from the formation energies listed in Fig. 1-3A, the anion (O) Frenkel defect is dominant in SiO_2, and the cation (Al) Frenkel defect is dominant in Al_2O_3 as illustrated in Fig. 1-3B. Each lattice defect assists ionic diffusion in the oxide at high temperature. As a result, as shown in Fig. 1-3C, the SiO_2 layer is formed on the Si surface by inner diffusion of anion (O), whereas the Al_2O_3 layer is formed on the Al surface by outer diffusion of cation (Al), in agreement with experiments [7]. Thus, to reduce the oxidation rate at high temperature, we may need to control the defect structure in the oxides in some way, for example, by alloying. Thus, we need to learn metal oxides as well as alloys to understand a whole feature of the oxidation behavior in a fundamental manner. In other words, wide knowledge and information are necessary to understand the alloy property.

In particular, in the case of structural alloys it is strongly requested to have an excellent combination of various properties such as mechanical properties (e.g., tensile strength, creep strength), corrosion resistance, and oxidation resistance. However, it is rather difficult to predict these alloy properties in a reasonable way. In this book, several trials are presented to resolve a part of such difficulties using electron theory, since most of physical and chemical properties of alloys are related closely to the electronic states.

In the next section, previous electronic approaches will be reviewed briefly to deepen our understanding of electron theory for metals and alloys.

1.2 Electronic Approaches to Metals and Alloys

Pure metals have been investigated extensively from both experiments and theories, and valuable information has been accumulated on the electronic states in pure metals. For example, the Fermi surfaces are determined for a variety of metals [9]. The Fermi surface is spherical in alkali metals such as Na and K, in which a free electron model is satisfied. A nearly free electron model is satisfied for the lightweight metals such as Al and Mg, in which the *s* and *p* electrons spread over the crystal space. On the other hand, for transition metals the *d* electrons are localized around the atom site and the localized *d* electrons cause the ferromagnetism in Fe group metals. For example, the atomic moment is about 2.2 μ_B for Fe, 1.7 μ_B for Co, and 0.6 μ_B for Ni, where μ_B is a Bohr magneton.

Also, the impurity state in metals has been investigated. For example, the magnetic impurity (e.g., Fe, Cr, Mn) in metals (e.g., Au, Ag, Cu) causes a residual resistance minimum at low temperature that is known as the Kondo effect [10]. Also, the Friedel oscillation appears in either the charge distribution or the spin polarization around an impurity atom embedded in metal [10].

By contrast, electronic information is very limited in the solute-rich alloys. In many cases alloy properties have been analyzed using the electrons-per-atom ratio, *e/a*. This approach seems natural, since *e/a* is the number of outer electrons to form the chemical bonds in alloys and compounds. For example, there are the so-called electron compounds proposed by Hume-Rothery. The crystal structures of the intermetallic compounds change depending on the specific *e/a* values [10]. A rigid band model has been used together with the *e/a* value to represent the electronic structures of transition metal–based alloys. For example, the atomic moments for binary alloys consisting of the iron group elements are plotted against the group number (i.e., *e/a* value) of the periodic table of elements; this is well known as the Slater–Pauling curve [11,12].

Even in the field of alloy design, the electron vacancy number in the *d* band, *Nv*, which is expressed approximately as $Nv = 10.66\text{-}(e/a)$, is used in the PHACOMP method as explained earlier [13]. PHACOMP is a method to predict the formation of harmful and brittle phases (e.g., the σ phase) in nickel-based superalloys. Besides the *e/a*, the electronegativity and the atomic radius have been employed for treating the solid solubility problem of alloys [14,15]. These parameters are assigned to pure metals and any alloying effects are not involved in them.

To the author's limited knowledge, we do not have any alloying parameters despite a long history of metallurgy and metal science. Therefore, for alloy design it is first necessary to determine alloying parameters of elements under the same chemical circumstance as in real alloy. The molecular orbital method is very convenient to calculate local electronic structures around alloying elements in alloy. Thus, a theory for alloy design is constructed on the basis of the molecular orbital calculations.

The scope and outline of each chapter in this book will be given briefly in the next section.

1.3 Scope and Outline of Each Chapter

This book consists of 12 chapters.

Chapter 1, Introduction

The necessity of developing a theory for alloy design is explained concisely.

Chapter 2, Theory for Alloy Design

In case of transition metal alloys, two alloying parameters are determined for various alloying transition metals by using the DV-Xα molecular orbital method. One parameter is the *d* orbital energy level, *Md*, of alloying element, M, in a base metal, X, and another parameter is the bond order, *Bo*, between M and X atoms. A theory for alloy design is developed with these alloying parameters. The physical and chemical meanings of alloying parameters are explained together with the DV-Xα molecular orbital method, which is a first-principle calculation method.

Chapter 3, Nickel Alloys

Alloying parameters, *Md* and *Bo*, relevant to nickel alloys are determined and applied to the phase stability problem of the alloys. For example, new PHACOMP is developed to solve various contradictions inherent in the PHACOMP. Also, a single crystal Ni-based superalloy is developed following the theory.

Chapter 4, Iron Alloys

Two alloying parameters, *Md* and *Bo*, are determined for both bcc Fe and fcc Fe, and used for predicting the formation of the harmful δ-ferrite phase in ferritic steels. The evolutionary process of commercially available heat-resistant ferritic steels is well traced on the $\overline{Bo}$-$\overline{Md}$ diagram. Here, $\overline{Bo}$ and $\overline{Md}$ are the compositional averages of the *Bo* and the *Md* parameters in the alloy, respectively. Heat-resistant high-Cr ferritic steel for use of rotors in power plants is developed following the theory.

Chapter 5, Titanium Alloys

Alloying parameters, *Bo* and *Md*, relevant to titanium alloy are determined and used for predicting phase stability and alloy properties. For example, it is shown that any titanium alloys can be classified into either the α, or $\alpha+\beta$, or β type from the alloy composition by using the alloying parameters. Here, α is the hcp-Ti phase and β is the bcc-Ti phase. A concrete way of alloy design is explained using an example of high strength β-type Ti alloys. The corrosion resistance is also treated along this approach. Recent progress in this approach is presented to design titanium alloys for biomedical applications.

Chapter 6, Aluminum alloys and Magnesium Alloys

Not only the transition metal–based alloys, but also simple metal–based alloys such as aluminum alloys and magnesium alloys are treated theoretically. Big data of the mechanical properties are put in order using a new alloying parameter (s orbital energy level, *Mk*). Also, the mechanical properties are treated in an energy scale for the first time.

Chapter 7, Hydrogen Overpotential of Alloys and Intermetallic Compounds

A new concept is presented for understanding the catalytic activity for the hydrogen evolution reaction on transition metal–based binary alloys and also on intermetallic compounds of TiAl, FeAl, and NiAl containing alloying 3*d* transition elements.

Chapter 8, Crystal Structural Maps for Intermetallic Compounds

New crystal structure maps are presented for aluminide, silicide, and some transition metal compounds. They are constructed using *Bo* and *Md* parameters. There is a clear separation of the crystal structures on the *Bo–Md* structure maps. In addition, the brittleness of TiAl compound is discussed in view of the chemical bond between atoms in compounds.

Chapter 9, A Universal Relation in Electron Density Distribution in Materials

There is a universal relation between electron density minima and atomic or ionic radii. This is discovered from a series of calculations of electronic structures in matter over 130 species, including gases, water, and solids.

Chapter 10, Atomization Energy Approach to Alloys and Metal Compounds

Another electronic approach is newly developed to express the chemical bond between atoms in materials in an energy scale. The atomization energy is calculated for every constituent atom in the material. For example, metal hydrides, complex hydrides, and hydrocarbons can be located on the atomization energy diagram, despite the significant differences in the nature of the chemical bond among them. One of the applications of this approach is the catalyst design. For example, the catalytic activities of metal oxides (e.g., Nb_2O_5) are evaluated quantitatively on the dehydrogenation reaction of magnesium hydride (MgH_2), $MgH_2 \rightarrow Mg + H_2$.

Chapter 11, Local Lattice Strains Around Alloying Elements in Metals

Local strains are introduced to the lattice around the solute atom due to the size mismatch between solute and solvent atoms in alloy. Local lattice strains around various alloying elements, M, are calculated in a systematic way in Mg, Ti, and Fe metals using the pseudopotential method. Such local lattice strains are found to influence the martensitic transformation start temperature (*Ms*) in both titanium alloys and iron alloys.

Chapter 12, Conclusions

A theory for alloy design is summarized, and a perspective on the alloy design is presented.

Thus, this book covers various topics of not only alloys and intermetallic compounds, but also metal oxides, hydrides, and even hydrocarbons. The important properties such as phase stability, mechanical strength, corrosion resistance, hydrogen storage, and catalysis are treated in a consistent way in view of electron theory.

References

[1] T. Saito, T. Furuta, J.H. Hwang, S. Kuramoto, K. Nishino, N. Suzuki, et al., Multifunctional alloys obtained via a dislocation-free plastic deformation mechanism, Science 300 (2003) 464–467.

[2] R.Z. Valiev, Y. Estrin, Z. Horita, T.G. Langdon, M.J. Zehetbauer, Y.T. Zhu, Producing bulk ultrafine-grained materials by severe plastic deformation, JOM 58 (2006) 33–39.

[3] Y. Iwahashi, J. Wang, Z. Horita, M. Nemoto, T.G. Langdon, Principle of equal-channel angular pressing for the processing of ultra-fine grained materials, Scripta Materialia 35 (1996) 143–146.

[4] Y. Saito, H. Utsunomiya, N. Tsuji, T. Sakai, Novel ultra-high straining process for bulk materials-development of the accumulative roll bonding (ARB) process, Acta Mater. 47 (1999) 579–583.

[5] N. Tsuji, Y. Ito, Y. Saito, Y. Minamino, Strength and ductility of ultrafine grained aluminum and iron produced by ARB and annealing, Scripta Materialia 47 (2002) 893–899.

[6] H. Huang, N. Hansen, N. Tsuji, Hardening by annealing and softening by deformation in nanostructured metals, Science 312 (2006) 249–251.

[7] M. Yoshino, Y.Shinzato, M. Morinaga, Energetics of Native Defects in Al_2O_3 and SiO_2, The 3rd International Symposium on Designing, Processing and Properties of Advanced Engineering Materials, Eds. S. G. Kang, T. Kobayashi, Jeju, South Korea, Nov. 5-8, 2003; Materials Science Forum 449–452, 2004, 713–716.

[8] W.D. Kingery, H.K. Bowen, D.R. Uhlmann, Introduction to Ceramics, 2nd Edition, John Wiley & Sons, New York, 1976.

[9] A.P. Cracknell, K.C. Wong, The Fermi Surfaces of Metals : Its Concept, Determination, and Use in the Physics of Metals, Clarendon Press, Oxford, 1973.

[10] J.M. Ziman, Principles of the Theory of Solids, Second Edition, Cambridge University Press, Cambridge, 1972.

[11] C. Kittel, Introduction to Solid State Physics, Fifth Edition, John Wiley & Sons, Inc, New York, 1976.

[12] J.E. Goldman, Atomic moments in alloys, Rev. Mod. Phys. 25 (1953) 108–113.

[13] W.J. Boesch, J.S. Slaney, Preventing sigma phase embrittlement in nickel base superalloys, Metal Progress 86 (1964) 109–111.

[14] W. Hume-Rothery, G.V. Raynor, Structure of Metals and Alloys, Institute of Metals, London, 1954.

[15] L. Darken, R.W. Gurry, Physical Chemistry of Metals, Institute of Metals, London, 1954.

2 Theory for Alloy Design

2.1 Electronic Approach to Alloy Design

It is well known that physical and chemical properties of metals and alloys are related closely to the electronic state in them. Most commercially available alloys consist of multiple components. For example, nickel-based superalloys consist of about 10 alloying elements. For these complex alloy systems, it is very difficult to calculate the electronic state accurately, and even if the calculation could be performed in some way, it might be meaningless for alloy design, because the number of combinations of alloying species and compositions are infinitely large in the multiple-component alloys.

For alloy design, it is more practical to use alloying parameters that exhibit explicitly the character of each element in a mother metal. However, despite a long history of metallurgy and metal science, we do not have any alloying parameters. For this reason, most alloy properties have been evaluated by using the parameters assigned to pure metals (e.g., the electrons-per-atom-ratio (e/a), atomic radius, electronegativity) [1,2]. The same value of the parameter is always used for the same element even if the alloy system is varied, resulting in very poor prediction of alloy properties along this approach. To solve this problem, it is first necessary to determine alloying parameters theoretically using molecular orbital methods [3–7]. To make the problem clear, the present approach is compared with the previous approach as shown in Fig. 2-1.

To this end, the DV-Xα cluster method [8–17] is utilized. This is the first-principle calculation method, which is more appropriate to the simulation of the local electronic structure around the alloying element, as compared with the conventional band calculation method. For convenience, the DV-Xα cluster method will be explained later.

2.2 Metallic Bonds

It is well known that the nature of the chemical bond between atoms is very different between simple metals (e.g., Al, Mg) and transition metals (e.g., Fe, Ni, Ti). This difference is illustrated schematically in Fig. 2-2A for simple metals and Fig. 2-2B for transition metals. The *s* and *p* electrons in simple metals behave as if they are nearly free electrons, resulting in high electrical conductivity. In contrast to this, in case of transition metals, the *d* electrons are rather localized around atom site and the strong *d-d* covalent interactions are operating between atoms. As a result, transition metals have higher cohesive energy than simple metals, which leads to higher melting temperature for transition metals than for simple

A Quantum Approach to Alloy Design. DOI: https://doi.org/10.1016/B978-0-12-814706-1.00002-9

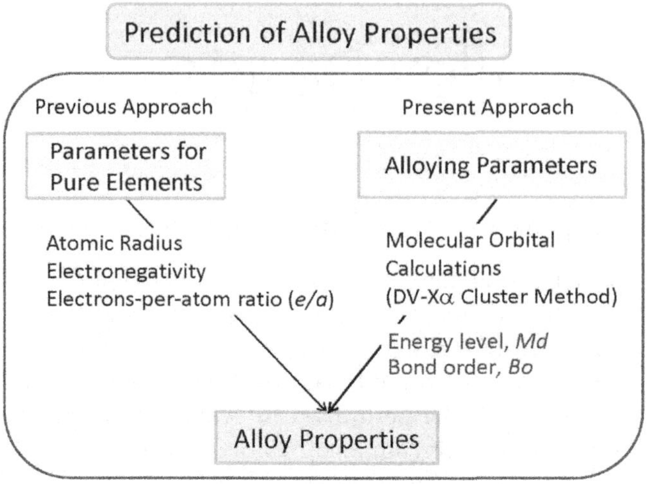

FIGURE 2-1 Comparison between the previous approach and the present approach.

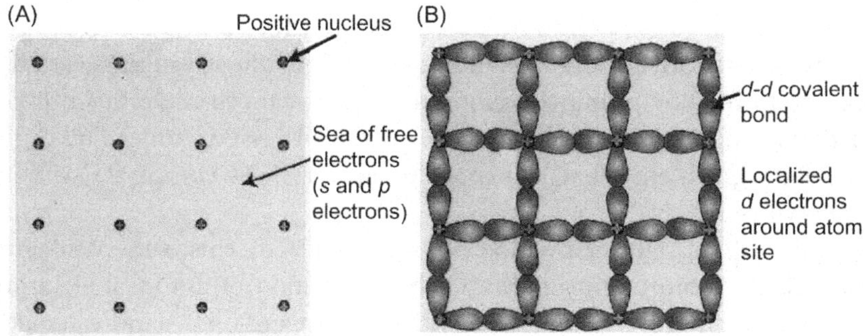

FIGURE 2-2 Schematic comparison in the chemical bond between atoms in (A) simple metals and (B) transition metals.

metals. The localization of the *d* electrons also causes magnetism, for example, the ferromagnetism emerges in pure iron due to the partial occupancy of *d* electrons in the 3*d* band. In this book, attention is mainly directed toward the transition metal–based alloys, except for Chapter 6 where both aluminum and magnesium alloys are treated to show a feature of simple metals in view of electron theory.

The transition metal–based alloys such as Fe alloys and Ti alloys are so common in our daily life. As mentioned previously, the chemical bond is characterized by *d* electrons in the transition metals. Needless to say, depending on the use of alloys, we have to select the most suitable alloying parameters. In the case of structural use, two alloying parameters are useful for the design of transition metal–based alloys [3–7].

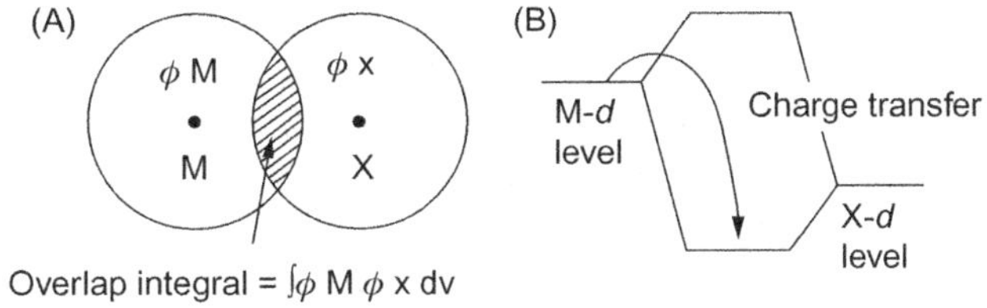

FIGURE 2-3 Alloying parameters: (A) bond order, *Bo* and (B) *d*-orbital energy level, *Md*. In (A), ϕ_M and ϕ_X are the atomic orbitals of M and X, and *Bo* is proportional to the overlap population between them.

2.3 Alloying Parameters

2.3.1 Bond order, *Bo*

One alloying parameter is the bond order (hereafter referred to *Bo*). As shown in Fig. 2-3A, *Bo* is a parameter to show the overlapping of the electron clouds between alloying element (M) and mother metal (X). So, this is a measure of the strength of the covalent bond between M and X atoms. If both M and X are transition metals, the *d-d* covalent bond is most important between them. As the *Bo* value increases, the stronger chemical bond is operating between M and X atoms.

2.3.2 *d* orbital energy level, *Md*

Another alloying parameter is the *d* orbital energy level (hereafter referred to *Md*). As shown in Fig. 2-3B, when M atom makes a molecule with X atom, the bonding level at the lower energy and the antibonding level at the higher energy are formed by hybridizing the *d* orbitals of isolated M and X atoms. As shown in the figure, if the *d* level of an isolated M atom is higher than that of an isolated X atom, the electrons move from M to X so as to reduce the energy. As a result of such a charge transfer, the effective charge becomes positive for M, and negative for X. Thus, the energy level controls the direction for charge transfer, and hence it is related to the electronegativity [18]. Higher electronegativity element has the lower *d* orbital energy level (*Md*).

Also, *Md* correlates with the atomic radius. Any atom with the larger atomic radius has the larger average radius of the *d* orbital. In such a case, the average distance between *d* electrons and the nucleus at the center increases, which weakens the attractive Coulomb force operating between them. As a result, the *d* orbital level (*Md*) becomes high as the atomic radius increases.

It is stressed here that both *Md* and *Bo* are new alloying parameters obtained for the first time from the molecular orbital calculation. These two electronic parameters change following the order of elements in the periodic table. As most physical and chemical properties change following the periodic table, these are indeed suitable parameters for treating alloy properties. Furthermore, these two parameters represent the characteristics of the two-dimensional periodic table. The use of only one parameter (for example, Mendeleev number) never expresses any changes following the periodic table.

Both *Bo* and *Md* parameters are determined for various alloying elements, M, in several mother metals, X, such as Ni, Fe and Ti, as explained in Chapters 3 to 5.

2.4 Alloy Design Based on Molecular Orbital Method

As already shown in Fig. 2-1, the present approach is a great contrast to the previous approach. A whole scope of the present approach is illustrated in Fig. 2-4.

First, using the alloy cluster two alloying parameters of *Bo* and *Md* relevant to each alloy system are determined using the DV-Xα molecular orbital method. Second, a new guideline to the design of novel alloys is derived by analyzing current experimental data in terms of the calculated alloying parameters. Third, alloying elements and compositions are selected on the basis of the guideline, and candidate alloys are prepared and tested experimentally to examine alloy properties. If the experimental results of candidate alloys are better than the required properties, the alloy design is completed, and a new alloy is developed in a reasonable way without relying on many trial-and-error experiments.

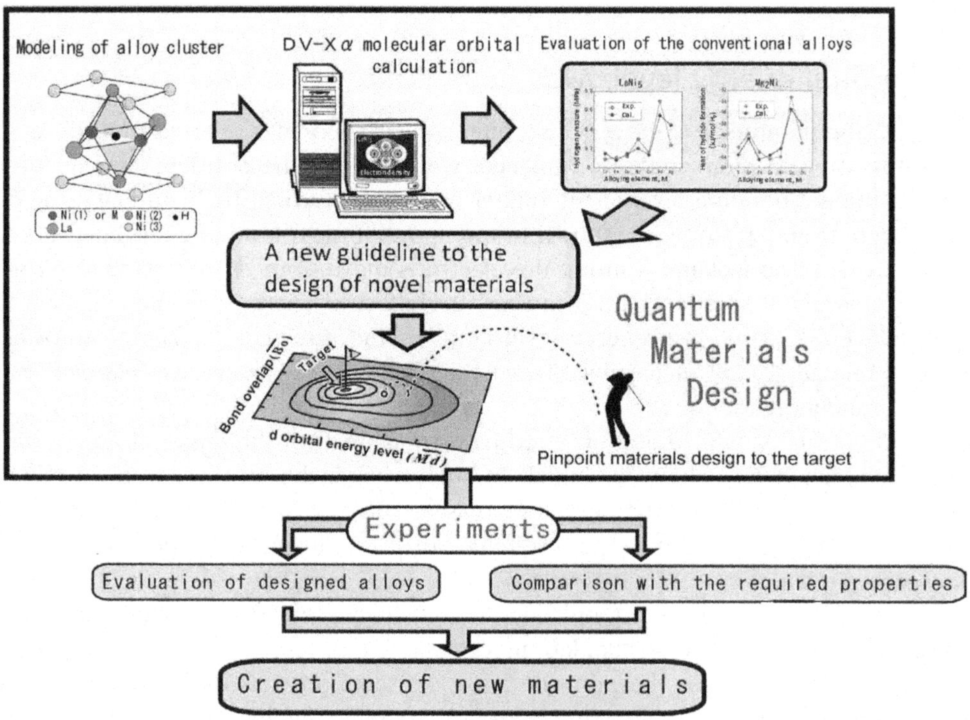

FIGURE 2-4 Alloy design based on the molecular orbital method.

2.5 DV-Xα Molecular Orbital Method

The discrete variational (DV)-Xα molecular orbital method is used for the calculation of electronic structures of alloys [8–17]. This is one of the first-principle molecular orbital calculation methods. A series of papers [13–17] and review articles [19,20] by Adachi et al. have been published on this method. Here, for convenience, only the main framework of this method will be presented.

2.5.1 Hartree–Fock–Slater Approximation

As shown in Fig. 2-5, the Schrödinger equation for one electron is solved under the Hartree–Fock–Slater approximation. The exchange-correlation interaction between electrons is expressed as the Slater's Xα potential [8], and it is proportional to one-third of the density of the up-spin electron at the position r_1, $\rho_\uparrow(r_1)$. Also, α is a scaling parameter that is usually fixed at 0.7. The use of this Slater's Xα potential greatly reduces the computational time.

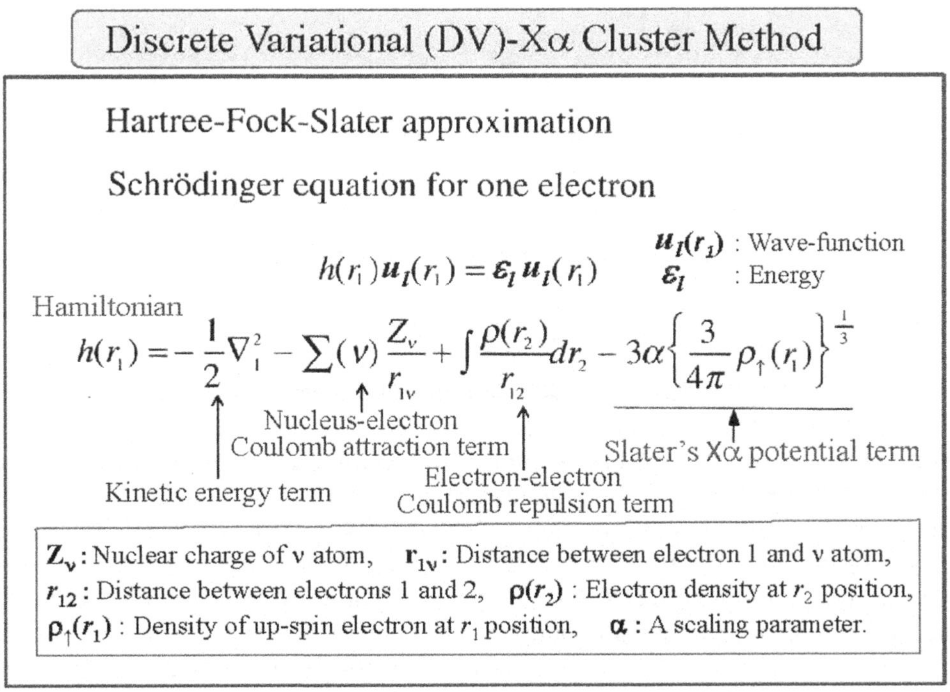

FIGURE 2-5 DV-Xα cluster method, in which the Slater's Xα potential is employed to express the exchange-correlation interaction between electrons.

2.5.2 Linear Combination of Atomic Orbitals (LCAO) Approximation

The wave function of molecule, u_l, is constructed by a linear combination of atomic orbitals (LCAO), and it is expressed as,

$$u_l = \Sigma_{(i)} C_{il} \phi_i. \tag{2.1}$$

Here, C_{il} is the coefficient for the atomic orbital i in the LCAO wave function of the lth molecular orbital. This LCAO expression is commonly used in the molecular orbital method. However, in the DV-Xα cluster method real atomic orbital for the basis function ϕ_i is calculated without using analytical functions such as the Gaussian-type orbital (GTO) and the Slater-type orbital (STO). The basis function ϕ_i is generated numerically in the molecule by solving the Schrödinger equation [20]. So, "atom in molecule" is employed as a basis function, ϕ_i. The ϕ_i is renewed after each self-consistent field (SCF) iteration by taking into account the modification of the charge density during iteration [13]. Thus, the basis function is not the same as that of an isolated atom.

2.5.3 Resonance Integral *Hij* and Overlap Integral *Sij*

The molecular orbital energy ε_l and the wave function u_l are obtained by solving the secular equation,

$$\mathbf{HC} = \varepsilon \mathbf{SC} \tag{2.2}$$

Here, the matrix element of **H** and **S** are evaluated by the DV numerical integration [9,10]. Namely, the matrix elements of the resonance integral *Hij* and the overlap integral *Sij* in the secular equation are calculated numerically as follows,

$$Hij = \Sigma_{(k)} w(r_k) \phi_i(r_k) h(r_k) \phi_j(r_k), \tag{2.3}$$

$$Sij = \Sigma_{(k)} w(r_k) \phi_i(r_k) \phi_j(r_k). \tag{2.4}$$

Here, $h(r_k)$ is the one-electron Hamiltonian for the molecule as shown in Fig. 2-5. In the DV-Xα cluster method, these matrix elements are calculated by a random sampling method, so that the integration is replaced by a summation over a large number of sample points in the molecule. The value $w(r_k)$ is a weighting factor of the sample points. This treatment is different from the conventional molecular calculation method in which the analytical integration for *Hij* and *Sij* is feasible using analytical functions for ϕ_i and ϕ_j as mentioned earlier. The secular equation of Eq. (2.2) that is composed of these matrix elements is solved by an ordinary matrix diagonalization to obtain the wave function, u_l (i.e., C_{il}) and the energy, ε_l. The self-consistent charge (SCC) scheme is used to obtain an approximate SCF potential [11,13]. A series of calculations is repeated until the output charge agrees well with the input charge in the molecule.

2.6 Analysis of Chemical Bond between Atoms

The nature of the chemical bond between atoms is analyzed following the Mulliken population analysis [21].

2.6.1 Spatial Distribution of Electron Density

The electron density, $\rho(r)$, is expressed as the square of the wave function, u_l.

$$\begin{aligned}\rho(r) &= \Sigma_{(l)}\, u_l(r)^2 \\ &= \Sigma_{(l)} \sum\nolimits_{(i,j)} C_{il}C_{jl}\phi_i(r)\phi_j(r).\end{aligned} \tag{2.5}$$

Here, $\Sigma_{(l)}$ means the summation over the occupied orbitals. The spatial distribution of electron density calculated by Eq. (2.5) helps us to understand a whole feature of the chemical bond between atoms in the molecule.

2.6.2 Bond Order

The overlapping between ϕ_i and ϕ_j atomic orbitals, n_{ij}, is defined as,

$$n_{ij} = \Sigma_{(l)}C_{il}C_{jl}\int \phi_i(r)\phi_j(r)\mathrm{d}r. \tag{2.6}$$

The summation of n_{ij} between atoms is denoted as the overlap population or the bond order, which is a measure to show the covalent bond strength between atoms. In this book, the bond order is mainly used instead of the overlap population.

2.6.3 Ionic Value of Atom

Another parameter, n_i, is defined as,

$$n_i = \Sigma_{(j)}n_{ij}. \tag{2.7}$$

This n_i expresses the electron number in the atomic orbital ϕ_i. The summation of all the atomic orbitals in the atom gives the electron number belonging to the atom. The ionic value of the atom is obtained by subtracting this electron number from the electron number in a neutral atom. This is used as a measure of the charge transfer between atoms, and hence it is a sort of parameter to show the ionic interaction between atoms.

2.6.4 Energy Level Structure and the *d* Orbital Energy Level, *Md*

The energy level structure is obtained from the energy, ε_l of the molecular orbital, u_l . For example, the *d* orbital energy level, *Md*, is obtained from the energy level structure. The local electron density of states is also calculated by overlapping Gaussian functions with a proper half-width (e.g., 0.5 eV) and with the centers located at each energy, ε_l, of the molecular orbital, u_l. This is comparable with the result of the band calculation. The alloying behavior of elements in metals is evaluated quantitatively through these analyses.

Finally, it is summarized that the theory for alloy design has been developed on the basis of the molecular orbital calculations of electronic structures of alloys [22]. Two electronic parameters of the bond order, *Bo*, and the *d* orbital energy level, *Md* reflect the alloying behavior of elements in transition metals, while changing the magnitudes following the order of elements in the periodic table. The usefulness of these new parameters for alloy design will be presented in the following chapters.

References

[1] W. Hume-Rothery, G.V. Raynor, Structure of Metals and Alloys, Institute of Metals, London, 1954.

[2] L. Darken, R.W. Gurry, Physical Chemistry of Metals, Institute of Metals, London, 1954.

[3] M. Morinaga, N. Yukawa, H. Adachi, H. Ezaki, New PHACOMP and its applications to alloy design, in: M. Gell et al. (Eds.), Superalloys 1984, Proc. of the 5th International Symposium on Superalloys, Seven Springs Mountain Resort, Warrendale, PA, The Metallurgical Society of AIME, 1984, pp. 523–532.

[4] M. Morinaga, N. Yukawa, H. Adachi, Alloying effect on the electronic structure of Ni_3Al (γ'), J. Phys. Soc. Jpn 53 (1984) 653–663.

[5] M. Morinaga, N. Yukawa, T. Maya, K. Sone, H. Adachi, Theoretical design of titanium alloys, in: Proc. Sixth World Conf. on Titanium, Cannes, France, 6–9 June 1988, Paris, Société Française de Métallurgie, 1988, pp. 1601–1606.

[6] M. Morinaga, M. Kato, T. Kimura, M. Fukumoto, I. Harada, K. Kubo, Theoretical design of β-type titanium alloys, in: Proc. 7th World Conference on Titanium, San Diego, California, June 29–July 2, 1992, Warrendale, Pennsylvania, TMS, 1992, pp. 217–224.

[7] M. Morinaga, N. Yukawa, H. Ezaki, H. Adachi, Solid solubilities in transition-metal-based FCC alloys, Philos. Mag. A 51 (1985) 223–246.

[8] J.C. Slater, Quantum Theory of Molecules and Solids, vol. 4, McGraw-Hill, New York, 1974.

[9] D.E. Ellis, G.S. Painter, Discrete variational method for the energy-band problem with general crystal potentials, Phys. Rev. B 2 (1970) 2887–2898.

[10] F.W. Averill, D.E. Ellis, An efficient numerical multicenter basis set for molecular orbital calculations: application to $FeCl_4$, J. Chem. Phys. 59 (1973) 6412–6418.

[11] A. Rosén, D.E. Ellis, H. Adachi, F.W. Averill, Calculations of molecular ionization energies using a self-consistent-charge Hartree-Fock-Slater method, J. Chem. Phys 65 (1976) 3629–3634.

[12] D.E. Ellis, H. Adachi, F.W. Averill, Molecular cluster theory for chemisorption of first row atoms on nickel (100) surfaces, Surf. Sci. 58 (1976) 497–510.

[13] H. Adachi, M. Tsukada, C. Satoko, Discrete variational $X\alpha$ calculations. 1. Application to metal clusters, J. Phys. Soc. Jpn 45 (1978) 875–883.

[14] C. Satoko, M. Tsukada, H. Adachi, Discrete variational $X\alpha$ cluster calculations. II. Application to the surface electronic structure of MgO, J. Phys. Soc. Jpn 45 (1978) 1333–1340.

[15] H. Adachi, S. Shiokawa, M. Tsukada, Discrete variational $X\alpha$ cluster calculations. III. Application to transition metal complexes, J. Phys. Soc. Jpn 47 (1979) 1528–1537.

[16] M. Tsukada, H. Adachi, C. Satoko, Theory of electronic structure of oxide surfaces, Progress in Surface Science 14 (1983) 113–173.

[17] H. Adachi, Electronic state calculation of transition metal cluster, Adv. Quant. Chem. 29 (1997) 49–81.

[18] L. Pauling, The Nature of the Chemical Bond, third ed, Cornell University, Ithaca, 1960.

[19] H. Adachi, Theoretical analysis of X-ray and electron spectra by DV-Xa method, Adv. Quant. Chem. 37 (2000) 1–29.

[20] H. Adachi, Electronic state calculation of transition metal cluster, Adv. Quant. Chem. 29 (2000) 49–81.

[21] R.S. Mulliken, Electronic population analysis on LCAO–MO molecular wave functions. I, II, III, IV, J. Chem. Phys. 23 (1955) 1833–1840, 1841–1846, 2338–2342, 2343–2346.

[22] M. Morinaga, Alloy design based on molecular orbital method, Mater. Trans. 57 (2016) 213–226.

3 Nickel Alloys

3.1 Molecular Orbital Calculation for Alloyed Ni_3Al

Nickel alloys are strengthened by the precipitation of the Ni_3Al (γ') phase in the fcc Ni (γ) matrix. The volume fraction of the γ' phase is as high as about 60% in advanced Ni-based superalloys. The high temperature strength is attributable to the Ni_3Al (γ') phase precipitated in the γ matrix [1–6]. The Ni_3Al has a unique property suitable for the high-temperature alloy, since the yield strength increases with temperature [6,7]. A variety of alloying elements are soluble in either the γ or the γ' phase, and nickel alloys are further strengthened by the solution hardening.

3.1.1 Level Structures and Electron Density of States

It is known that the electronic structure of pure Ni_3Al [8,9] resembles that of pure Ni. For example, the Fermi level lies in the Ni 3*d* band, and the covalent interaction between Ni *d* electrons is responsible for the cohesive energy of both Ni_3Al and pure Ni [10]. However, no calculations have been carried out in the Ni_3Al alloyed with various transition metals. So, the alloying effect on the electronic structure of Ni_3Al is first calculated using the DV-Xα cluster method [11].

In Fig. 3-1A, the crystal structure of Ni_3Al is shown [12]. An Al atom is surrounded by 12 Ni atoms in the first-nearest-neighbor sites, and by 6 Al atoms in the second-nearest-neighbor sites. So, a cluster model, [$MNi_{12}Al_6$], is constructed as shown in Fig. 3-1B and used for the calculation. In this model, every alloying element, M, is substituted for an Al atom at the center. As M is surrounded by 12 Ni atoms, it is in the same chemical circumstance as in fcc Ni. The experimental lattice parameter used is 0.3570 nm, which close to the value of pure Ni, 0.3524 nm [13].

The calculated level structures are shown in Fig. 3-2 for pure and alloyed Ni_3Al with 3*d* transition metals [12]. Here, the energy of the Fermi level (E_f) in pure Ni_3Al is set to be zero and used for reference. The E_f level for alloyed Ni_3Al is indicated by an arrow ($\leftarrow$) in the figure. In pure Ni_3Al, the levels of $13a_{1g}$ to $15e_g$ are mainly originated from Ni 3*d* orbitals and form the Ni 3*d* band where the E_f level lies. The lower energy level of $12a_{1g}$ to $9t_{2g}$ and the upper energy levels of $15a_{1g}$ to $14t_{2u}$ are the mixed states of Al 3*s*, 3*p* and Ni 4*s*, 4*p*, 3*d*. For alloyed Ni_3Al with 3*d* transition metals, new energy levels due mainly to the M-3*d* orbitals appear above the E_f. For example, $16e_g$ and $14t_{2g}$ levels correspond to these levels, as indicated by a dotted level in the figure. The height of these levels changes monotonously with the order of elements in the periodic table. Similarly, new energy levels due to the M-4*d* or 5*d* orbitals appear in the alloyed Ni_3Al with 4d or 5d transition metals.

A Quantum Approach to Alloy Design. DOI: https://doi.org/10.1016/B978-0-12-814706-1.00003-0

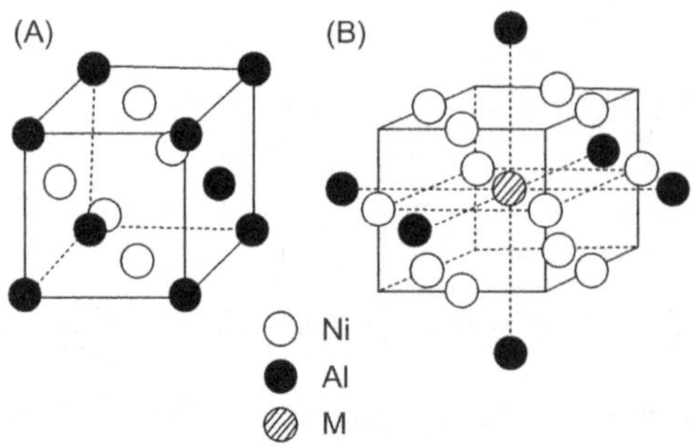

FIGURE 3-1 (A) Crystal structure of Ni_3Al and (B) cluster model used for the calculation.

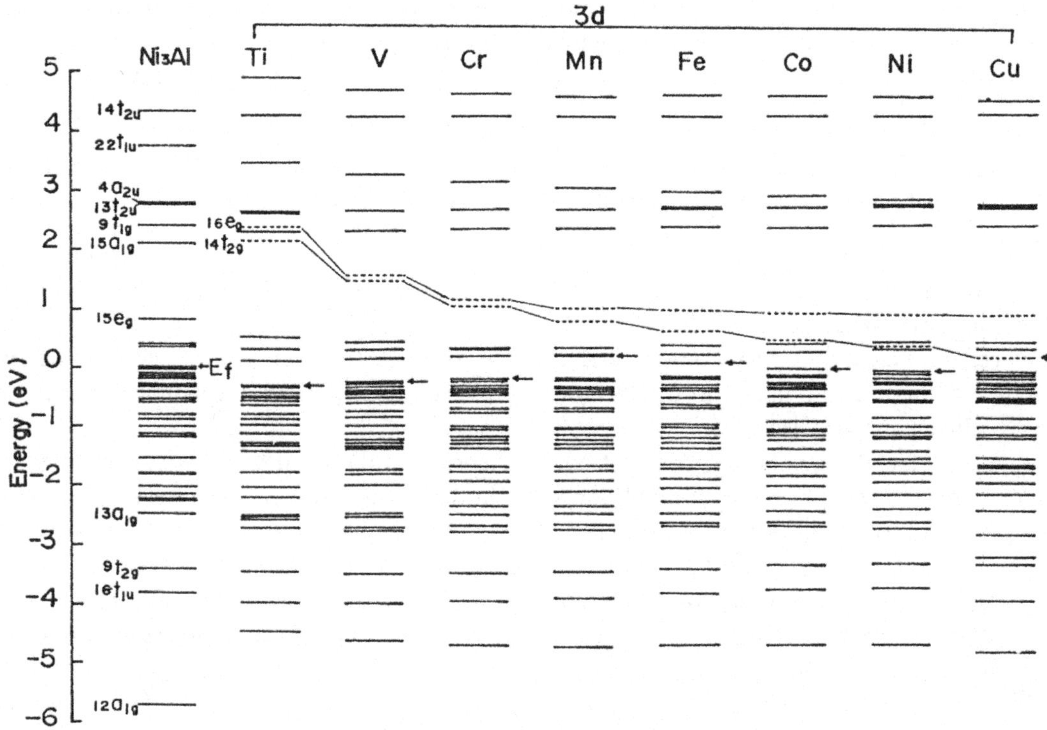

FIGURE 3-2 Energy level structures of pure and alloyed Ni_3Al with 3*d* transition metals.

In Fig. 3-3 the electron density of states is shown for (A) M = Al, (B) M = Ti, and (C) M = Mo. These are obtained by overlapping Gaussian functions with a width of 0.5 eV, and with the centers located at each cluster energy level [14]. The main features shown in Fig. 3-3A for pure Ni_3Al quite resemble the results obtained from the band calculation [8–10] despite the use of a small cluster model shown in Fig. 3-1B. It is evident from Fig. 3-3A–C that the electron density of states for pure Ni_3Al is modified by the Ti or Mo substitution for Al. This modification occurs mainly due to the appearance of the M-*d* levels above the E_f, as explained previously.

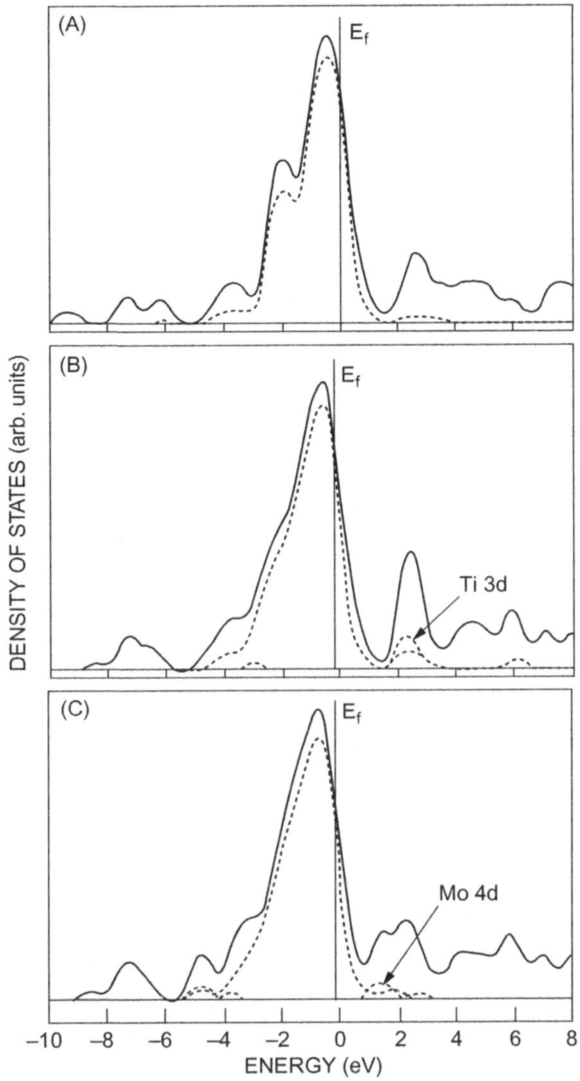

FIGURE 3-3 Density of states for (A) M = Al, (B) M = Ti and (C) M = Mo. *Dashed line*, Ni 3*d*; *dotted-dashed line*, Ti 3*d* or Mo 4*d*; *solid line*, total.

3.1.2 *Md* Level

The *d* levels of alloying elements, M-*d*, for example, $16e_g$ and $14t_{2g}$ levels for the 3*d* transition metals shown in Fig. 3-2, correlate with the electronegativity. Following Mulliken [15], eigenvalues of energies obtained by the DV-Xα calculation may represent electronegativity, although the covalent interaction between the neighboring atoms somehow modifies this idea for a free atom. In Fig. 3-4, the M-*d* levels are plotted against the electronegativity proposed by Watson and Bennett [16]. The M-*d* levels decrease with increasing electronegativity of M.

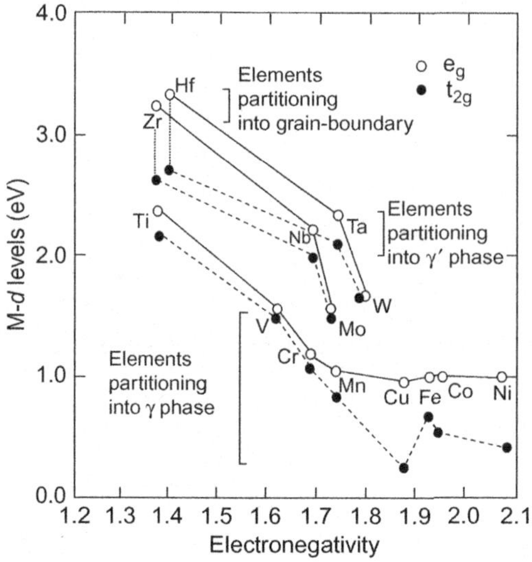

FIGURE 3-4 M-*d* levels (e_g and t_{2g}) vs. electronegativity.

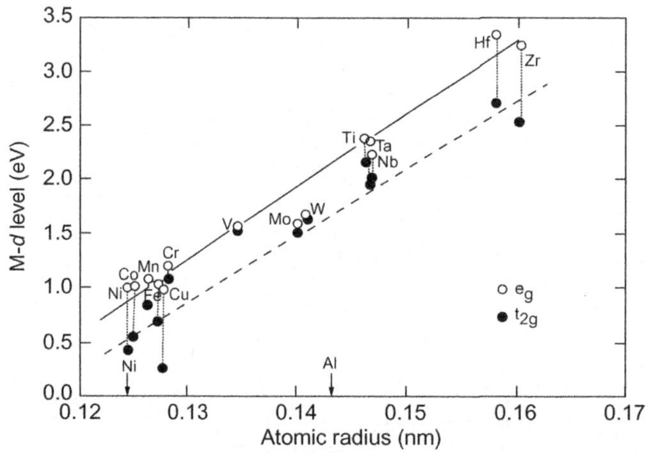

FIGURE 3-5 Correlation of M-*d* levels, e_g and t_{2g} with atomic radius.

Following the classification of alloying elements in nickel-based superalloys by Decker [17,18], Zr is an alloying element partitioning into the grain boundary. Hf is probably a similar element to Zr. Also, Ti, Nb, and Ta are elements mainly partitioning into the γ' phase. The other transition elements largely partition into the γ phase. As shown in Fig. 3-4, these three groups are distinguishable from their M-*d* energy levels.

In Fig. 3-5, the M-*d* levels are also plotted against the atomic radius. Here, the atomic radii are taken from the values of closed packed metals with CN = 12, reported by Teatum et al. [19]. As explained in Chapter 2, there is a trend that the M-*d* level increases with increasing atomic radius.

Table 3-1 *Md* and *Bo* Values for Various Elements in Ni_3Al

Element		*Md* (eV)	*Bo*
	Al	1.900	0.533
	Si	1.900	0.589
3d	Ti	2.271	1.098
	V	1.543	1.141
	Cr	1.142	1.278
	Mn	0.957	1.001
	Fe	0.858	0.857
	Co	0.777	0.697
	Ni	0.717	0.514
	Cu	0.615	0.272
4d	Y	3.817	1.279
	Zr	2.944	1.479
	Nb	2.117	1.594
	Mo	1.550	1.611
	Tc	1.191	1.535
	Ru	1.006	1.314
	Rh	0.898	1.068
5d	Hf	3.020	1.518
	Ta	2.224	1.670
	W	1.655	1.730
	Re	1.267	1.692
	Os	1.063	1.500
	Ir	0.907	1.256

The *Md* level of each alloying element, M, is defined simply by taking the average of these e_g and t_{2g} levels. The values of *Md* level are listed in Table 3-1 for various alloying elements.

3.1.3 Wave Functions of Bonding Orbitals

To clarify the bonding mechanism between M and the surrounding Ni atoms, the wave functions of the bonding orbitals are examined. Fig. 3-6 shows the contour map of the wave function for the $12e_g$ orbital, which exists below the E_f and contributes substantively to the M-Ni bonding. In case of pure Ni_3Al (i.e., M = Al), the Al $3d\gamma$ orbital participates in the bonding to make π type bond with the Ni 3*d* orbitals, although the Al 3*d* component in this orbital is small because the atomic Al 3*d* level is situated at the higher energy region. In the case when Ti is substituted for a central Al (i.e., M = Ti), the Ti 3*d* component is remarkably high in the bonding orbital. Then, the covalent bond between atoms increases significantly by the substitution of transition metals for Al.

3.1.4 Bond Order, *Bo*, and Ionicity

In Fig. 3-7A–D, both the bond order between atoms and the ionicity, evaluated from the Mulliken population analysis [20], are summarized together with (E) the Fermi energy level (E_f)

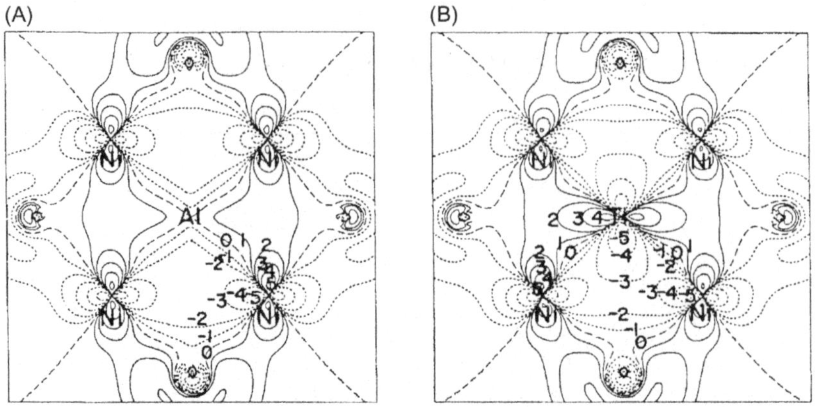

FIGURE 3-6 Contour maps of the wave function for the $12e_g$ orbital in (A) $[AlNi_{12}Al_6]$ cluster and (B) $[TiNi_{12}Al_6]$ cluster. The label of the curve, 0, ±1, ±2, ±3, ±4, ±5 corresponds to the curve of wave function, 0.00, ±0.01, ±0.02, ±0.04, ±0.08, ±0.16, respectively.

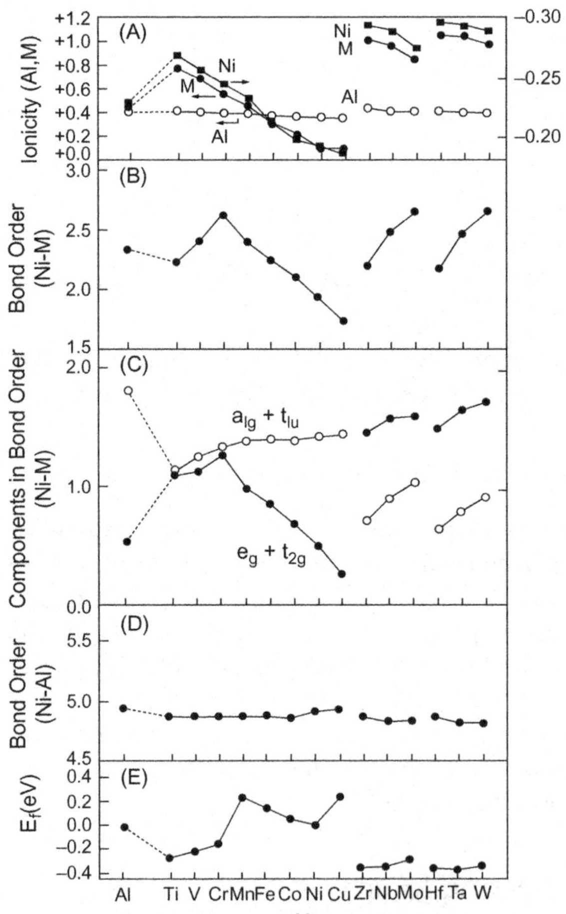

FIGURE 3-7 (A) Ionicity and (B–D) bond order obtained for various elements, M. The change in the Fermi level (E_f) with M is also shown in (E).

of the alloy cluster. As shown in (A), the ionicity of Al is about +0.4, and the value scarcely changes with alloying elements, M. However, the iconicity of Ni depends largely on M. Namely, Ni has a negative ionicity, but M has a positive ionicity. This means that the charge transfer takes place from M to Ni following the electronegativity difference between them. The ionicity of Ni becomes negative, as Ni is most electronegative among various elements M, shown in the horizontal axis in the figure. Thus, the ionic bond is operating between M and Ni, although it is weaker as compared with the covalent bond.

The bond order shows the overlap population between atoms and hence it is a measure of the covalent bond strength between atoms, as explained in Chapter 2. As shown in (D), the bond order between Ni and Al scarcely changes with M, whereas that of Ni-M changes largely with M as shown in (B). A peak appears at Cr in the 3*d* elements. Such a high bond order is seen for Mo in the 4*d* elements and for W in the 5*d* elements. All these elements belong to the 6A group in the periodic table. In the 3*d* series elements, the occupancy of electrons at the antibonding levels begins at Mn, resulting in the decrease of the bond order with the atomic number of elements larger than Cr.

As shown in (C), there are four components in the bond order between Ni and M. The components of a_{1g}, t_{1u}, e_g, and t_{2g} contribute to the bond order through the covalent bonding of M-*s*, M-*p*, M-*d*, and M-*d* electrons with the Ni-*s*, *p*, *d* electrons, respectively. Therefore, among them the ($e_g + t_{2g}$) contributions to the bond order appear to be the most essential parts of the covalent bond between transition metals, because they are associated with the bonding due to the M-*d* electrons. As shown in (C), the plots of the ($e_g + t_{2g}$) bond order again show a peak at Cr, Mo, and W of the 6A group elements. It is also apparent that the ($e_g + t_{2g}$) value is higher in the 5*d* elements than in the 4*d* elements than in the 3*d* elements. Here, the ($e_g + t_{2g}$) value is defined as the bond order, *Bo*, and the values are listed in Table 3-1 for various alloying elements, M, together with the *Md* parameter as explained earlier.

3.1.5 Definition of $\overline{Md}$ and $\overline{Bo}$ Parameters for Alloy

For an alloy, the average values of *Md* and *Bo* are defined simply by taking the compositional average, then $\overline{Md}$ and $\overline{Bo}$ are defined as [4],

$$\overline{Md} = \sum_{i=1}^{n} X_i \cdot (Md)_i, \tag{3.1}$$

$$\overline{Bo} = \sum_{i=1}^{n} X_i \cdot (Bo)_i. \tag{3.2}$$

Here, X_i is the atomic fraction of component *i* in the alloy, and $(Md)_i$ and $(Bo)_i$ are the respective values for component *i*, which are listed in Table 3-1. The summation extends over the components, $i = 1, 2, \ldots, n$. $\overline{Md}$ is the average value of the *d* level of each alloying element, so it represents a center of gravity in the *d*-band of the alloy. The values of $\overline{Md}$ and $\overline{Bo}$ are calculated readily from the alloy composition using Eqs. (3.1) and (3.2).

3.2 New PHACOMP

The solid solubility problem of alloy is one of the important problems in physical metallurgy. According to the classical approach by Hume-Rothery [21] and Darken–Gurry [22], this problem has been treated using the atomic radius and the electronegativity. However, this approach is still difficult in the alloy when both solute and solvent atoms are transition metals.

As described earlier, the phase computation (PHACOMP) method [1,2] was developed in 1964 to predict the solid solubility limit of elements in the Ni matrix (γ) to suppress the formation of brittle phases (e.g., the σ phase) in the γ matrix. The electron vacancy number, Nv, is used for the prediction in this method. Here, Nv is the number of electron vacancies or holes existing above the Fermi energy level in the d band, and expressed approximately as $Nv = 10.66\text{-}(e/a)$, where e/a is the electrons-per-atom ratio. For example, for the 4A group elements (i.e., $e/a = 4$) such as Cr, Mo, and W, the Nv value is 6.66. However, the atomic radius is smaller in Cr than in Mo or W. Thus, the atomic size concept is missing in the Nv parameter. In this sense, this method is very different from the classical approach by Hume-Rothery and Darken–Gurry [21,22], and inevitably the prediction is poor along this Nv method. Nevertheless, PHACOMP has been used for the design and the quality control of nickel-based superalloys.

New PHACOMP was proposed in 1984 [4]. This is a method for predicting the solid solubility limit by using the Md parameter. As explained earlier, the Md parameter is concerned with the electronegativity and the atomic radius of elements. It is stressed here that this parameter is determined using the alloy cluster that has the same chemical circumstance as in the Ni alloy. Therefore, the Md parameter has a great potential for dealing with the solid solubility problem of alloys in which both solute and solvent atoms are transition metals.

We assume that when the $\overline{Md}$ value is larger than a certain value, the phase instability takes place and the second phase appears in the γ matrix [4]. In other words, such a critical $\overline{Md}$ determines the solid solubility limit of the γ phase. The critical $\overline{Md}$ value depends on a type of the second phase, because the solubility limit is determined by the tangent to the Gibbs free energy vs. composition curves of the γ matrix phase and the second phase so as to make the chemical potential of elements equal between the two phases. As the solid solubility changes with temperature, it depends on temperature, too. Such a critical value for $\overline{Md}$ is determined by fitting the solubility line in phase diagrams by a constant $\overline{Md}$ line.

3.2.1 Solid Solubility in Ternary Alloys

To verify new PHACOMP, four types of the phase boundaries are mainly examined here. They are (1) $\gamma/\gamma+\sigma$, (2) $\gamma/\gamma+\mu$, (3) $\gamma/\gamma+\gamma'$ (Ni_3Al), and (4) $\gamma/\gamma+\beta$ (NiAl). Here, the σ and μ phases are called the topologically close-packed (TCP) phase, and γ'(Ni_3Al) is called the geometrically close-packed (GCP) phase.

3.2.1.1 $\gamma/\gamma + \sigma$ Phase Boundaries

Typical phase diagrams of Ni-Co-Cr and Ni-Cr-Mo at 1477K are shown in Fig. 3-8A–B. In each phase diagram, the $\gamma/\gamma + \sigma$ phase boundary is traced using the iso-$\overline{Md}$ line of 0.925. For simplicity, the unit of $\overline{Md}$ (eV) is omitted in the figure. For comparison, the iso-$\overline{Nv}$ line and iso-$\overline{R}$ line are also shown in each phase diagram. Here, $\overline{Nv}$ and $\overline{R}$ are the compositional averages of Nv and atomic radius, R, of elements, respectively. The iso-$\overline{Nv}$ line of $\overline{Nv} = 2.49$, which is often used for the prediction of the $\gamma/\gamma + \sigma$ phase boundary, is far away from the boundary. On the other hand, the iso-$\overline{R}$ line traces well the boundary shown in (A), but it is far away from the boundary shown in (B). Compared to these, the iso-$\overline{Md}$ line is close to the boundary in both (A) and (B) [23].

Another phase diagram of Fe-Ni-Cr at 1073K is given in Fig. 3-8C. The critical $\overline{Md}$ value changes with temperature. It is 0.900 (eV) at 1073K, lower than 0.925 (eV) at 1477K shown in Fig. 3-8A–B [23].

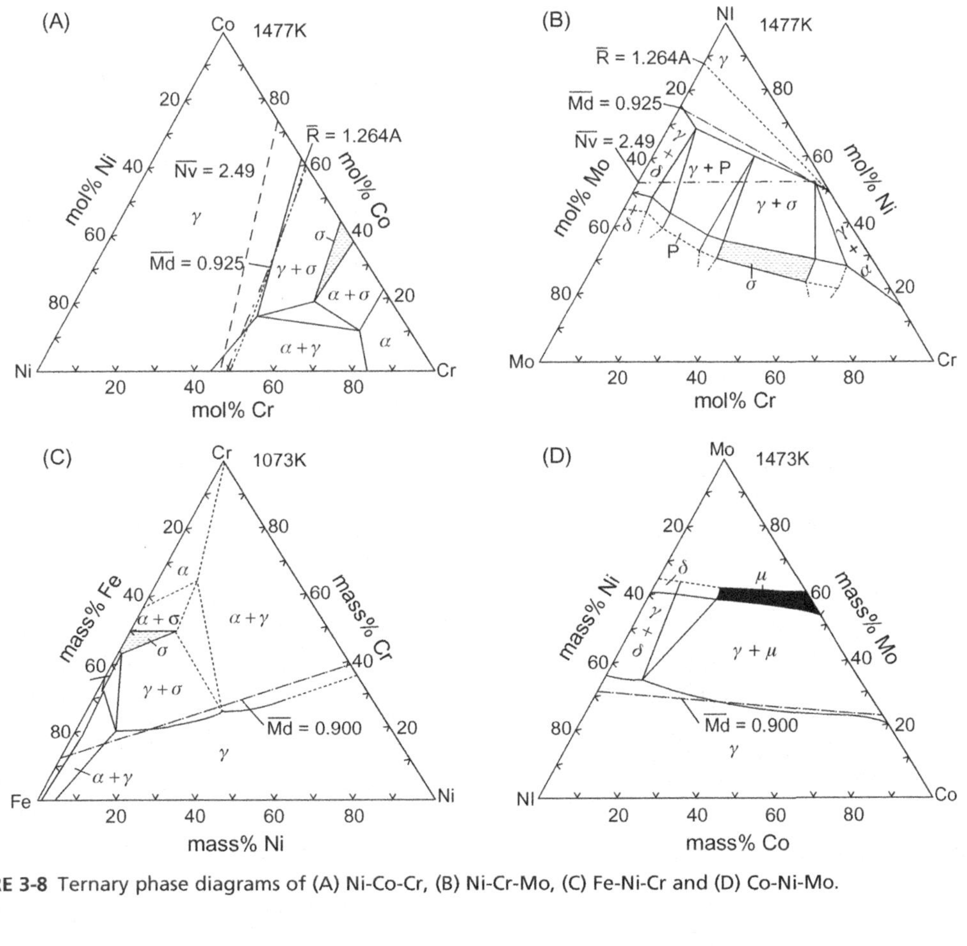

FIGURE 3-8 Ternary phase diagrams of (A) Ni-Co-Cr, (B) Ni-Cr-Mo, (C) Fe-Ni-Cr and (D) Co-Ni-Mo.

3.2.1.2 $\gamma/\gamma + \mu$ Phase Boundaries

The phase diagram of Co-Ni-Mo at 1473K is shown in Fig. 3-8D. The critical $\overline{Md}$ is 0.900 (eV) for the μ phase, which is smaller than 0.925 (eV) for the σ phase. This is due to the stability difference of the σ and μ phases relative to the γ phase, mainly arising from the crystal structure difference between them. The σ phase consists of 30 atoms per unit cell of a complex body-centered tetragonal lattice and has high coordination numbers [24]. The μ phase contains 13 atoms per unit cell of a rhombohedral lattice with the formula, A_7B_6.

Also, as shown in Fig. 3-9A–B, the μ phase appears in the phase diagrams of Co-Fe-Mo and Fe-Ni-Mo at 1473K [23]. The critical $\overline{Md}$ is 0.900 (eV) as is the same as the value of Co-Ni-Mo at 1473K shown in Fig. 3-8D.

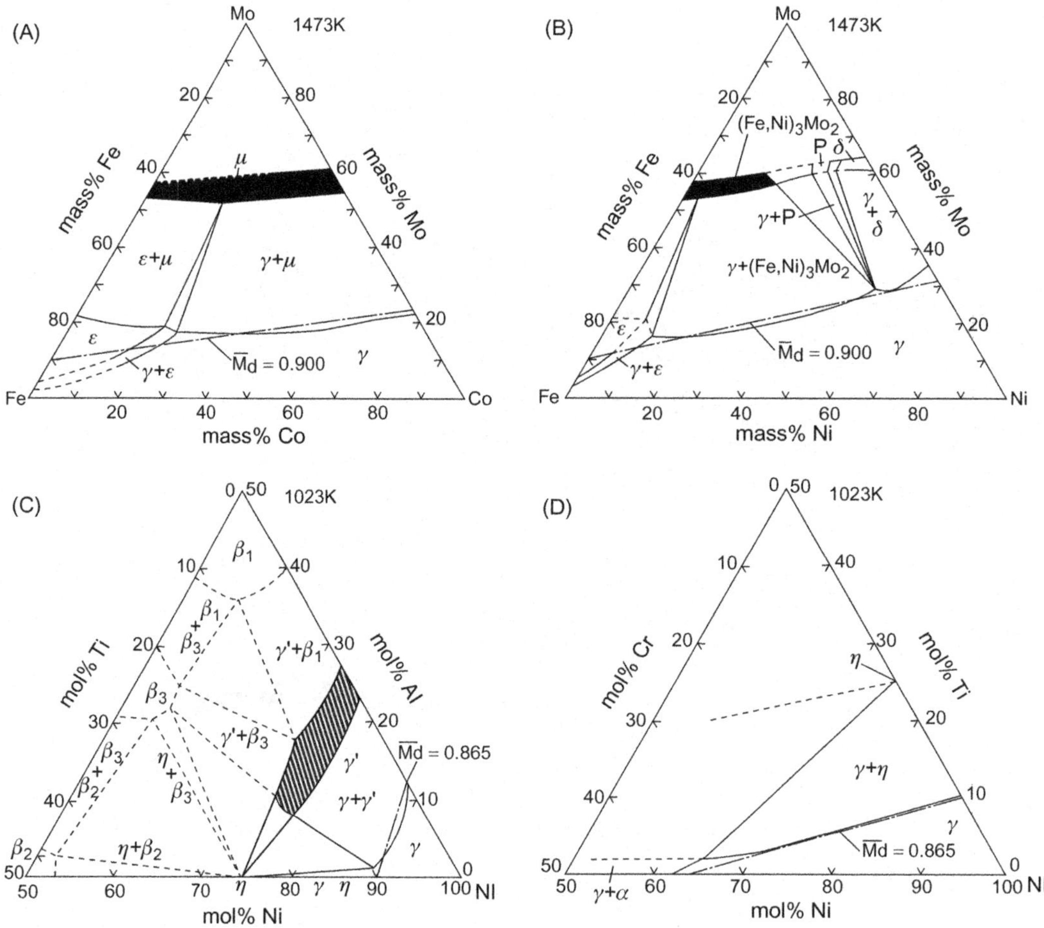

FIGURE 3-9 Ternary phase diagrams of (A) Co-Fe-Mo, (B) Fe-Ni-Mo, (C) Ni-Al-Ti and (D) Ni-Ti-Cr.

3.2.1.3 $\gamma/\gamma + \gamma'$ Phase Boundaries

The phase diagrams of Ni-Al-Ti and Ni-Ti-Cr at 1023K are given in Fig. 3-9C–D [23]. The γ' phase is Ni_3Al and the η phase is Ni_3Ti [24]. The phase boundaries for these two phases are traced using the same $\overline{Md}$ value of 0.865 (eV), probably owing to the resemblance in their crystal structures. The validity of the present method is further confirmed through the evaluation of the $\gamma/\gamma + \gamma'$ phase boundaries in various ternary alloys [25].

3.2.1.4 $\gamma/\gamma + \beta$ (NiAl) Phase Boundary

The $\gamma/\gamma + \beta$ (NiAl) phase boundary in Ni-Fe-Al ternary alloys is also traced by the iso-$\overline{Md}$ line as shown in Fig. 3-10A–C. The critical $\overline{Md}$ is 0.950 (eV) at 1523K, 0.930 (eV) at 1323K, and 0.900 (eV) at 1023K [23].

In addition, the present method is found to be applicable to the TCP phases (e.g., Laves phase) other than the σ and μ phases. The critical $\overline{Md}$ value seems to decrease in the order, σ phase $>$ μ phase $>$ Laves phase [26]. Also, the volume fraction of the α-W phase precipitated in Ni-Cr-W alloys increases monotonously with the $\overline{Md}$ values. The α-W free alloy is obtained when the $\overline{Md}$ value is smaller than about 0.885 (eV) [26]. Thus, the validity of this *Md* method is proved firmly through the examination of more than 30 ternary phase diagrams [23,25,26].

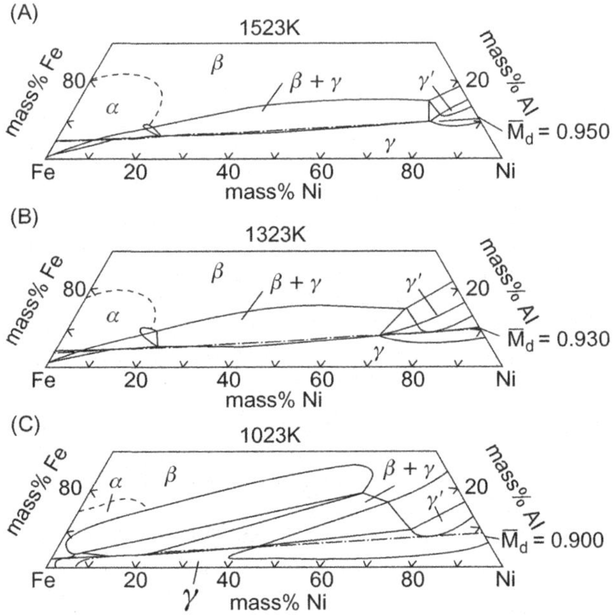

FIGURE 3-10 Partial phase diagrams of Ni-Fe-Al at (A) 1523K, (B) 1323K and (C) 1023K.

3.2.1.5 Temperature Dependence of Critical $\overline{Md}$ Values

The temperature dependence of the critical $\overline{Md}$ for the σ phase is shown in Fig. 3-11 [23]. It is expressed approximately as

$$\sigma \text{ phase: Critical } \overline{Md} = 6.25 \times 10^{-5}\text{T} + 0.834. \tag{3.3}$$

Here, T is the temperature (K), and the temperature coefficient can be compared with the coefficient of thermal energy, $k_B\text{T} = 8.62 \times 10^{-5}$ T (eV), where k_B is the Boltzmann constant. This temperature dependence of the critical $\overline{Md}$ is applicable to the design of the σ-free alloy at service temperatures.

Also, the temperature dependence of the critical $\overline{Md}$ for the μ phase [26] and for the γ′ phase [26] is expressed approximately as,

$$\mu \text{ phase: Critical } \overline{Md} = 1.54 \times 10^{-4}\text{T} + 0.675, \tag{3.4}$$

$$\gamma' \text{ phase: Critical } \overline{Md} = 1.41 \times 10^{-4}\text{T} + 0.727. \tag{3.5}$$

The temperature coefficient for these phases is about twice as large as that for the σ phase. The temperature dependence of the critical $\overline{Md}$ corresponds to the solid solubility change with temperatures. The critical $\overline{Md}$ changes with the type of the second phases such as the σ, μ and γ′ phases, because the solid solubility limit depends on the second phases.

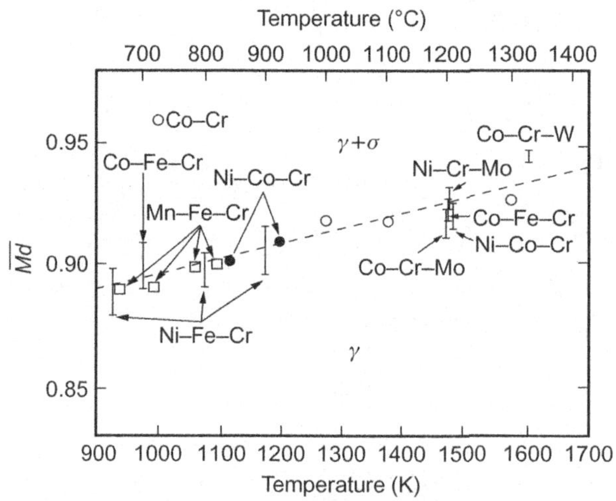

FIGURE 3-11 Temperature dependence of critical $\overline{Md}$ for the σ phase in various alloys.

3.3 Applications to New PHACOMP to Practically Used Austenite (fcc) Alloys

New PHACOMP is applicable not only to ternary phase diagrams, but also to practically used alloys with multiple components, and the formation of the TCP phases is predictable in them [4,23].

3.3.1 Ni-Based Alloys

The σ phase usually precipitates in the γ matrix. The prediction for the σ phase formation is carried out using the analyzed results of the γ phase compositions in the alloys [27,28]. Such results of the compositional analyses are summarized in the paper by Barrows and Newkirk [29]. About 10 elements exist in the γ phase in the practically used alloys. The calculated results are shown in Fig. 3-12 for (A) $\overline{Md}$ and (B) $\overline{Nv}$. As reported by Barrows and Newkirk [29], the $\overline{Nv}$ method does not give a right prediction in IN 713C and IN 713LC as shown in (B). On the other hand, the $\overline{Md}$ method predicts rightly even in these alloys as shown in (A). All the σ-prone alloys have a higher $\overline{Md}$ value than about 0.915 (eV), which is a value to correspond to the temperature of about 1300K, following the estimation from Eq. (3.3). But the reason for the wrong prediction for TRW 1900 by both $\overline{Md}$ and $\overline{Nv}$ methods is still unknown.

By the way, the *Md* parameters used here are calculated using the Ni alloy cluster model shown in Fig. 3-1. Nevertheless, the calculated results appear to be useful even for Co-based alloys and Fe-based alloys, as long as they are in the fcc (austenite) state. This is seen in various ternary phase diagrams, for example, Ni-$\underline{\text{Co}}$-Cr shown in Fig. 3-8A, $\underline{\text{Fe}}$-Ni-Cr shown in Fig. 3-8C, $\underline{\text{Co}}$-Ni-Mo shown in Fig. 3-8D, $\underline{\text{Co}}$-$\underline{\text{Fe}}$-Mo shown in Fig. 3-9A, and $\underline{\text{Fe}}$-Ni-Mo shown in Fig. 3-9B. According to the classical classification, Fe, Ni, and Co all belong to the 8 group in the periodic table, and their *Md* values resemble each other as shown in Table 3-1. This point will be further discussed in Chapter 4, using the calculated results for fcc Fe. In this chapter, both the Co-based and the Fe-based alloys are treated in a similar way as in the Ni-based alloys.

3.3.2 Co-Based Alloys

The tendency for the formation of the TCP phases in six Co-based alloys is examined as shown in Fig. 3-12 for (C) $\overline{Md}$ and (D) $\overline{Nv}$. In the figure, open circle means the TCP-free alloy and solid circle means the TCP-prone alloy. The TCP (μ and Laves) phases appear in L-605, and the σ phase appears in both UMCO 50 and FSX-414, and the other alloys are free from the TCP phases in them [23]. Sims [3] assigned about 2.70 to the critical $\overline{Nv}$ for Co-based alloys. However, TCP (μ and Laves) phases appear in L-605 around $\overline{Nv} = 2.48$, which is considerably lower than the critical $\overline{Nv}$. In the $\overline{Md}$ method there is a clear

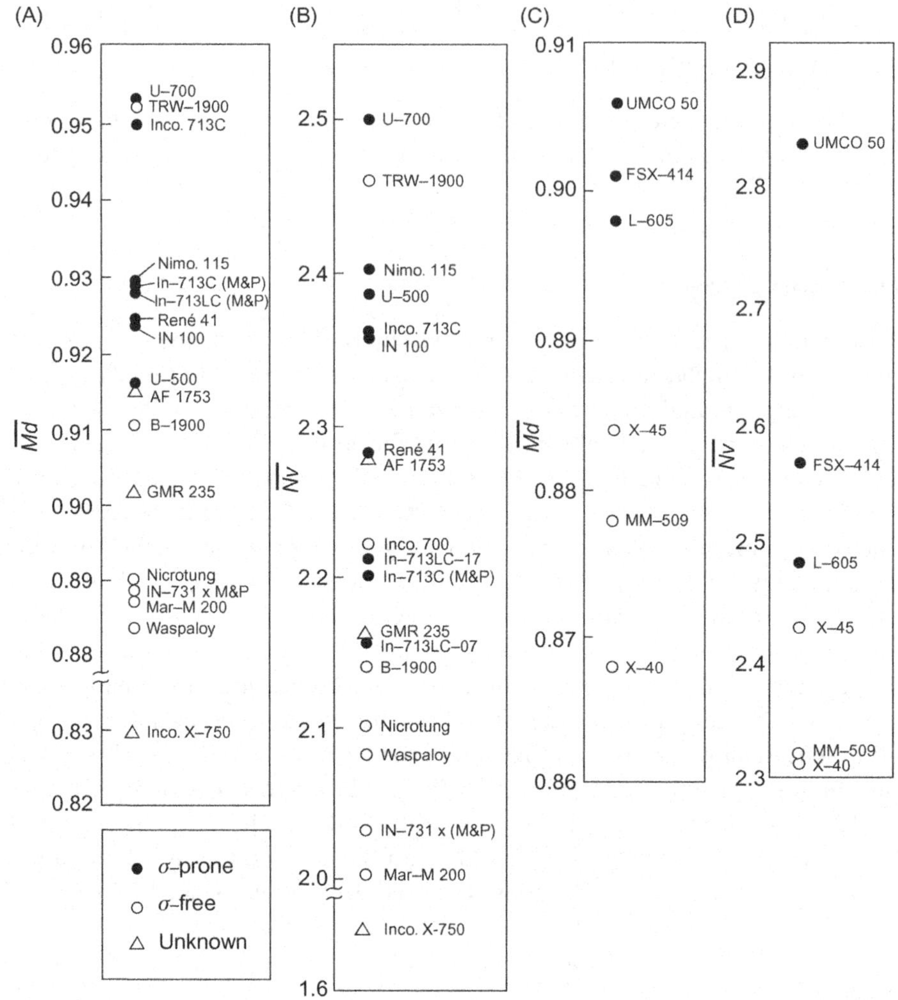

FIGURE 3-12 Estimation for the occurrence of TCP phases in (A,B) Ni-based superalloys and (C,D) Co-based superalloys.

separation in the alloys denoted by open and solid circles, and L-605 lies near $\overline{Md} = 0.90$(eV), being close to the critical $\overline{Md}$ for the μ phase shown in Figs. 3-8D and 3-9A–B.

3.3.3 Fe-Based Alloys

HK40 is one of the heat-resisting Fe-based alloys. Ohta and Saori examined the σ phase formation with the 53 HK40-type alloys with various compositions [30]. The compositional

Table 3-2 The Range of Alloy Compositions (Mass%) of HK40: (A) Total Alloys and (B) A-Group and B-Group Alloys

(A)	Element	C	Si	Mn	Cr	Ni	Fe	Total alloy number
	Specimen	0.02–0.63	0.05–2.16	0.05–2.95	15.23–31.55	18.76–31.50	Bal.	53
(B)	Alloy group	C	Si	Mn	Cr	Ni	Fe	Alloy number of each group
	A	0.02–0.59	0.05–0.98	0.05–2.95	22.87–25.37	18.76–22.63	Bal.	22
	B	0.08–0.63	1.49–2.16	0.93–1.13	15.23–31.55	19.25–31.50	Bal.	31

range of the alloys is listed in Table 3-2A. The σ-phase formation is very sluggish in them. So, to accelerate the formation reaction, specimens are first cold-worked to reduce the thickness by 40% and then aged at 1073K for 54 Ms.

In HK40, the $M_{23}C_6$-type carbide precipitates in the early stage of aging, and then the σ-phase formation follows. By assuming that the carbide is a $Cr_{23}C_6$ and all carbon atoms form this carbide, both the Cr and C contents necessary for the carbide formation are subtracted from the total alloy composition, and the remaining elements in the matrix are scaled so as to total 100%.

Then, by using $\overline{Md}$ and $\overline{Nv}$ methods, it is examined whether the σ phase is prone or free in each alloy [31]. The results are shown in Fig. 3-13 for (A) $\overline{Nv}$ and (B) $\overline{Md}$. In this analysis the alloys are divided into group A and group B, according to the Si content in them. The compositional range of each group is listed in Table 3-2B. The A-group alloys contain less than 0.98 mass% Si, and the critical $\overline{Md}$ is 0.90 (eV) as shown in Fig. 3-13B. Below this value no σ phase precipitates. This value agrees with the critical $\overline{Md}$ value for the $\gamma/\gamma + \sigma$ phase boundary in the Fe-Cr-Ni phase diagram at 1073K shown in Fig. 3-8C. However, in the B-group alloys containing more than 1.49 mass% Si, the σ phase does not appear despite the larger $\overline{Md}$.

In general, Si has been considered to be a σ-forming element, but the result that the σ phase does not appear in the higher Si-content alloys conflicts with this trend. So, it is presumed that a certain Si-rich compound precipitates in the B-group alloys, and causes a compositional change in the residual matrix. In accordance with this presumption, the existence of a Si-rich compound, $Cr_5Ni_3Si_2(C)$, is confirmed in the B-group alloys. Its secondary electron image and Si-Kα X-ray image are shown in Fig. 3-13B. It contains about 10 mol% C according to the EPMA experiment. A similar Si-rich compound, $(Cr,Si)_3Ni_2Si$, exists in Cr-Ni-Si ternary alloys when the C content is higher than 1 mass% [32]. The Si-rich compound observed in the B-group alloys probably corresponds to this type of compound.

The discrepancy of the B-group alloys in Fig. 3-13B can be removed if the precipitation of the Si-rich compound is taken into account. By assuming that Si is soluble in the matrix up to 1 mass% and the balanced Si forms a $Cr_5Ni_3Si_2(C)$ compound, the composition of the residual matrix is recalculated for the B-group alloys. Here, the 1 mass% solubility limit of Si is simply chosen, as it is an upper limit of the Si content in the A-group alloys (0.98 mass%). The partition of carbon atoms into the $Cr_5Ni_3Si_2(C)$ as well as $Cr_{23}C_6$ is counted in this

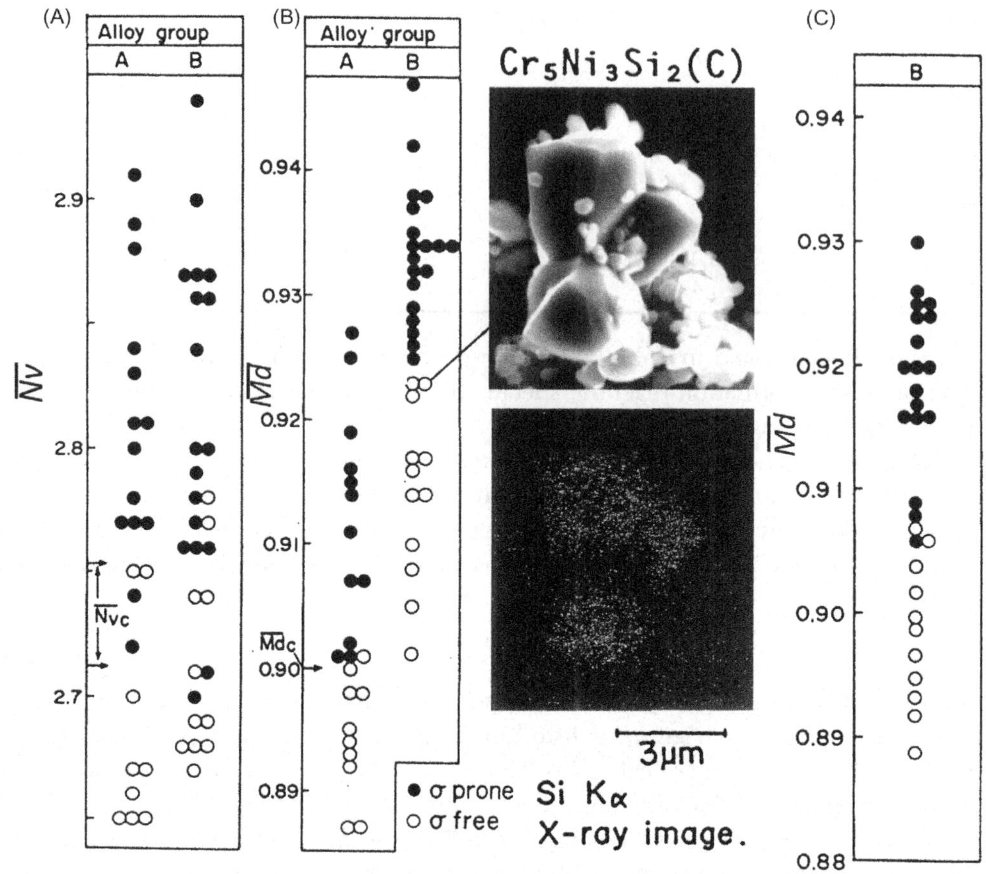

FIGURE 3-13 Estimation for the occurrence of the σ phase in HK40: (A) $\overline{Nv}$, (B) $\overline{Md}$, and (C) re-calculated $\overline{Md}$ for the B group.

recalculation. The partitioning ratio is estimated by assuming the Si-rich compound to be $Cr_5Ni_3Si_2C_1$. The reestimated $\overline{Md}$ is shown in Fig. 3-13C. The critical $\overline{Md}$ is about 0.905 (eV), in reasonable agreement with the critical value for the A-group alloys.

On the other hand, in the $\overline{Nv}$ method, there are no clear boundaries between the A-group alloys and B-group alloys as shown in Fig. 3-13A. Its prediction for the σ phase formation is poor, and the existence of a Si-rich compound is not able to be predicted by this method.

Thus, new PHACOMP predicts the appearance of various phases such as the σ, μ, and γ′, Laves phases, and α-W phase as well [4,23,25,26,31]. Various contradictions inherent in the *Nv*-PHACOMP are solved in the new PHACOMP.

Cieslak et al. [33] used this method to predict the formation of TCP phases in various Ni-based weld alloys. Caron and Khan [34] employed new PHACOMP to predict the μ phase

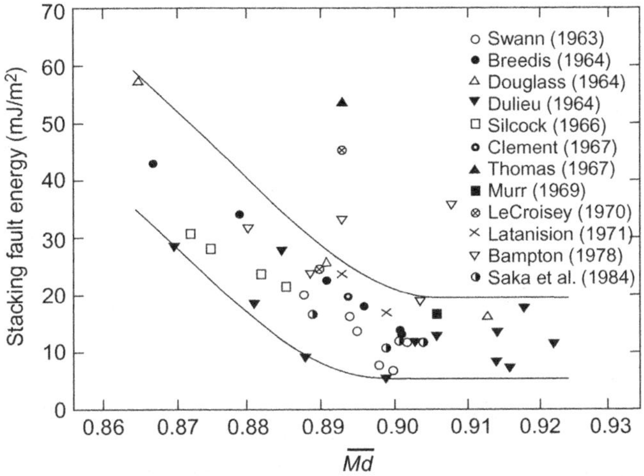

FIGURE 3-14 Correlation of $\overline{Md}$ with the stacking fault energy of Fe-Cr-Ni and its related alloys.

formation in the superalloys with high Mo and W contents. Also, as shown in Fig. 3-14, Yukawa et al. [35] reported that the stacking fault energies of Fe-Cr-Ni and its related alloys keep decreasing with increasing $\overline{Md}$ values and then become nearly constant around $\overline{Md} = 0.90$(eV), although the data are rather scattered in the figure. When a stacking fault is introduced to the C layer in the fcc lattice with the ABCABC stacking sequence, then it is equivalent to the local formation of the ABAB stacking sequence in a hcp lattice. Therefore, the stacking fault energy may be considered to be a measure to show the relative stability between the fcc and the hcp, although the strain energy around the stacking fault is not counted in this explanation. It is, however, interesting to note that commercial austenitic stainless steels have the compositions near $\overline{Md} = 0.90$(eV), where the $\gamma/\gamma+\sigma$ phase boundary exists at 1073K, and the stacking fault energy takes a lowest value of about 12 mJ/m^2.

3.3.4 Correlation of the *Md* and *Bo* Parameters With Thermodynamic Data

There may be a certain correlation of the *Md* and *Bo* parameters with thermodynamic data of the Ni alloys. For binary Ni-10mol% M alloys, the experimental values of enthalpy of mixing for liquid alloys, ΔH^L, and the enthalpy of formation for solid solutions, ΔH^S, are taken from Ref. [36]. Such values of ΔH^L and ΔH^S for various M are plotted in Fig. 3-15 against (A) *Md* and (B) *Bo*. For each element M, ΔH^L is very similar to ΔH^S, although experimental temperatures are very different between them. For example, in case of M = Mn, ΔH^L is - 4.7 kJ/mol, which is measured at 1743K, and ΔH^S is −5.0 kJ/mol, which is measured at 1050K. So, as shown in Fig. 3-15, the ΔH^L data, indicated by the ● symbol, are overlapped with the ΔH^S data indicated by the ○ symbol for various elements, M (=Cu, Co, Fe, Mn, and Pd).

As shown in Fig. 3-15A, ΔH^L and ΔH^S change linearly with the *Md,* and ΔH^L is expressed approximately as, $\Delta H^L = 6.91 - 9.89Md$, with the correlation coefficient of 0.99. So, the iso-$\overline{Md}$

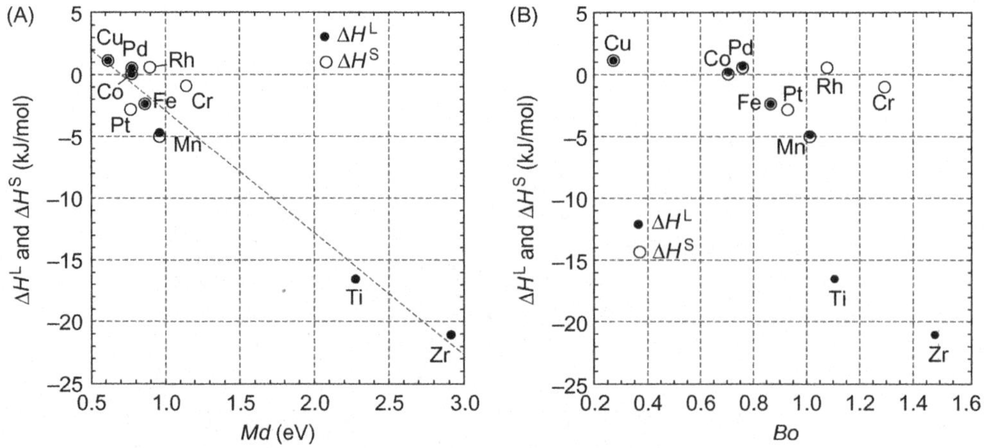

FIGURE 3-15 Correlation of (A) *Md* and (B) *Bo* with the enthalpy of mixing for liquid alloys, ΔH^L and the enthalpy of formation for solid solutions, ΔH^S of Ni-10mol%M alloys.

line shown in Fig. 3-8 may be a line to express a constant enthalpy of formation for solid solutions of the alloys, as ΔH^S is nearly equal to ΔH^L. On the other hand, as shown in Fig. 3-15B, ΔH^L and ΔH^S become more negative with increasing *Bo*, since the chemical bond between M and Ni atoms becomes stronger with the *Bo* as described before [37].

3.4 Target Region for Alloy Design

3.4.1 $\overline{Bo}$-$\overline{Md}$ Diagram

In Fig. 3-16, 19 conventionally cast superalloys are plotted on the $\overline{Bo}$-$\overline{Md}$ diagram [38]. The values of $\overline{Bo}$ and $\overline{Md}$ for each alloy are calculated from the alloy composition by using Eqs. (3.1) and (3.2). The contour lines of the 0.2% yield stress at 1255K [39] are also drawn by the solid curves in the figure. The 0.2% yield stress shows a maximum near the position of $\overline{Md} = 0.98$(eV) and $\overline{Bo} = 0.67$. Also, most single crystal Ni-based superalloys are located in the shadow region near the maximum position. The 100-hours (3.6 Ms) creep rupture strength measured at 1255K under the applied stress of 137.9 MPa also shows the maximum in the same region as the 0.2% yield stress [39]. Thus, a target region for alloy design is specified on the $\overline{Bo}$-$\overline{Md}$ diagram.

The value of $\overline{Md} = 0.98$(eV) in the target region is higher than the critical $\overline{Md}$ value (e.g., 0.925 (eV)) for the $\gamma/\gamma + \sigma$ phase as shown in Fig. 3-8A–B. This is simply due to the difference in the calculation method between them, because the present $\overline{Md}$ value is calculated from the total alloy composition, but not from the composition of the γ phase. However, in any alloy the γ' phase coexists with the γ phase, and the σ phase precipitates in the γ phase. The $\overline{Md}$ value for the $\gamma'(Ni_3Al)$ phase may be calculated to be about 1.013 (eV), since $0.75 \times 0.717 + 0.25 \times 1.90 \fallingdotseq 1.013$ by using *Md* values of 0.717 (eV) for Ni and

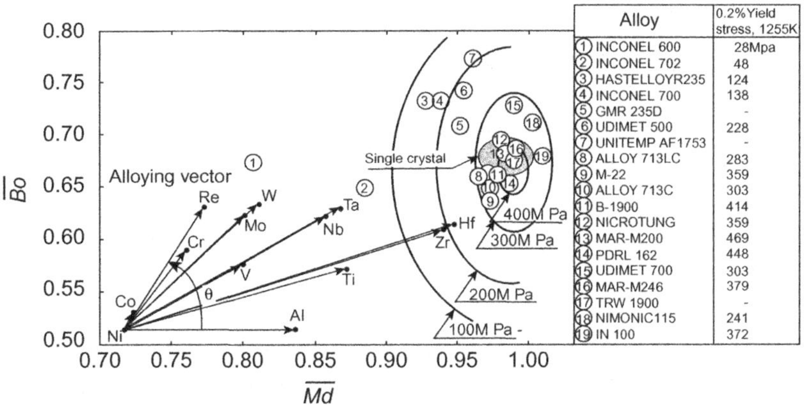

FIGURE 3-16 $\overline{Bo}-\overline{Md}$ diagram showing the locations of conventionally cast superalloys and the contour lines of 0.2% yield stress. Alloying vectors are also drawn, and the angle θ is counted from the Ni-Al direction.

1.90 (eV) for Al (see Table 3-1). If the volume fraction of the γ' phase is assumed to be about 60%, we may get about 0.98 (eV) for a whole alloy from the calculation, $0.60 \times 1.013 + 0.40 \times 0.925 \fallingdotseq 0.98$ (eV). In other words, any alloy in the target region seems to be composed of a large amount of alloying elements that are nearly close to the critical $\overline{Md}$ level for the σ phase formation in it, and also its volume fraction of the γ' phase is as high as 60% to increase the creep rupture strength of the alloy.

3.4.2 Alloying Vectors

The concept of alloying vectors is also important for alloy design. As shown in Fig. 3-16, an alloying vector starts at the position of pure Ni and ends at the position of Ni-10mol% M alloy on the $\overline{Bo}$-$\overline{Md}$ diagram, where M is an alloying element in the alloy. As the M composition increases, the alloy position moves away from the position of pure Ni along the alloying vector. The magnitude and direction of the alloying vectors vary depending on M. For example, the magnitude is small for Co because of the resemblance in the *Bo* and *Md* values between Ni and Co, but it is large for Re because of the large differences between Ni and Re. It is noted here that the direction is very similar among the elements in the same group of the periodic table, for example, among Ti, Zr, Hf (4A group elements), V, Nb, Ta (5A group elements), and Cr, Mo, W (6A group elements). This is simply due to the fact that both *Bo* and *Md* parameters change following the order of elements in the periodic table. Thus, alloying behavior reflects clearly the alloying vector and hence the $\overline{Bo}$-$\overline{Md}$ diagram.

For example, as shown in Fig. 3-17, the γ'solvus temperatures are plotted on the $\overline{Bo}$-$\overline{Md}$ diagram by using the experimental data of 35 alloys in the compositional range of Ni-11mol %Cr-14mol%Al-(1.0-4.0)mol%Ta-(1.3-3.5)mol%W-(0-7.5)mol%Co [39]. Here, the γ' solvus temperature means an upper limit temperature below which the γ' phase is stable, but above which it becomes unstable and disappears in the alloy. Following the experimental

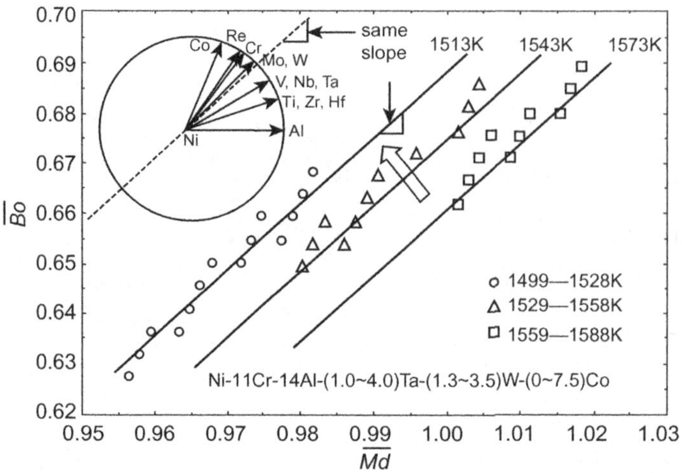

FIGURE 3-17 Representation of the γ′solvus temperatures of Ni-11Cr-14Al-(1.0–4.0)Ta-(1.3–3.5)W-(0–7.5)Co alloys on the $\overline{Bo}$-$\overline{Md}$ diagram.

data, three parallel iso-γ′ solvus temperature lines are drawn in Fig. 3-17. As the γ′ solvus temperatures decrease from 1573K to 1513K, the iso-γ′ solvus temperature lines shift toward the higher $\overline{Bo}$ and the lower $\overline{Md}$ region, as indicated by a white arrow in the figure. Also, in an inset circle, a dotted line of the same slope as the iso-γ′ solvus temperature line is drawn. The line lies in between Ni-Mo (or W) and Ni-V (or Nb, Ta) vectors. This means that the γ′ solvus temperature increases with the addition of V, Nb, Ta, Ti, Zr, Hf, and Al in the alloys. So, these elements except for Zr and Hf are the γ′phase stabilizing elements that occupy preferentially in the γ′phase. As discussed before, Zr and Hf are elements to be partitioned mainly into the grain boundaries. By contrast, the γ′ solvus temperature decreases with the addition of Co, Re, Cr, Mo, and W. So, these elements are not the γ′ phase stabilizing elements, but the γ phase stabilizing elements, and occupy preferentially in the γ phase. The alloying elements are classified using the *Md* value in Fig. 3-4, but also possible using the alloying vectors as shown in Fig. 3-17.

The concept of alloying vectors is useful for alloy modification. To keep the alloy position on the $\overline{Bo}$-$\overline{Md}$ diagram nearly unchanged with the modification, alloy compositions may be adjusted among those elements that have similar vector directions. For example, Re is an element to increase the strength and corrosion resistance of alloys. The Re addition may require to reduce the Cr content in the alloy, since their vectors point to a similar direction as shown in Fig. 3-16. Such a compositional change is actually seen in the modification from PWA 1400 (the first-generation superalloy) to PWA 1484 (the second-generation superalloy). Namely, the Cr content of PWA 1484 decreases by 5 mass%, whereas the Re content increases by 3 mass% as compared with the respective values of PWA 1400. The further decrease in Cr content and the attendant increase in the Re content are seen in René N6 (the third-generation superalloy). According to the evolutionary process of superalloys, the first-, second-, and third-generation superalloys are classified by the Re content, 0%, 3%, and 6% in mass%, respectively.

3.4.3 Relation Between γ' Volume Fraction and Alloying Vectors

The stability of the γ' phase in the alloy is associated with the direction of alloying vector. As shown in Fig. 3-16, the Ni-Al direction is set to be zero, and any vector direction is presented using the θ angle to be counted from the Ni-Al direction. Following the previous discussion, the γ' phase tends to become more stable with decreasing θ angle. So, the volume fraction of the γ' phase is expected to increase with decreasing θ angle of the alloy. As shown in an inset in Fig. 3-18, the θ angle of the alloy is defined as the angle between the Ni-Al direction and the alloy direction, which starts at the pure Ni position and ends at the alloy position on the $\overline{Bo}$-$\overline{Md}$ diagram. The θ angles are determined for various binary and quaternary alloys as well as the multiple-component superalloys, and the results are plotted against the measured γ' volume fractions in Fig. 3-18. Although the data are somewhat scattered in this plot, there is an approximately linear relationship between them in a wide volume fraction range of 20% to 75%. As might be expected from the lower θ angles for the alloying vectors of the γ'-partitioning elements (e.g., Al, Ti, Ta), the volume fraction increases with decreasing θ angle of the alloy. The volume fraction of the γ' phase in the superalloy is predictable using an equation, γ' vol% $= -2.3\ \theta + 132$ [40]. The θ angle is about 30° for the alloys in the target region shown in Fig. 3-16, so that the γ' vol% is estimated to be about 63% using the previously mentioned equation. In fact, the γ' volume fractions are as high as 60%–65% in advanced single crystal superalloys.

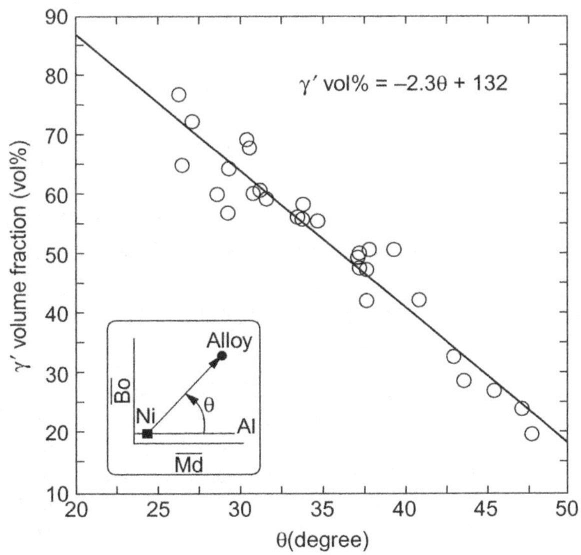

FIGURE 3-18 Correlation between the γ' volume fraction and the θ angle of the alloys.

3.5 Design of Single Crystal Superalloys

Ni-based superalloys are classified into the conventionally cast (CC) superalloys, directionally solidified (DS) superalloys, and single crystal (SC) superalloys. The creep strength increases in the order, CC < DS < SC. In the advanced industrial gas-turbine system in power plants, there is a great demand for new single crystal superalloys with an excellent combination of high-temperature creep strength, hot corrosion resistance, and oxidation resistance. As mentioned before, there is a clear trend of lowering Cr content and increasing Re content in the evolution of the first-, second-, and third-generation superalloys. However, the high-temperature oxidation resistance tends to decrease with increasing Re content. Therefore, it is required to improve the oxidation resistance of the third-generation superalloys in some ways. Standing on this background, we have developed a nickel-based single crystal superalloy containing 4.4 mass% Re following the present approach [41–43].

3.5.1 Designed Alloy

A new alloy, A1, named NKH-202, has been designed and the alloy composition is shown in Table 3-3 together with four reference alloys, KR, LR, MR, and IR. Here, KR and LR are typical commercially available second-generation superalloys containing 3 mass% Re. Each superalloy has been widely used in practical plants. MR is a third-generation single crystal superalloy developed in Japan and IR corresponds to the CC superalloy developed in the United States and currently used in practical gas turbines, so it is included here as a reference for the corrosion resistance and the oxidation resistance at high temperatures.

The alloy A1 contains 11% Co and 0.12% Hf, since the Co addition increases the creep rupture strength, and the Hf addition prevents the surface oxide layer (e.g., Al_2O_3) from peeling off from the alloy. Also, it does not contain Mo, because Mo decreases the oxidation resistance, as might be expected from the formation of volatile MoO_3 in pure Mo. Also, the Re content is not 6 mass%, but 4.4 mass %, because its excess addition may be detrimental to the oxidation resistance, even though the creep strength may decrease by lowering the Re content in the alloy. All the alloys listed in Table 3-3 lie in the range of $0.97 < \overline{Md} < 0.99$ and $0.66 < \overline{Bo} < 0.67$, so they are located in the target region shown in Fig. 3-16.

The designed alloy A1 is heat-treated following the conditions: solution treatment (1573K/4 hours + 1583K/6 hours + 1588K/12 hours), first step aging (1413K/4 hours, air

Table 3-3 Chemical Compositions (Mass%) of Designed Alloy and Reference Alloys and Their $\overline{Bo}$ and $\overline{Md}$ Values

	Ti	Cr	Co	Ni	Mo	Hf	Ta	W	Re	Al	$\overline{Md}$	$\overline{Bo}$
A1	1.40	6.00	11.00	Bal.	–	0.12	6.80	5.50	4.40	5.40	0.987	0.668
KR	1.00	6.50	9.00	Bal.	0.60	0.10	6.50	6.00	3.00	5.60	0.984	0.664
LR	0. 70	6.60	9.20	Bal.	0.50	1.40	3.20	8.50	3.00	5.70	0.983	0.664
MR	–	3.00	12.00	Bal.	2.00	0.10	6.00	6.00	5.00	5.70	0.971	0.666
I R	4.90	14.00	9.50	Bal.	1.50	–	2.80	3.80	–	3.00	0.971	0.666

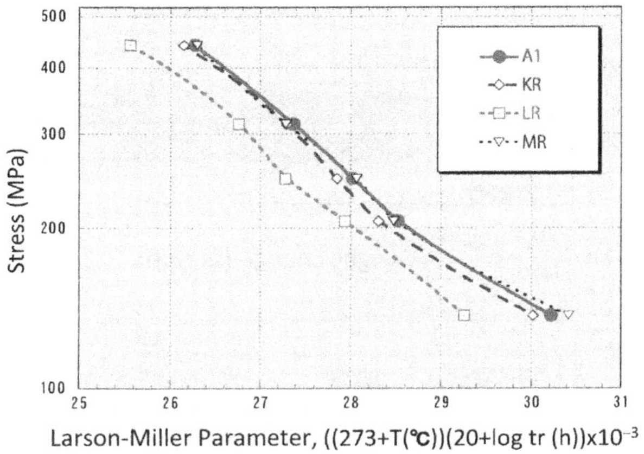

FIGURE 3-19 Creep rupture strength of A1 (NKH-202) and other Ni-based superalloys.

cool), and second step aging (1144K/20 hours, air cool). The size of the γ' phase is about 0.5 μm after heat treatments.

3.5.2 Creep Rupture Strength

The experimental results of the creep rupture strength are shown in Fig. 3-19. In the horizontal axis, the Larson–Miller parameter is used, in which T (°C) is the test temperature and tr (h) is the creep rupture life. As this parameter increases, the alloy has high creep strength. The curve for the designed alloy, A1 (NKH-202), is nearly overlapped with that for MR (the third SC superalloy), and it is stronger than KR (the second-generation superalloy). The temperature capability is estimated to be 20K higher than that of second-generation superalloy.

3.5.3 Hot Corrosion Resistance

The hot corrosion resistance is examined using the burner rig test at 1173K for 35 hours (126 ks), by spraying about 80 ppm NaCl solution with the light fuel oil containing 0.04 mass % S. The experimental results are shown in Fig. 3-20A–B. The weight change of A1 (NKH-202) is lowest among the alloys. Also, IR that is practically used in the 1300°C class gas turbines shows a negative weight change, indicating that the corrosion scaled layer is peeled off from the alloy surface.

3.5.4 Oxidation Resistance

The oxidation test is performed at 1313K and the results are shown in Fig. 3-21. A1 (NKH-202) exhibits very good oxidation resistance as compared with the other alloys. This A1(NKH-202) does not contain Mo, but IR and MR contain 1.5–2.0 mass% Mo. Thus, the design of Mo-free or nearly-free alloys is probably one way to achieve the high oxidation resistance. It is also noted that the weight change is small in KR and LR, both of which

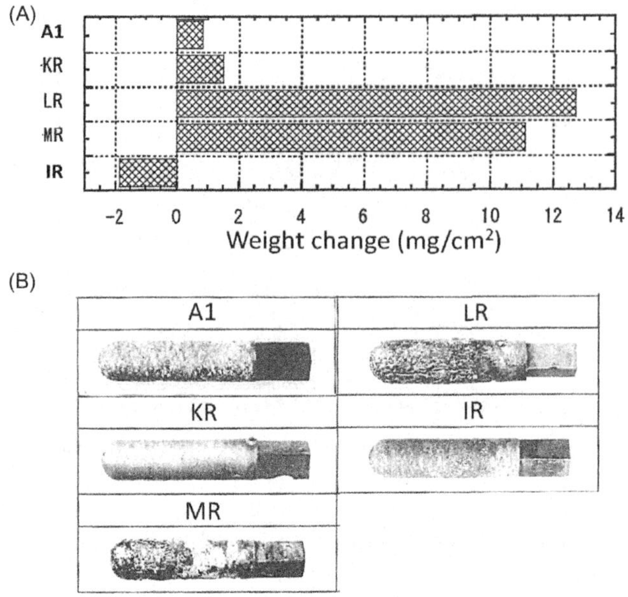

FIGURE 3-20 Results of the burner rig test at 1173K for 35 hours (126 ks) in atomized 80 ppm NaCl; (A) weight changes and (B) appearance of specimens after tests.

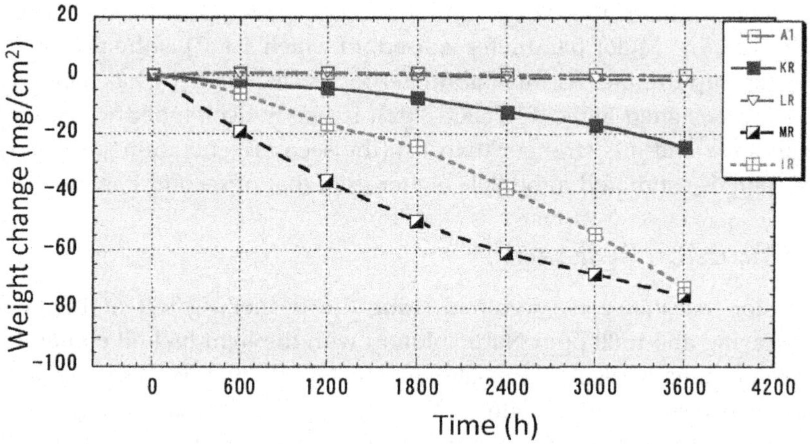

FIGURE 3-21 Oxidation resistance at 1313K for A1(NKH-202) and other Ni-based superalloys.

contain 3.0 mass% Re. The Re addition up to 4.4 mass% content level (i.e., NKH-202) is allowable without having any detrimental effect on the oxidation resistance of the alloy.

3.5.5 Single Crystal Growth

A model blade is made to prove a potential of growing single crystal. As shown in Fig. 3-22, a blade of about 170 mm in height is grown without any troubles. Thus, NKH-202 is a

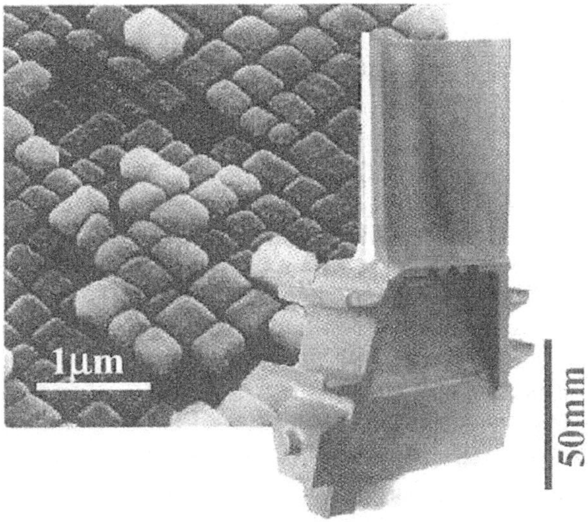

FIGURE 3-22 Ni-based single crystal superalloy for turbine blade in power plant. The cubes of the γ' phase are arranged three-dimensionally in the γ matrix.

promising single crystal superalloy with an excellent combination of creep rupture strength, hot corrosion resistance, and oxidation resistance.

Thus, a theory for alloy design based on the molecular orbital calculation is useful not only for the fundamental understanding of the phase stability of alloys, but also for the practical design of superalloys.

References

[1] W.J. Boesch, J.S. Slaney, Preventing sigma phase embrittlement in nickel base superalloys, Metal Progress 86 (1964) 109–111.

[2] C.S. Barrett, Some industrial alloying practice and its basis, J. Institute of Metals 100 (1972) 65–73.

[3] C.T. Sims, A contemporary view of cobalt-base alloys, J. Metals 21 (1969) 27–42.

[4] M. Morinaga, N. Yukawa, H. Adachi, H. Ezaki, New PHACOMP and its applications to alloy design, in: Superalloys 1984, M. Gell et al., (Eds.), Proc. of the 5th International Symposium on Superalloys, Seven Springs Mountain Resort, Warrendale, PA, The Metallurgical Society of AIME,1984, pp. 523–532.

[5] F. Schubert, in: V. Guttmann (Ed.), Phase Stability in High Temperature Alloys, Applied Science Publishers, London, 1981, pp. 119–149.

[6] R.C. Reed, The Superalloys : Fundamentals and Applications, Cambridge University Press, Cambridge, 2006.

[7] J.H. Westbrook, High Strength Materials, 1965, John-Wiley & Sons, New York, 1965, pp. 724–768.

[8] G.C. Fletcher, Electron bands and related magnetic properties of Ni_3Al, Physica 56 (1971) 173–184.

[9] G.C. Fletcher, Electron bands of Ni_3A1 and Ni_3Ga, Physica 63 (1972) 41–50.

[10] D. Hackenbracht, J. Kübler, Electronic magnetic and cohesive properties of some nickel-aluminium compounds, J. Phys. F 10 (1980) 427–440.

[11] H. Adachi, M. Tsukada, C. Satoko, Discrete variational Xα calculations. 1. Application to metal clusters, J. Phys. Soc. Jpn 45 (1978) 875–883.

[12] M. Morinaga, N. Yukawa, H. Adachi, Alloying effect on the electronic structure of Ni_3Al (γ'), J. Phys. Soc. Jpn. 53 (1984) 653–663.

[13] R.W. Guard, J.H. Westbrook, Alloying behavior of Ni_3Al (gamma-phase), Trans. Met. Soc. AIME 215 (1959) 807–814.

[14] C. Satoko, M. Tsukada, H. Adachi, Discrete variational Xα cluster calculations. II. Application to the surface electronic structure of MgO, J. Phys. Soc. Jpn 45 (1978) 1333–1340.

[15] R.S. Mulliken, Quelques Aspects de La Theorie des Orbitales Moleculaires, Journal de chimie physique et de physico-chimie biologique 46 (1949) 497–542.

[16] R.E. Watson, L.H. Bennett, Transition metals: d-band hybridization, electronegativities and structural stability of intermetallic compounds, Phys. Rev. B 18 (1978) 6439–6449.

[17] R.F. Decker, Alloy Design, Using Second Phases, Metallurgical Transactions A 4 (1973) 2495–2518.

[18] R.F. Decker, J.R. Mihalisi, Coherency strains in γ' hardened nickel alloys, ASM Transactions quarterly 62 (1969) 481.

[19] E.T. Teatum, K.A. Gshneidner Jr., J.T. Waber, LA-2345, U.S. Department of Commerce, Washington D. C., 1968.

[20] R.S. Mulliken, Electronic population analysis on LCAO–MO molecular wave functions. I, II, III, IV, J. Chem. Phys 23 (1955) 1833–1840, 1841–1846, 2338–2342, 2343–2346.

[21] W. Hume-Rothery, G.V. Raynor, Structure of Metals and Alloys, Institute of Metals, London, 1954.

[22] L. Darken, R.W. Gurry, Physical Chemistry of Metals, Institute of Metals, London, 1954.

[23] M. Morinaga, N. Yukawa, H. Ezaki, H. Adachi, Solid solubilities in transition-metal-based F.C.C. alloys, Phil. Mag. A 51 (1985) 223–246.

[24] W.B. Pearson, The Crystal Chemistry and Physics of Metals and Alloys, Wiley-Interscience, New York, 1972.

[25] M. Morinaga, N. Yukawa, H. Ezaki, H. Adachi, Solid solubilities in nickel-based F.C.C. alloys, Phil. Mag. A 51 (1985) 247–252.

[26] H. Ezaki, M. Morinaga, N. Yukawa, H. Adachi, Solid solubilities of Molybdenum and Tunguten in F.C.C. Transition metal alloys, Phil. Mag. A 65 (1992) 1249–1260.

[27] O.H. Kriege, J.M. Baris, Chemical partitioning of elements in gamma prime separated from precipitation-hardened, high-temperature nickel-base alloys, ASM Transactions Quarterly 62 (1969) 195.

[28] J.R. Mihalisin, D.L. Pasquine, Phase transformation in nickel-base superalloys, Proc. of Inter. Symposium on Structural Stability in Superalloys Vol.1 (1968) pp. 134–171. Seven Springs, Pa.

[29] R.G. Barrows, J.B. Newkirk, A modified system for predicting σ formation, Metall. Trans. A 3 (1972) 2889–2893.

[30] S. Ohta, M. Saori, T. Yoshida, Special Report on Superalloys, Tokyo, Nippon Gakujutsu Shinkokai, Vol. 16, 1975, p. 197.

[31] H. Ezaki, M. Morinaga, N. Yukawa, H. Adachi, Prediction of the occurrence of the σ phase in Fe-Cr-Ni alloys, Phil. Mag. A 53 (1986) 709–716.

[32] E.I. Gradyshevskii, L. Borusevic, Z. neorg. Khim 8 (1963) 1915.

[33] M.J. Cieslak, G.A. Knorovsky, T.J. Headley, A.D. Romig Jr., The use of new PHACOMP in understanding the solidification microstructure of nickel base alloy weld metal, Metall. Trans. A 17 (1986) 2107–2115.

[34] P. Caron, T. Khan, Design of superalloys for single crystal blade applications: A 20-year experience, in: J.-C. Zhao, M. Fahrmann, T.M. Pollock (Eds.), Materials Design Approaches and Experiences, TMS, Warrendale, PA, 2001, pp. 1–14.

[35] N. Yukawa, M. Morinaga, K. Nishiyama, Y. Matsumoto, Y. Murata, H. Ezaki, Design of low activation austenitic steels by the *d*-electron concept, Standard Technical Publication 1047, American Society for Testing and Materials, Philadelphia, PA, 1990, pp. 30–46.

[36] B. Predel, Phase Equilibria, Crystallographic and Thermodynamic Data of Binary Alloys, Landolt-Bömstein, Numerical Data and Functional Relationships in Science and Technoloy, New Series/Editors in Chief: O. Madelung, W. Martienssen, Group IV: Macroscopic Properties of Matter, Volume 5, Springer-Verlag, Berlin, 1991.

[37] M. Morinaga, Alloy design based on molecular orbital method, Materials Transactions 57 (2016) 213–226.

[38] K. Matsugi, Y. Murata, M. Morinaga, N. Yukawa, Realistic advancement for nickel-based single crystal superalloys by the *d*-electrons concept, in Superalloys 1992, Eds. S. D. Antolovich et al, Proc. of the 7th International Symposium on Superalloys, Seven Springs Mountain Resort, Warrendale, PA, The Minerals, Metals and Materials Society 1992, pp. 307–316.

[39] Y. Murata, S. Miyazaki, M. Morinaga, Evaluation of the partitioning ratios of alloying elements in nickel-based superalloys by the d-electrons parameters, in Materials for Advanced Power Engineering 1994, Liege, Belgium, October 3–6, 1994.

[40] Y. Murata, S. Miyazaki, M. Morinaga, R. Hashizume, Realistic hot corrosion resistant and high strength nickel-based single crystal and directionally-solidified superalloys developed by the d-electrons concept, in Superalloys 1996, Eds. R. D. Kisshinger et al, Proc. of the 8th International Symposium on Superalloys, Seven Springs Mountain Resort, Warrendale, PA, The Minerals, Metals and Materials Society1996, pp.61–70.

[41] R. Hashizume, A. Yoshinari, T. Kiyono, Y. Murata, M. Morinaga, Development of novel Ni-based single crystal superalloys for power-generation gas turbines, Materials at High Temperatures 24 (2007) 163–172.

[42] Y. Murata, M. Morinaga, R. Hashizume, Oxidation behaviour of nickel-based alloys containing high Re contents comparable to the third generation nickel-based superalloys, Materials Science Forum 449-4 (2004) 349–352.

[43] Y. Murata, M. Moniruzzaman, M. Morinaga, Double bladed effect of Re on high-temperature oxidation and hot-corrosion of nickel-based superalloys, Materials Science Forum 426-4 (2003) 4561–4566.

4 Iron Alloys

4.1 Molecular Orbital Calculation of bcc Fe

The electronic structure of pure iron has been calculated by many authors [1–4], and there is general agreement among various calculations. Several experiments support the results of the band calculations [5]. On the other hand, information is very limited of the electronic structures of iron alloys. So, a series of calculations is carried out using the alloy cluster model and new alloying parameters relevant to bcc Fe are determined from the DV-Xα molecular orbital calculation [6].

4.1.1 Energy Level Structure and *Md* Parameter

In bcc Fe, each Fe atom is surrounded by eight first nearest neighbor Fe atoms, $Fe^{(1)}$, and by six second nearest neighbor Fe atoms, $Fe^{(2)}$. So, the cluster model used is MFe_{14}, as shown in Fig. 4-1 [6]. The alloying element, M, is substituted for the Fe atom at a center of the cluster model. A series of calculations is carried out with nonmagnetic Fe, since we have neither attempted to account for the ferromagnetism of iron, nor tried to predict the crystal structure of the lowest energy. The lattice parameter used is 0.2866 nm, the same as in the bulk. A volume change due to the magnetic interaction [7] does not substantially alter the results on the alloying behavior of various elements in bcc Fe.

The results of the electron density of states for pure bcc Fe is shown in Fig. 4-2A, B. The calculated result of the molecular orbital calculation with the Fe_{15} cluster (M = Fe) is shown in Fig. 4-2B. It agrees reasonably well with the result of the band calculation shown in Fig. 4-2A, despite the use of a relatively small cluster model. For example, the Fermi energy level (E_f) lies at a peak of the electron density of states in the 3*d* band in both cases.

The energy level structures of pure and alloyed Fe with 3*d* transition metals are shown in Fig. 4-3 [6]. The E_f for pure Fe (i.e., M = Fe) is set to be zero and used as a reference. The position of the E_f level is indicated by arrows ($\leftarrow$) in the figure. In pure Fe, the levels of the $8e_g$ to the $16t_{2g}$ are mainly originated from the Fe 3*d* orbitals, and form the 3*d* band where the E_f level lies. The higher levels, $17a_{1g}$ and $9t_{2u}$, are the mixed states of the 3*d*, 4*s*, and 4*p* components. The 4*s* and 4*p* components extend to the wide energy range, so that these electrons behave as if they are nearly free electrons in the crystal.

In the alloyed Fe with the 3*d* transition metals, new energy levels derived mainly from the *d* orbitals of the alloying metal, appear above the Fermi level, E_f. For example, the $13e_g$ and $16t_{2g}$ levels drawn by broken lines correspond to these new levels. The height of these levels changes monotonously with the order of elements in the periodic table. The appearance of these metal *d* levels in the energy level structure is very similar to the result of Ni_3Al [8], as

A Quantum Approach to Alloy Design. DOI: https://doi.org/10.1016/B978-0-12-814706-1.00004-2

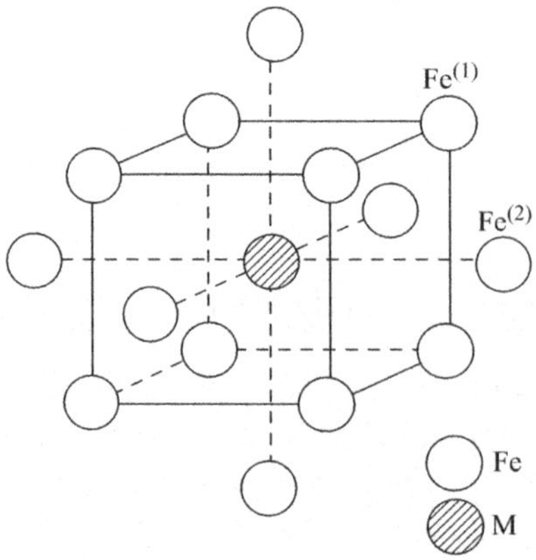

FIGURE 4-1 Cluster model, $[MFe_{14}]$, used for the calculation of bcc Fe.

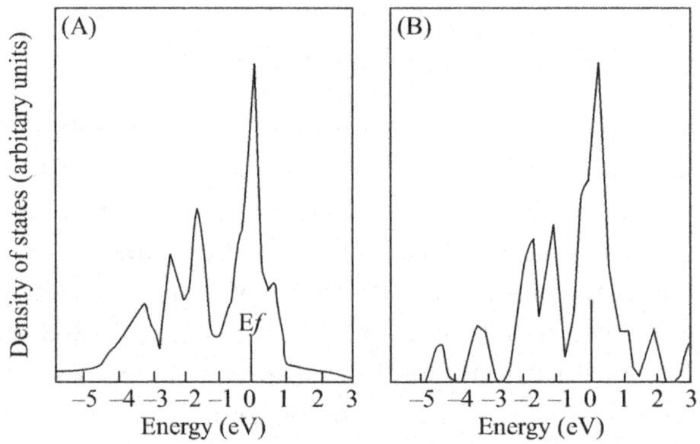

FIGURE 4-2 Electron density of states obtained from (A) the band calculation and (B) the molecular orbital calculation.

explained in Chapter 3. Such new levels also appear in the alloyed Fe with the 4*d* and 5*d* transition metals. These levels are related to the electronegativity and atomic radius of M [6].

The *Md* level is defined simply by taking the average of these new e_g and t_{2g} levels of M. The values of *Md* levels are listed in Table 4-1 for a variety of alloying elements, M, in bcc Fe. In this table the *Md* values for C, N, and Si are also included, because all these elements are important in Fe alloys. The *Md* values are empirically determined for these nontransition elements of C, N, and Si, to count them properly in the framework of transition elements [9].

4.1.2 Ionicity and *Bo* Parameter

Charge transfer takes place between atoms in the alloy cluster. The ionicity of each atom is evaluated from the Mulliken population analysis [10] and the results are shown in Fig. 4-4A. The ionicity of the second nearest neighbor Fe atom, $Fe^{(2)}$, is about +0.05 and shows little change with M. On the other hand, the ionicity of the first nearest neighbor Fe atom, $Fe^{(1)}$,

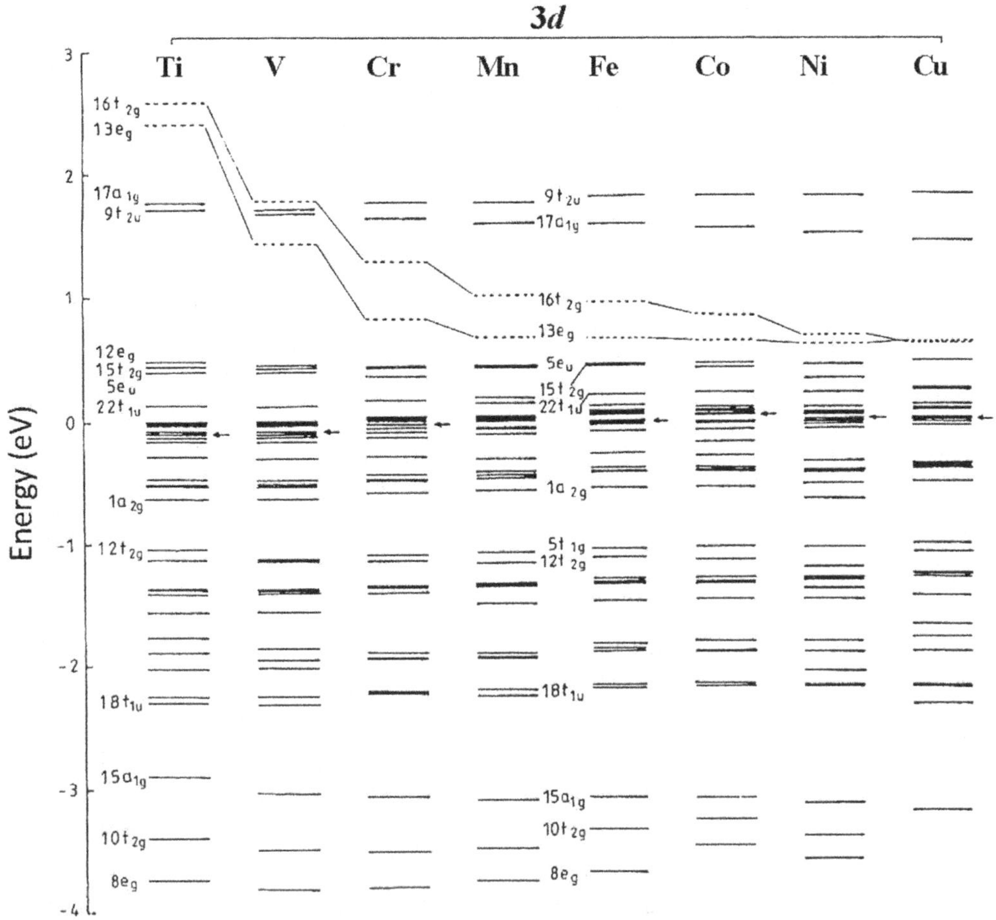

FIGURE 4-3 Energy level structures of pure and alloyed Fe with 3*d* transition metals.

varies largely depending on M. Namely, the ionicity of M gradually changes following the electronegativity of elements. For example, as Ti is the most electropositive element in the 3*d* series, it has a positive ionicity of +0.15. Also, as Ni is the most electronegative element in the 3*d* series, it has a negative ionicity of −0.23. It is evident from Fig. 4-4A that there is a strong correlation in the ionic values between $Fe^{(1)}$ and M. The weak ionic interaction is operating between them.

The bond order is a measure to show the covalent bond strength between atoms. The covalent *d-d* bond between atoms contributes largely to the cohesive energy of transition metals. Therefore, the bond order associated with the *d-d* electrons is calculated as shown in Fig. 4-4B. In the figure, "Total" means the sum of the three components, $Fe^{(1)}$-$Fe^{(2)}$, M-$Fe^{(1)}$, and M-$Fe^{(2)}$ bond orders. It is apparent from this figure that the bond order depends largely on M. In the 3*d* series, the element M, located to the left side of Fe, has a higher total bond order. Total bond order is larger in 4*d* and 5*d* elements than in 3*d* elements.

Table 4-1 *Md* and *Bo* Values for Alloying Elements in bcc Fe

Element		*Md* (eV)	*Bo*
3d	Ti	2.497	2.325
	V	1.610	2.268
	Cr	1.059	2.231
	Mn	0.854	1.902
	Fe	0.825	1.761
	Co	0.755	1.668
	Ni	0.661	1.551
	Cu	0.637	1.361
4d	Zr	3.074	2.551
	Nb	2.335	2.523
	Mo	1.663	2.451
5d	Hf	3.159	2.577
	Ta	2.486	2.570
	W	1.836	2.512
	Re	1.294	2.094
Others	C	−0.230	0
	N	−0.400	0
	Si	1.034	0

It is noticed that the second nearest neighbor bond order, M-Fe$^{(2)}$, is smaller than the first nearest neighbor bond order, M-Fe$^{(1)}$, but it is definitely present. This is interpreted as due simply to the short second nearest neighbor distance in bcc metal, because it is only about 15% longer than the first nearest neighbor distance. As compared with the bcc metal, this distance is about 41% longer in fcc metal, and hence a second nearest neighbor interaction is less significant in it [11]. In fact, according to the calculation of fcc Fe using a Fe_{19} cluster model as explained later, the second nearest neighbor bond order is less than 4% of the first nearest neighbor bond order.

The ratio of the bond orders, M-Fe$^{(2)}$/M-Fe$^{(1)}$, is shown in Fig. 4-4C. In the 3*d* series, a large peak appears at Fe, indicating that the second nearest neighbor atomic interactions are most predominant in pure bcc Fe. With the addition of M to bcc Fe the bond order ratio decreases rather drastically. For Ni and Cu, the second nearest neighbor atomic interactions are nearly zero. A large ratio is also seen at Os that is the 5*d* element located below Fe in the periodic table.

The directional nature of the *d* (e_g and t_{2g}) orbitals is illustrated in Fig. 4-5A [6]. It is expected from this figure that a larger portion of M-Fe$^{(1)}$ bonds is attributable to the t_{2g} like orbitals, whereas the M-Fe$^{(2)}$ bonds arise mainly from the e_g like orbitals. The calculated result of the components in the M-Fe$^{(1)}$ and M-Fe$^{(2)}$ bond orders is shown in Fig. 4-5B. In accordance with this expectation, the t_{2g} components are dominant in the M-Fe$^{(1)}$ bond order, and the e_g components are dominant in the M-Fe$^{(2)}$ bond order.

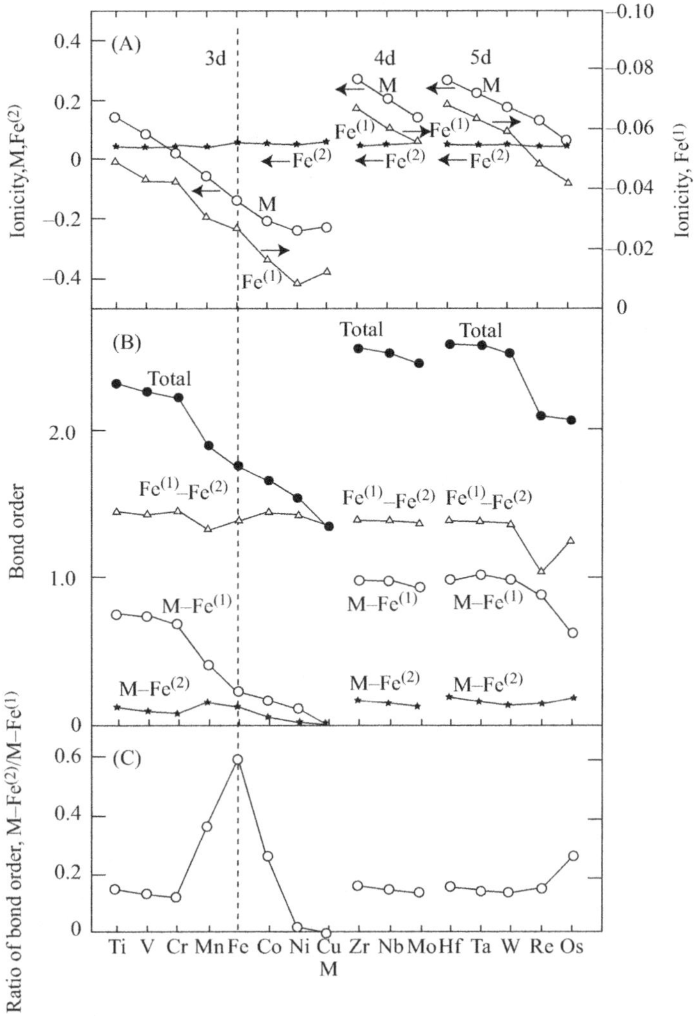

FIGURE 4-4 Changes of (A) ionicity, (B) bond order, and (C) ratio of bond order, M-Fe$^{(2)}$/M-Fe$^{(1)}$ with M.

The total bond order shown in Fig. 4-4B is defined as the *Bo* value for M. The *Bo* values are listed in Table 4-1, together with the *Md* values for various M. The *Bo* values are large for the elements, Ti, V, Cr (3*d*), Zr, Nb, Mo (4*d*) and Hf, Ta, W (5*d*), all of which belong to the 4A, 5A, and 6A groups in the periodic table. The commercially available Fe-based alloys often contain large bond order elements such as Cr, Mo, and W as the principal alloying elements. This trend that principal alloying elements have large bond orders is also found in Ni-based and Ti-based alloys. It is also interesting to note that the creep strength increases with the addition of M into bcc Fe in the order, Mo > Cr > Mn > Co > Ni [12]. This order is consistent with the order of the *Bo* value listed in Table 4-1.

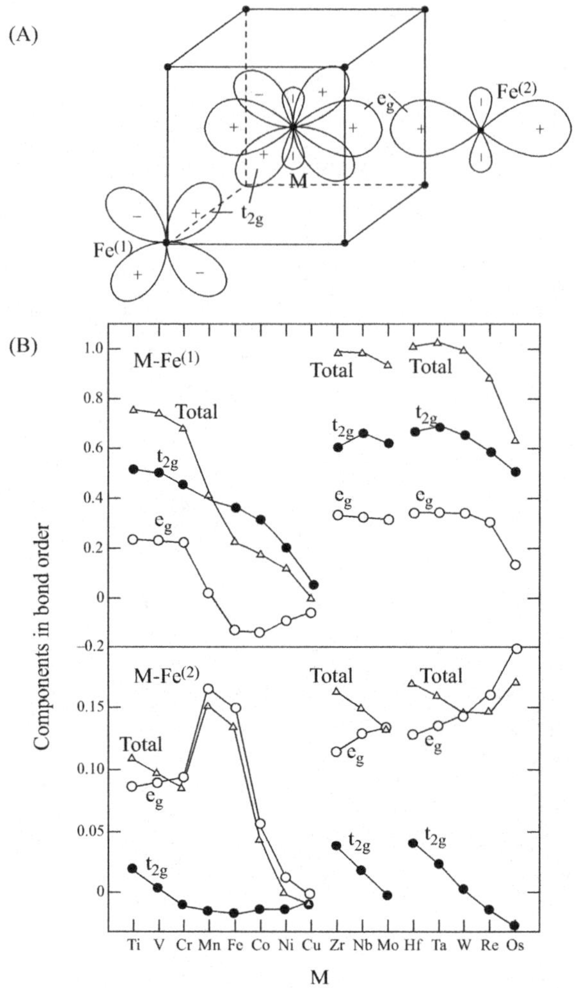

FIGURE 4-5 (A) Schematic illustration of e_g and t_{2g} orbitals in bcc Fe and (B) components in the bond order of M-Fe$^{(1)}$ and M-Fe$^{(2)}$.

4.1.3 Second Nearest Neighbor Atomic Interactions in bcc Metals

For comparison, the electronic structures of bcc Cr, V, Ti, Zr, Nb, and Mo are calculated to understand a general trend of the second nearest neighbor atomic interactions in bcc metals. For the calculation, Cr_{15}, V_{15}, Ti_{15}, Zr_{15}, Nb_{15}, and Mo_{15} cluster models are employed, which are the same as the Fe_{15} cluster model shown in Fig. 4-1. In Table 4-2 the calculated bond orders of these metals are given together with the result of bcc Fe. The ratio of the second nearest neighbor bond order to the first nearest neighbor bond order is very large for Fe, but small for Mo, Cr, Nb, and V. The ratios for Ti and Zr are intermediate between them.

Table 4-2 Comparison of the First and Second Nearest Neighbor Bond Orders Among bcc Metals

	Bond Order (First Neighbors)	Bond Order (Second Neighbors)	Ratio of Bond Order (Second Neighbors/First Neighbors)
[Fe-Fe] in Fe_{15}	0.230	0.136	0.591
[Cr-Cr] in Cr_{15}	0.851	0.072	0.085
[V-V]in V_{15}	0.922	0.134	0.145
[Ti-Ti] in Fe_{15}	0.556	0.160	0.287
[Zr-Zr] in Zr_{15}	0.770	0.197	0.255
[Nb-Nb] in Nb_{15}	1.254	0.138	0.110
[Mo-Mo] in Mo_{15}	1.303	0.098	0.075

It is well known that bcc Fe (α-Fe) transforms to fcc Fe (γ-Fe), bcc Ti (β-Ti) transforms to hcp Ti (α-Ti), and bcc Zr (β-Zr) transforms to hcp Zr (α-Zr), but such an allotropic transformation never takes place in bcc Mo, Cr, Nb, and V. Thus, the second nearest neighbor atomic interactions are important in all bcc metals, but they are significantly large in those bcc metals that transform to the fcc or hcp phase. In a geometrical point of view, an atom in the close-packed phase is surrounded by 12 first nearest neighbor atoms. In the bcc phase, an atom is surrounded by 14 atoms if the second nearest neighbor atoms are included as shown in Fig. 4-1. Therefore, the present calculation implies that attractive atomic interactions are operating over the first and second nearest neighbors even in the bcc phase of those metals that transform to the fcc or hcp phase [8]. In particular, this trend is remarkable in α-Fe.

4.1.4 Bond Order Change with Alloying Elements in Various bcc Metals

The bond order, *Bo*, is calculated for various bcc metals (Ti, Fe, V, Zr, Mo, and Nb) alloyed with transition elements. The results are shown in Fig. 4-6. For convenience, their *Bo* values are also listed in Table 4-3 [13]. Every curve shown in Fig. 4-6 varies with M in a very similar manner. A general trend of the bond order change with alloying elements appears almost independent of the mother metals. The *Bo* values are, however, larger for Nb, Mo, and V metals than Zr, Ti, and Fe metals. It is interesting to note that the former three metals with the larger bond orders are stable in the bcc state for a whole temperature range, whereas the latter three metals with the smaller bond orders undergo an allotropic transformation from bcc to fcc or hcp phase at a certain temperature. Thus, the bond strength between atoms reflects the structural stability of metals.

4.2 Molecular Orbital Calculation of fcc Fe

4.2.1 *Bo* and *Md* Parameters

In fcc Fe, each Fe atom is surrounded by twelve first nearest neighbor Fe atoms, and by six second nearest neighbor Fe atoms. So, the cluster model used is MFe_{18} as shown in Fig. 4-7.

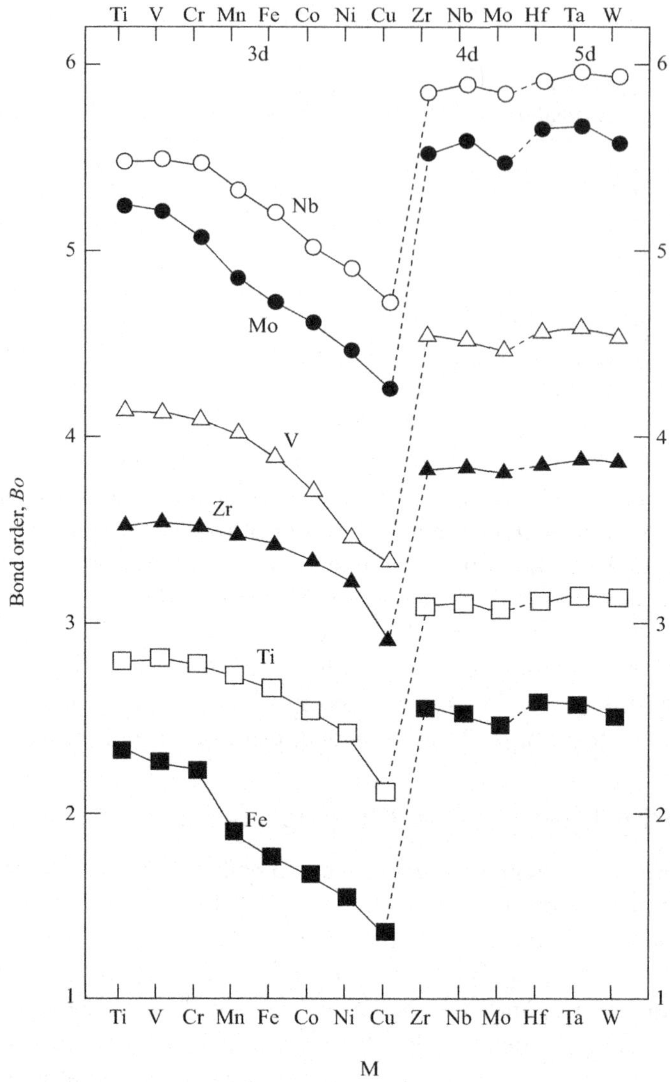

FIGURE 4-6 Bond order change with alloying elements in various bcc metals.

The alloying element, M, is substituted for the Fe atom at a center of the cluster model. Using this cluster model the electronic structures are calculated and both the *Bo* and *Md* parameters are determined in a similar way as explained in Ni_3Al and bcc Fe. Their values are listed in Table 4-4. Also, the *Bo* values are plotted against various elements, M, in Fig. 4-8. Total bond order is the sum of the Fe-Fe and M-Fe bond orders. For comparison, the results of the bcc Fe are also shown using solid circles. There is a certain resemblance in the bond order change with M between fcc Fe and bcc Fe.

Table 4-3 Bond Order, *Bo*, of Alloying Elements, M, in Various bcc Metals, Ti, Fe, V, Zr, Mo, and Nb

		Ti	Fe	V	Zr	Mo	Nb
M	Ti	2.790	2.325	4.142	3.523	5.238	5.470
	V	2.805	2.268	4.135	3.538	5.212	5.481
	Cr	2.779	2.231	4.095	3.515	5.068	5.463
	Mn	2.723	1.902	4.019	3.468	4.849	5.319
	Fe	2.651	1.761	3.886	3.417	4.716	5.198
	Co	2.529	1.668	3.708	3.328	4.614	5.012
	Ni	2.412	1.551	3.459	3.223	4.459	4.901
	Cu	2.114	1.361	3.332	2.911	4.248	4.709
	Zr	3.086	2.551	4.538	3.817	5.511	5.828
	Nb	3.099	2.523	4.506	3.829	5.578	5.867
	Mo	3.063	2.451	4.449	3.797	5.453	5.825
	Hf	3.110	2.577	4.557	3.839	5.630	5.890
	Ta	3.144	2.570	4.567	3.877	5.642	5.930
	W	3.125	2.512	4.532	3.864	5.554	5.905

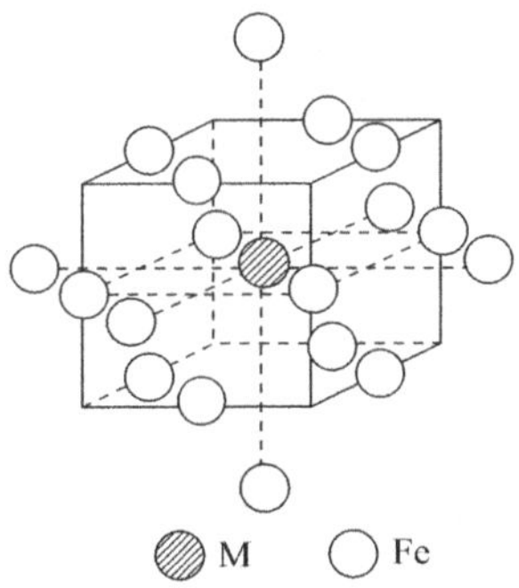

FIGURE 4-7 Cluster model, $[MFe_{18}]$, used for the calculation of fcc Fe.

4.2.2 Comparison of *Md* Parameter Between fcc Fe and bcc Fe

In Fig. 4-9A the *Md* values are compared between fcc Fe and bcc Fe. There is almost a linear relationship between them. According to the linear regression analysis, it is expressed as $y = 0.862\,x - 0.025$, where y is the *Md* value for fcc Fe and x is the *Md* value for bcc Fe, and the correlation coefficient, R, is as high as 0.994. Thus, each *Md* value is different between fcc Fe and bcc Fe, but the variation with M quite resembles between them. It may be said that the alloying behavior of each element in the γ (fcc)-Fe and the α (bcc)-Fe appears to be nearly the same and almost independent of crystal structures of Fe, as far as the *Md* parameter is concerned.

Table 4-4 *Md* and *Bo* Values for Alloying Elements in fcc Fe

Element		*Md* (eV)	*Bo*
3*d*	Ti	2.181	2.242
	V	1.491	2.270
	Cr	0.916	2.166
	Mn	0.623	2.072
	Fe	0.558	1.952
	Co	0.541	1.749
	Ni	0.513	1.374
	Cu	0.555	1.214
4*d*	Zr	2.571	2.467
	Nb	2.060	2.518
	Mo	1.496	2.452
5*d*	Hf	2.552	2.555
	Ta	2.104	2.571
	W	1.622	2.527
	Re	1.173	2.410
	Os	0.813	2.190
Others	C	− 0.985	0
	N	− 1.273	0
	Si	0.674	0

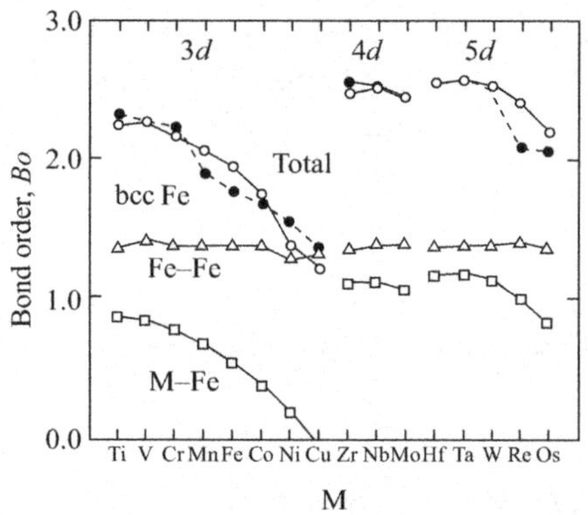

FIGURE 4-8 Bond order, *Bo*, for various M in fcc Fe. For comparison, the results of bcc Fe is also shown by solid circles.

4.2.3 Comparison of *Md* Parameter Between fcc Fe and Ni_3Al

As explained in Chapter 3, new PHACOMP is applied not only to the Ni-based alloys, but also to the Co-based and Fe-based alloys, by using the same *Md* parameter obtained from the calculation of alloyed Ni_3Al. The results shown in Chapter 3 are reasonable and satisfactory, but this approach is still the first approximation to the Co-based and Fe-based alloys. It is more reasonable to use new alloying parameters relevant to the respective alloy. So, it will

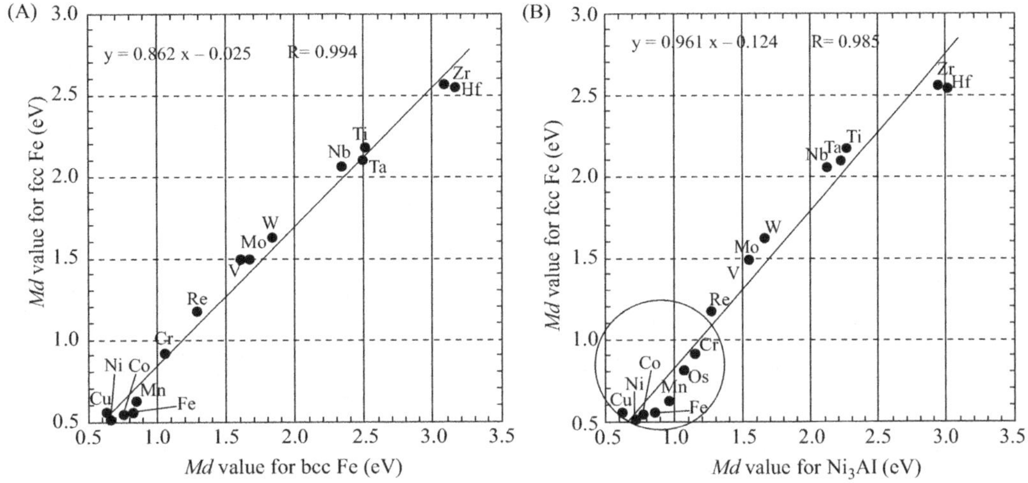

FIGURE 4-9 Comparison of the *Md* values (A) between fcc Fe and bcc Fe, and (B) between fcc Fe and Ni_3Al.

be discussed here how the previous analyses of the phase diagrams do change when the *Md* parameter for fcc Fe is used instead of that for Ni_3Al.

First, the *Md* values for fcc Fe are compared with those for Ni_3Al as shown in Fig. 4-9B. There is an approximately linear relationship in the *Md* values between fcc Fe and Ni_3Al. Their correlation is expressed as $y = 0.961\ x - 0.124$, where y is the *Md* value for fcc Fe and x is the *Md* value for Ni_3Al, and the correlation coefficient, R, is as high as 0.985. As a whole, this correlation may allow us to co-use the *Md* value in both the Ni-based and Fe-based alloys. However, there is still a clear deviation from a straight line around M = Fe and Mn as shown in a circle inset in the figure. So, the further examination is continued by tracing the $\gamma/\gamma + \sigma$ phase boundary using the iso-$\overline{Md}$ line.

The $\gamma/\gamma + \sigma$ phase boundary of Fe-Ni-Cr alloy at 1073K was previously shown in Fig. 3-8C. For comparison, this $\gamma/\gamma + \sigma$ phase boundary is traced here using the $\overline{Md}$ (Fe) values obtained from the calculation of fcc Fe. The result is shown in Fig. 4-10A together with the previous result using the $\overline{Md}$ (Ni_3Al) value. As the σ phase exists in the Fe-rich side, the iso-$\overline{Md}$ (Fe) line of $\overline{Md} = 0.635$(eV) appears to trace the $\gamma/\gamma + \sigma$ phase boundary better than the iso-$\overline{Md}$ (Ni_3Al) line of $\overline{Md} = 0.900$(eV). However, if the $\gamma/\gamma + \alpha$ phase boundary in the Ni-rich side is included, the iso-$\overline{Md}$ (Ni_3Al) line may be better. In Fig. 4-10B, another phase diagram of Fe-Cr-Mn alloy at 923K [14] is also shown. As Ni is not contained in this alloy system, the $\gamma/\gamma + \sigma$ phase boundary is well traced using the iso-$\overline{Md}$ (Fe) line of $\overline{Md} = 0.595$(eV) rather than the iso-$\overline{Md}$ (Ni_3Al) line of $\overline{Md} = 0.890$(eV).

On the other hand, no calculations have been done in fcc Co. As Co is an element located intermediately between Fe and Ni in the periodic table, two ternary phase diagrams are reexamined using both $\overline{Md}$ (Fe) and $\overline{Md}$ (Ni_3Al) values. They are Ni-Co-Cr at 1477K previously shown in Fig. 3-8A and Co-Ni-Mo at 1473K previously shown in Fig. 3-8D. The results are

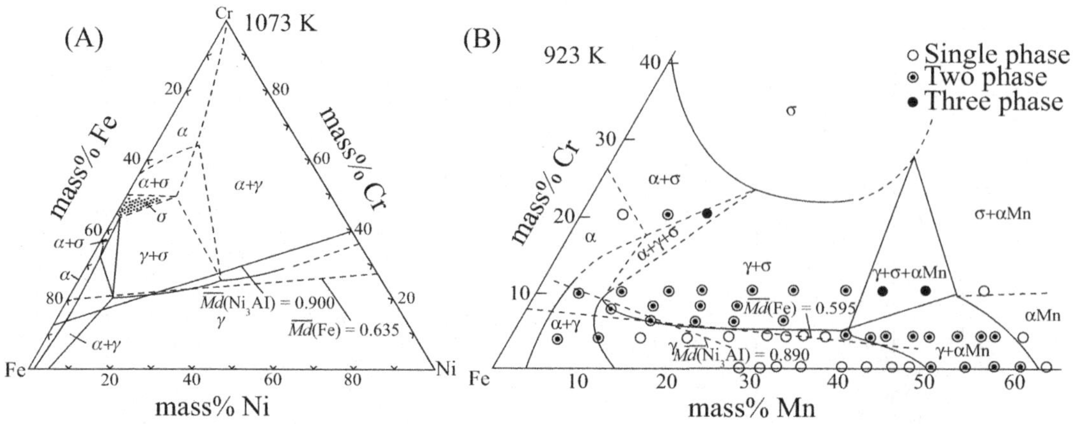

FIGURE 4-10 Ternary phase diagram of (A) Fe-Ni-Cr at 1073K and (B) Fe-Cr-Mn at 923K.

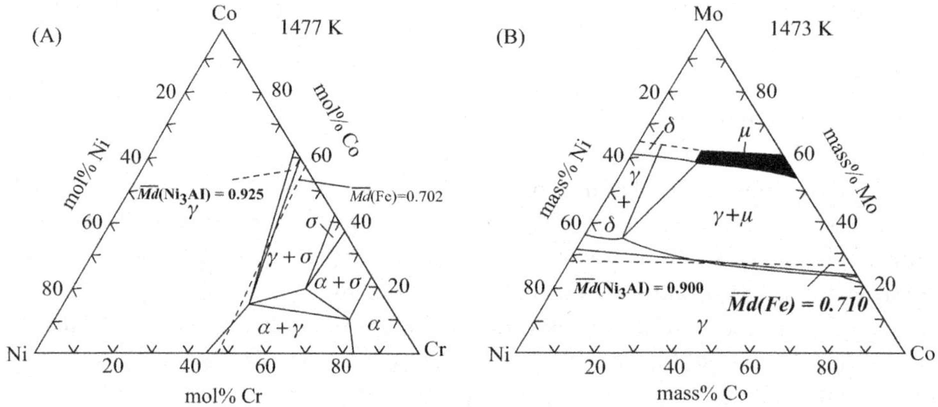

FIGURE 4-11 Ternary phase diagrams of (A) Ni-Co-Cr at 1477K and (B) Co-Ni-Mo at 1473K.

shown in Fig. 4-11A for Ni-Co-Cr and Fig. 4-11B for Co-Ni-Mo. Both the $\gamma/\gamma+\sigma$ and the $\gamma/\gamma+\mu$ phase boundaries are traceable using the iso-$\overline{Md}$ (Ni_3Al) line and the iso-$\overline{Md}$ (Fe) line. But, in either alloy, the iso-$\overline{Md}$ (Ni_3Al) line seems slightly closer to the phase boundaries than the iso-$\overline{Md}$ (Fe) line.

Thus, the *Md* parameter for Ni_3Al is probably valid for the Co-based alloys, but the use of new *Md* parameter listed in Table 4-4 is more appropriate for the fcc Fe-based alloys, in particular, for the Ni-free alloys.

4.3 Electronic State of Interstitial Atoms (C, N, O) in fcc Fe

4.3.1 Cluster Model

The interstitial atoms such as C, N, and O play important roles in iron alloys. So, local electronic structures around the interstitial atoms C, N, and O in fcc Fe are calculated using a cluster model, $Fe_{14}X$ (X = C, N, O), shown in Fig. 4-12 [15]. The C, N, and O atoms occupy

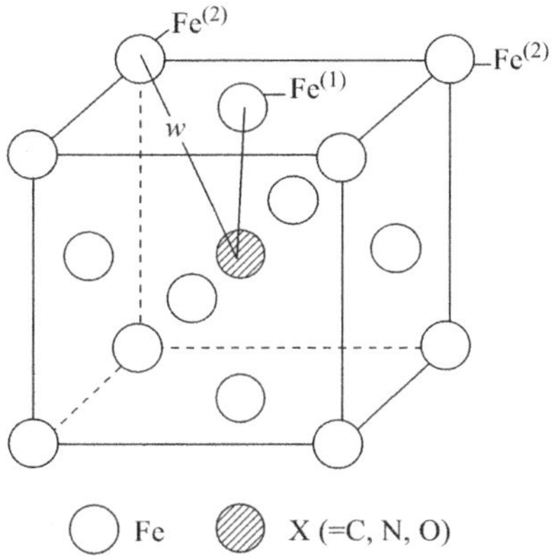

FIGURE 4-12 Cluster model, $Fe_{14}X$ (X = C, N, O) used for the calculation. $Fe^{(1)}$ and $Fe^{(2)}$ are the first and second nearest neighbor Fe atoms from X, respectively.

the octahedral interstitial site [16,17], so that each X atom is surrounded by six first nearest neighbor Fe atoms, $Fe^{(1)}$, and eight second nearest neighbor Fe atoms, $Fe^{(2)}$. Also, l and w denoted in the figure are the X-$Fe^{(1)}$ distance and the X-$Fe^{(2)}$ distance respectively, and the respective distances in a nondistorted fcc lattice are named l_0 and w_0. Bell and Owen [18] reported that the lattice expands by about 5% when one interstitial C or N atom exists in the fcc unit cell. Also, in the Fe-N system, there is a compound, Fe_4N, which has the exactly same crystal structure as the cluster model shown in Fig. 4-12. Its lattice parameter is 0.3797 nm, about 4% larger than that of pure fcc Fe, 0.3647 nm. For these reasons, the calculation is performed with varying l/l_0 from 1.00 to 1.04 and w/w_0 from 1.00 to 1.04 [15].

4.3.2 Bond Order

The bond order is a measure of the covalent bond strength between atoms. Using a nondistorted cluster model (i.e., $l = l_0$, $w = w_0$), the bond order is calculated and the result is shown in Fig. 4-13. In this figure, for example, $Fe^{(1)}$ *s,p,d*- X *p* and $Fe^{(1)}$ *s,p,d*- X *s* mean the bond order between $Fe^{(1)}$ *s,p,d* and X *p* electrons and between $Fe^{(1)}$ *s,p,d* and X *s* electrons, respectively. Similarly, $Fe^{(2)}$ *s,p,d*- X *p* and $Fe^{(2)}$ *s,p,d*- X *s* mean the respective bond orders between the corresponding electrons. Total means the sum of these four components.

The bond order depends largely on the interstitial atom. The total bond order changes in the order, C > N > O, and hence the covalent interactions change in the order, Fe-C > Fe-N > Fe-O. No significant change is observed in this trend, even in the calculation using a 4% distorted cluster (i.e., $l/l_0 = 1.04$ and $w/w_0 = 1.04$). Similar results are also reported in bcc Fe [19]. The further analysis reveals that the Fe-C chemical bond arises mainly from the covalent interactions between C 2*s*, 2*p*, and Fe 3*d* electrons in fcc Fe, which is also the case in bcc Fe [19].

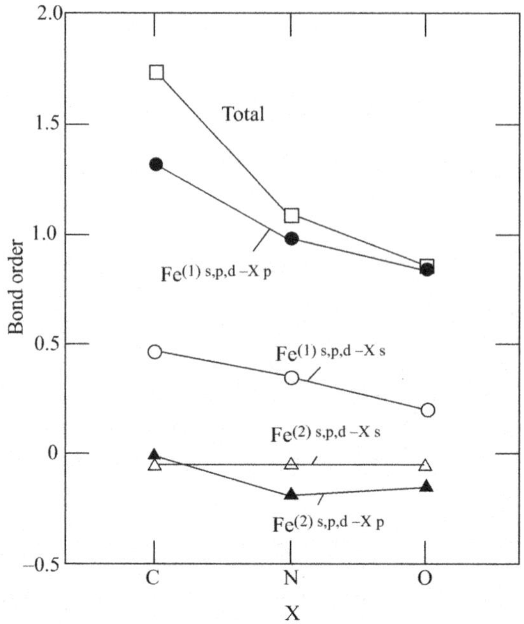

FIGURE 4-13 Bond order change with interstitial atoms, C, N, and O.

4.3.3 Difference Electron Density Maps

To examine the change of electron distributions with the addition of C, N, and O into fcc Fe, the difference electron density maps are calculated on the (100) atomic plane. The results are shown in Fig. 4-14 [15]. Here, the difference electron density is defined as,

$$\Delta\rho_X = \rho(\mathrm{Fe_{14}X}) - \rho(\mathrm{Fe_{14}}), \tag{4.1}$$

where $\rho(\mathrm{Fe_{14}X})$ and $\rho(\mathrm{Fe_{14}})$ are the respective electron densities for the corresponding cluster model without any distortion. Therefore, $\Delta\rho_X$ represents the change of electron distributions with the addition of X into fcc Fe. The broken curves in Fig. 4-14 indicate the electron deficient region and the full curves indicate the electron rich region. The interstitial site is denoted by a cross in each map. Large positive contour lines around this site are simply due to the existence of the interstitial atom in the octahedral site.

In Fig. 4-14(A), the region of the positive contour lines is extended over the C and $Fe^{(1)}$ atoms. This indicates that there is a strong covalent bonding between these atoms. As the excess electrons are distributed mainly along the C-$Fe^{(1)}$ direction, this covalency is probably attributed to the *p-d* σ-type bonding. On the other hand, in Fig. 4-14(C) the region of the negative contour lines exists along the O-$Fe^{(1)}$ direction. Instead, positive contour regions are located near the $Fe^{(1)}$ atom, as indicated by the symbol, P. Their lobes of excess electrons are directed towards the O atom, while keeping some angles inclined to the O-$Fe^{(1)}$ direction.

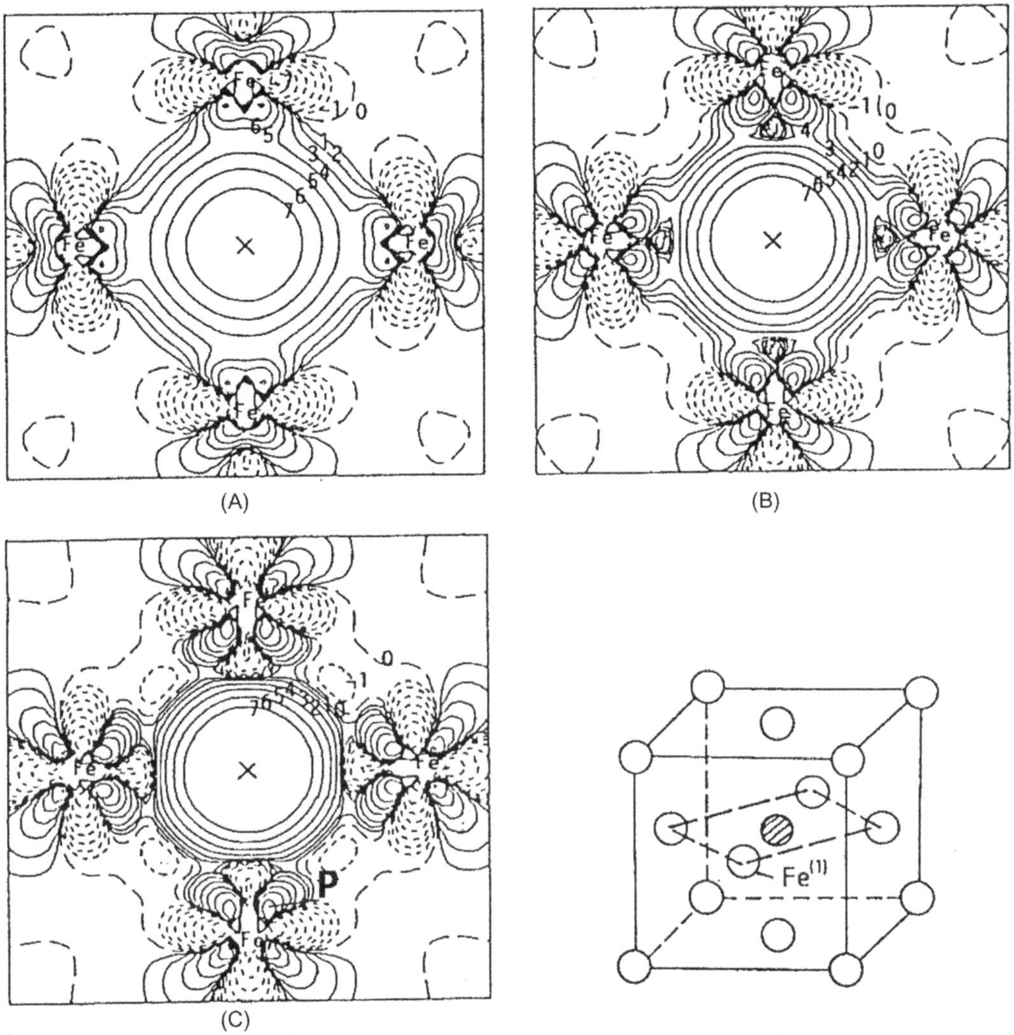

FIGURE 4-14 The difference electron density maps on the (100) atomic plane for (A) $\Delta\rho_C$, (B) $\Delta\rho_N$, and (C) $\Delta\rho_O$. The labels of the curves, 0, ±1, ±2, ±3, ±4, ±5, ±6, ±7 correspond to the values of electron density, 0, ±0.002, ±0.004, ±0.008, ±0.016, ±0.032, ±0.064, ±0.128 electrons/(a.u.)^3, respectively (1 a.u. = 0.0529 nm). The cross at the center of each figure indicates the interstitial site.

Such off-directional distributions of electrons are probably attributable to the *p-d* π-type bonding between O and $Fe^{(1)}$ atoms. From Fig. 4-14(B), it is evident that the bonding between N and $Fe^{(1)}$ atoms is intermediate between them. The σ-type bonding is, in general, stronger than the π-type bonding, and hence the change in the electron distributions is the one expected from the bond order calculation shown in Fig. 4-13.

4.3.4 Ionicity of Interstitial Atom

The charge transfer takes place in the direction from the Fe atom to the interstitial atom (C, N, O), following the electronegativity difference between them. From the Mulliken population analysis [10], the effective ionicity of the interstitial atom is estimated to be about (−0.33 to −0.35) for C, (−0.41 to −0.43) for N, and (−0.23 to −0.26) for O. The ionic values change slightly with the distortion. When the atom is perfectly ionized (i.e., 100% ionic), the ionicity is −4 for C, −3 for N, and −2 for O. Therefore, simply by taking the ratio of the effective ionic value to the perfect one, the ionic percentages are estimated to be 8.3%–8.7% for C, 13.5%–14.4% for N, and 11.4%–13.1% for O. The C atom is less ionic, and the N atom has the ionicity comparable to the O atom. Demangeat et al. [20] reported the ionicity of the interstitial atom in bcc Fe to be −1.22 for C, −1.80 for N, and −1.44 for O. The trend of the ionicity change with X is similar to the present result, but the magnitude is much larger than the present estimation for fcc Fe.

4.3.5 Bond Order Change with Lattice Distortion

The bond order changes with the lattice distortion around the interstitial atom. The results of the Fe-N system are shown in Fig. 4-15. Shown in the figure are the Fe-Fe bond order and

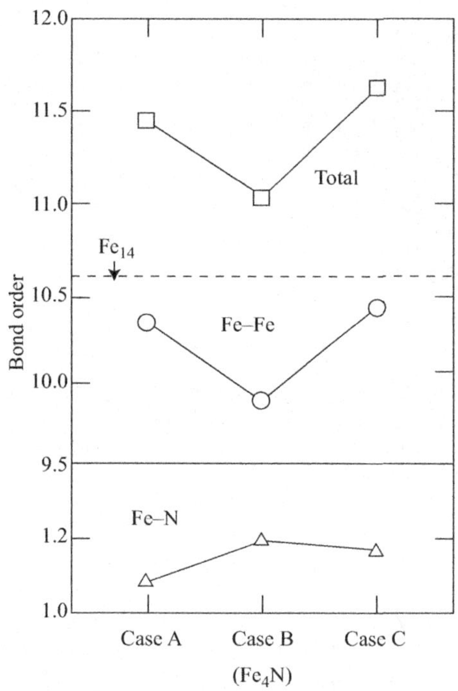

FIGURE 4-15 Bond order change with the local lattice distortion in the Fe-N system. The sizes of cluster $Fe_{14}N$ are as follows: case A, $l/l_0 = 1.00$, $w/w_0 = 1.00$; case B, $l/l_0 = 1.04$, $w/w_0 = 1.04$; case C, $l/l_0 = 1.04$, $w/w_0 = 1.00$.

the Fe-N bond order, and total bond order is the sum of them. The three cases are calculated here. Case A is the calculation using the nondistorted cluster with $l/l_0 = 1.00$, $w/w_0 = 1.00$. Case B is obtained with the distorted cluster with $l/l_0 = 1.04$, $w/w_0 = 1.04$. In case C only the $Fe^{(1)}$-N distance is expanded by 4%, while keeping the $Fe^{(2)}$-N distance unchanged, i.e., $l/l_0 = 1.04$, $w/w_0 = 1.00$. The last case is examined since the $Fe^{(2)}$-N distance is about 1.73 times longer than the $Fe^{(1)}$-N distance (see Fig. 4-12), and according to elasticity theory, the distortion is expected to be small in the second nearest neighbor $Fe^{(2)}$ atoms, which are distant from N. So, for simplicity, the zero distortion is assumed for $Fe^{(2)}$ atoms in case C.

By comparing case A with case B in Fig. 4-15, it is evident that the Fe-N bond order is higher in case B than in case A, but the Fe-Fe bond order is higher in case A than in case B. This indicates that the 4% expansion of the fcc lattice is effective in strengthening the Fe-N bond, but instead weakening the Fe-Fe bond. As the result, the total bond order is lower for the cluster of the 4% lattice expansion. From this result it may be said that the Fe_4N compound is stabilized by the Fe-N bonding, which is well accommodated with the small loss of the Fe-Fe bonding. However, this occurs only for the high N content, 20mol% N, in which there are relatively a large number of Fe-N bonds. The solubility limit of N in a disordered fcc Fe is at most about 10mol% N [21], and hence the Fe-Fe bonding may be more important in such a low N-content system. When case C is compared with case B, it is evident that the Fe-Fe bond order is remarkably higher in case C than in case B, and its increase covers the small decrease in the Fe-N bond order. As the result, the total bond order becomes much higher in case C than in case B. Thus, the interstitial N atom induces relatively large displacements of the first nearest neighbor $Fe^{(1)}$ atom, but very small displacements of the second nearest neighbor $Fe^{(2)}$ atom, in reasonable agreement with the elasticity theory.

In this way the displacement mode around an interstitial atom is possibly speculated qualitatively from the calculation of the bond order between atoms. In the figure, the total bond order for pure Fe is indicated by a dotted horizontal line. In every case of A, B, and C, the total bond order is higher than in pure Fe. This means that the interstitial N atom can exist stably in the fcc Fe lattice. The proper lattice distortion leads to the further stabilization. The importance of lattice distortion on the electronic structure around impurity atoms has been suggested by Akai et al. [22]. Further discussion will be given in Chapter 11 on the local lattice distortion introduced around various alloying elements including C and N in both bcc Fe and fcc Fe.

4.3.6 Bond Nature of Interstitial Atoms in fcc Fe

The bond nature of interstitial atoms in fcc Fe is understood qualitatively using the schematic drawing of the energy level structure shown in Fig. 4-16 [15]. For the Fe-C system, as the energy level of Fe 3*d* is relatively close to that of C 2*p*, the covalent interaction acts between Fe and C atoms. Then, the molecular orbital consists of the considerable mixture of the Fe *d* and C *p* orbitals. For example, as shown in Fig. 4-16B, the C 2*p* level has the 28.1% *d* and 44.4% *p* components. On the other hand, as shown in Fig. 4-16A for the Fe-O system, the energy level of Fe 3*d* is quite different from that of O 2*p*, so that the ionic interaction

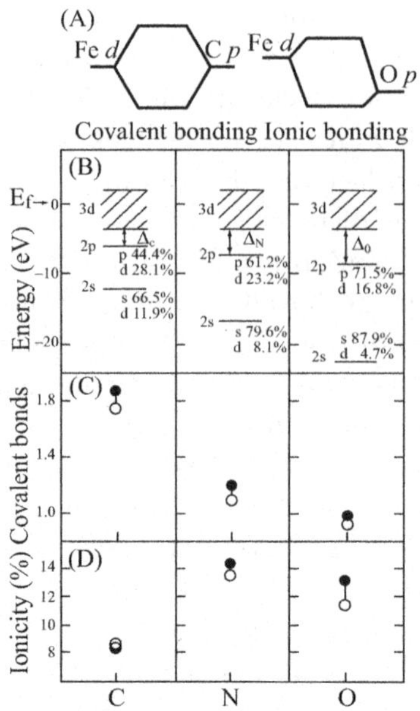

FIGURE 4-16 Schematic drawing for the feature of Fe-X interactions: (A) difference between covalent and ionic bonding, (B) energy level structure, (C) covalent bonds (i.e., bond order) and (D) ionicity (%). In (C) and (D), the open circle (○) and solid circle (●) refer to the results for nondistorted and 4% distorted clusters, respectively.

becomes more dominant between Fe and O atoms. In this case, the mixing of both atomic orbitals is small. For example, the O 2*p* level consists of 16.8% *d* and 71.5% *p*, and the *p* component is much higher in the O 2*p* level than in the C 2*p* level as explained previously. The nature of the covalent bonding decreases with increasing energy gap between the 2*p* and 3*d* band. As the order of the energy gap is $\Delta_C < \Delta_N < \Delta_O$, the covalent interaction becomes high in the order, C > N > O, in agreement with the result of the covalent bonds shown in Fig. 4-16C. The ionic bond character is the smallest for C but it is comparable between N and O as shown in Fig. 4-16D. Thus, as a whole, the electronic bond nature between Fe and interstitial atoms changes in the order, C, N, and O, following the order of elements in the periodic table. The same result is obtained in bcc Fe [19], so that the essential feature of the bonding between Fe and interstitial atoms is independent of the crystal structure.

4.4 Estimation of Alloying Behavior in bcc Fe from $\overline{Bo}$ and $\overline{Md}$ Parameters

Here, let's return to the topic of alloy design. Our attention will be directed mainly toward the high Cr ferritic steels, because they are important heat-resistant steels for use of turbines and boilers to operate power generating plants under ultrasupercritical conditions [23,24]. The steam condition is, for example, characterized by the temperature of 600°C (873K) and the pressure of 29 MPa [24]. The creep rupture strength is one of the most important properties of these heat-resistant steels, since they are serviced at high temperatures for many years. At elevated temperatures atomic diffusion may become significant and the creep deformation takes place gradually under the low external stress condition. As a result, the strain accumulates very slowly in the steel and finally reaches to the fracture strain of the steel. The long period of time until the steel is fractured is called creep rupture life. The creep rupture strength is usually represented by the allowable tensile stress. This is a measure of the creep rupture strength of the steel to guarantee a long service for 10^5 hours (about 11.4 years) under the external tensile stress condition at elevated temperatures. Recently great effort has been made to increase the steam temperature up to 620°C (893K) or 650°C (923K) [24]. Therefore, it has been requested strongly to develop advanced ferritic steels that are tolerable even in such severe operating conditions of power plants. Here, we will treat this problem in view of alloy design. The *Bo* and *Md* values for M in bcc Fe listed in Table 4-1 are used for the design of ferritic steels.

4.4.1 Alloying Vector

In Fig. 4-17 characteristics of alloying elements in bcc Fe are represented using the alloying vector, which starts from the position of pure Fe and ends at the position of binary Fe-1mol% M alloy on the $\overline{Bo}$-$\overline{Md}$ diagram. The alloying vectors point to the right and up for all the ferrite forming elements (e.g., Cr, Mo, W). On the other hand, the alloying vectors point to the left and down for the austenite forming elements (e.g., Co, Ni, Cu) except for Mn. The vector of Mn is small and exists near the pure Fe position.

It is known empirically that the elements with high *Bo* and low *Md* are principal alloying elements in most structural alloys [25]. The high *Bo* element will strengthen alloys by forming strong chemical bonds with the neighboring atoms, and the low *Md* element will increase the phase stability of alloys [25]. This is also the case in the ferritic steels. In other words, principal alloying elements have the high $\overline{Bo}/\overline{Md}$ ratio, namely, the large slope of

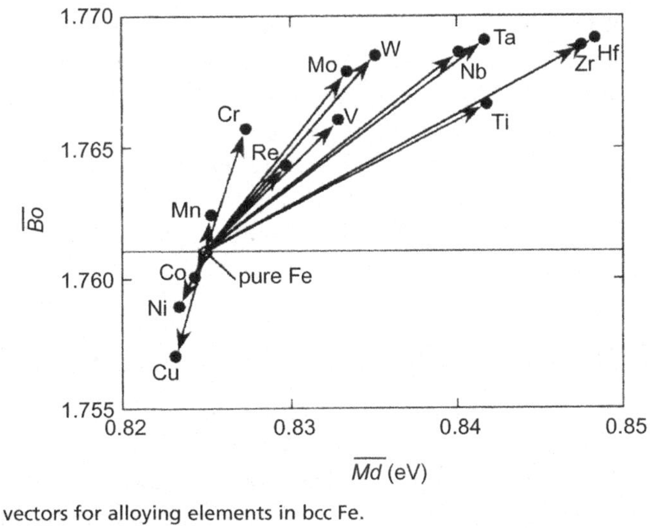

FIGURE 4-17 Alloying vectors for alloying elements in bcc Fe.

alloying vectors shown in Fig. 4-17. This ratio changes in the order, Cr > Mo > W > Re > V > Nb > Ta > Zr > Hf > Ti. As is consistent with this order, Cr, Mo, W, V, and Nb are indeed principal alloying elements in most ferritic steels. The only exception is Re, which is never used in ferritic steels despite the high $\overline{Bo}/\overline{Md}$ ratio. Therefore, special attention will be directed toward Re as explained later. As to the austenite stabilizing elements, the alloying vector has the opposite direction and the $\overline{Bo}/\overline{Md}$ ratio changes in the order, Co > Ni > Cu [26].

4.4.2 Prediction of δ-Ferrite Formation From $\overline{Md}$ Value

The Cr equivalent (Cr eq.) has been used as a parameter to predict the formation of the δ-ferrite in the ferritic steels. The Cr equivalent (Cr eq.) may be given by a Newhouse equation [9],

$$\begin{aligned} \text{Cr eq.} = {} & [\%\text{Cr}] + 6[\%\text{Si}] + 4[\%\text{Mo}] + 1.5[\%\text{W}] + 11[\%V] + 2.5[\%\text{Ta}] + 5[\%\text{Nb}] \\ & - (40[\%\text{C}] + 2[\%\text{Mn}] + 4[\%\text{Ni}] + 2[\%\text{Co}] + 30[\%\text{N}]), \end{aligned} \tag{4.2}$$

where [%M] is the mass% M in the steel.

However, the prediction for the δ-ferrite formation from the Cr equivalent is poorer, as compared to the $\overline{Md}$ method. For example, as shown in Fig. 4-18, the volume fractions of the δ-ferrite are measured with the 29 Cr-Mo ferritic steels normalized at 1323K, and the measured results are plotted against Fig. 4-18A the $\overline{Md}$ value and Fig. 4-18B Cr equivalent [9]. The formation of the δ-ferrite starts around $\overline{Md} = 0.852$(eV), and then the volume fraction increases with the $\overline{Md}$ value. On the other hand, for the prediction using the Cr equivalent, the data is scattered around the Cr eq. region of 12–16 (%) where the δ-ferrite starts forming in the steels.

The formation of δ-ferrite in ferritic steels has detrimental effect on their creep strength and the fracture toughness. The δ-ferrite in ferritic steels is a harmful phase like the σ phase in Ni-based superalloys. As another example, the measured volume fraction of the δ-ferrite existing in the steels normalized at 1323K [26] is plotted against the $\overline{Md}$ values in Fig. 4-19. For Ni-free and Co-free steels the δ-ferrite starts forming when the $\overline{Md}$ value exceeds about 0.852 (eV), and its volume fraction increases with the $\overline{Md}$ value. However, the presence of Ni and Co in the steel increases the critical $\overline{Md}$ value for the δ-ferrite formation. For example, the critical $\overline{Md}$ value is about 0.857 (eV) in the 1 mass% Ni containing steel. This is also the case of another austenite forming element, Co. So, the addition of Ni, Co, or Cu into the steel gives room to increase the amount of ferrite forming elements such as Mo and W without forming the δ-ferrite, resulting in the increase in creep rupture strength at high temperatures. In fact, recently developed high Cr ferritic steels contain both Mo and W together with the austenite forming element, Co and/or Ni (see Table 4-5).

4.5 Re Effect on the Creep Rupture Strength

As explained in Fig. 4-17, Re is an interesting element, since it has a high $\overline{Bo}/\overline{Md}$ ratio. So, the Re effect on the creep rupture strength is examined by preparing two experimental steels listed in Fig. 4-20A [27]. A2 and B2 contain nearly zero Re and 0.2 mass% Re, respectively, while keeping the other compositions constant, for example, about 10 mass% Cr, 4 mass%

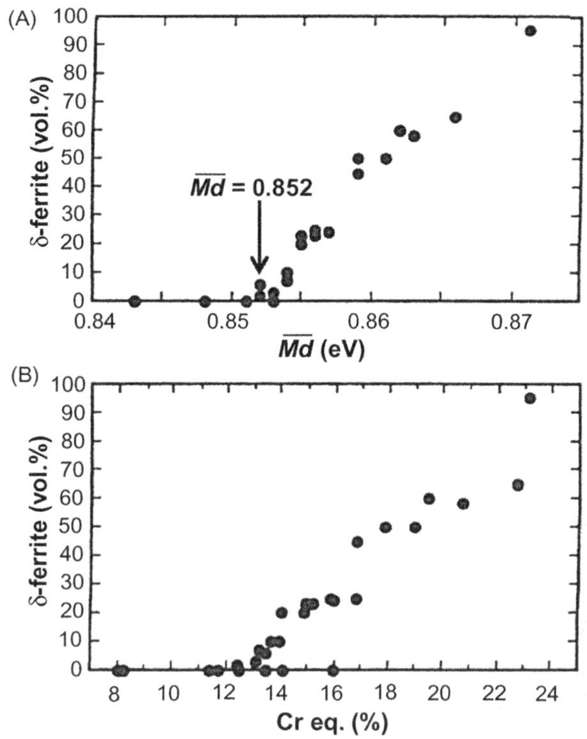

FIGURE 4-18 Prediction of the δ-ferrite formation in ferritic steels using (A) $\overline{Md}$ and (B) Cr equivalent.

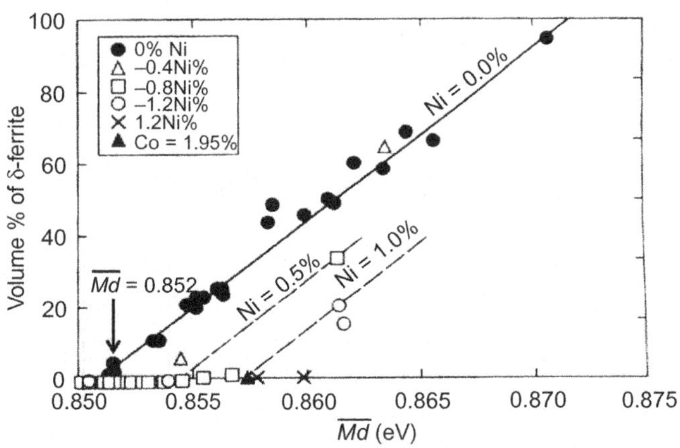

FIGURE 4-19 Prediction of the δ-ferrite formation in ferritic steels by the $\overline{Md}$ parameter.

Table 4-5 Chemical Compositions of 9%–12% Cr Ferritic Steels for Turbine Applications

Components, Steels	Chemical Compositions (Mass %)												Heat Treatment
Small Components	**C**	**Si**	**Mn**	**Ni**	**Cr**	**Mo**	**W**	**Co**	**V**	**Nb**	**B**	**N**	
H46 (12Cr-0.5MoVNbN)	0.15	0.40	0.60	—	12.0	0.5	—	—	0.30	0.25	—	0.050	1150°C–650°C
TAF (10.5Cr-1.5MoVNbB)	0.18	0.30	0.50	—	10.5	1.5	—	—	0.20	0.15	0.030	0.015	1150°C–700°C
TAF650 (11Cr-2.6W-3CoVNbB)	0.11	0.01	0.50	—	11.0	0.2	2.6	3.0	0.20	0.08	0.015	0.020	1100°C–750°C
Rotor													
GE (10.5Cr-1MoVNbN)	0.18	0.30	0.60	0.60	10.5	1.0	—	—	0.20	0.06	—	0.060	1050°C–620°C
TMK1 (10.3Cr-1.5MoVNbN)	0.14	0.05	0.50	0.60	10.3	1.5	—	—	0.17	0.06	—	0.040	1100°C–680°C
HR1100 (10.3Cr-1.2Mo-0.3WNbN)	0.15	0.03	0.60	0.60	10.3	1.2	0.3	—	0.20	0.05	—	0.050	1075°C–660°C
Mod. 12Cr (10Cr-1Mo-1WVNbN)	0.14	0.08	0.60	0.70	10.0	1.0	1.0	—	0.20	0.05	—	0.040	1050°C–660°C
TMK2 (10.2Cr-0.5Mo-2WVNbN)	0.14	0.05	0.05	0.50	10.5	0.5	1.8	—	0.17	0.06	—	0.040	1050°C–700°C
TR1200 (11Cr-0.2Mo-2.5WVNbN)	0.13	0.05	0.50	0.80	11.0	0.2	2.5	—	0.20	0.08	—	0.050	1050°C–710°C
HR1200 (11Cr-2.6W-3CoVNbB)	0.11	0.05	0.50	0.50	11.0	0.2	2.6	3.0	0.20	0.08	0.015	0.025	1050°C–720°C
New12Cr (10Cr-0.7Mo-1.8W-3CoVNbB)	0.11	0.08	0.10	0.20	10.0	0.7	1.8	3.0	0.20	0.05	0.010	0.020	1070°C–680°C
Casing													
MJC12 (9.5Cr-1MoVNbN)	0.10	0.70	0.70	0.50	9.5	1.0	—	—	0.15	0.06	—	0.040	1050°C–710°C

Source: By courtesy of Prof. F. Masuyama, History of power plants and progress in heat resistant steels, ISIJ International 41 (2001) 612–625.

(A) (mass%)

steel	C	Si	Mn	Ni	Cr	Mo	V	Nb	W	Co	B	Re	N	$\overline{Md}$	$\overline{Bo}$
A2	0.09	0.03	0.07	0.20	10.14	0.01	0.19	0.06	3.99	2.98	0.007	0.01	0.026	0.852	1.804
B2	0.09	0.03	0.07	0.21	10.20	0.01	0.19	0.06	3.97	3.00	0.007	0.23	0.025	0.855	1.811

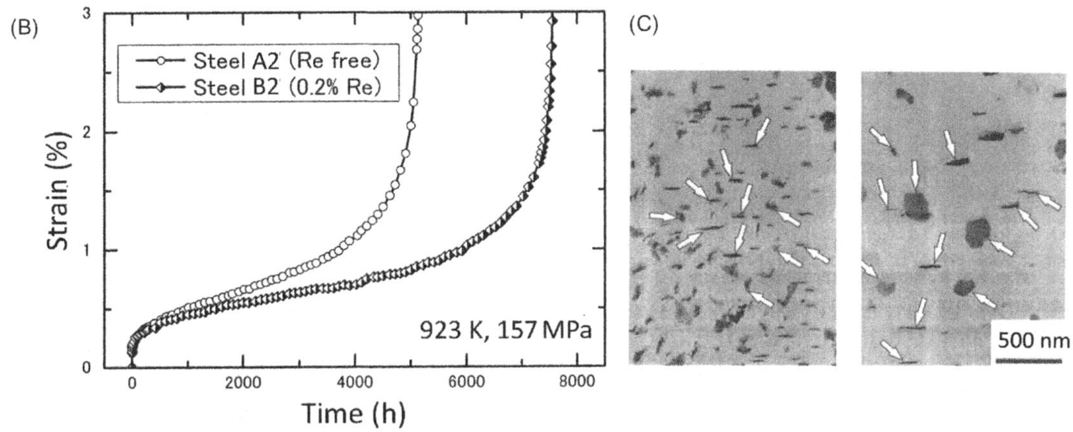

FIGURE 4-20 (A) Chemical compositions (mass%) for two experimental steels, (B) creep curves measured at 923K under the stress of 157 MPa, (C) transmission electron micrograph of B2 steel; the left and right pictures are taken before aging and after aging for 12,465 hours at 923K, respectively. The Laves phase is dispersed in the martensite lath as indicated by arrows.

W, and 3 mass% Co. The creep strain–rate curves measured at 973K under the stress of 157 MPa are shown in Fig. 4-20B. It is evident from this figure that the Re addition improves the creep rupture strength significantly. A similar result is found in the 9 mass% Cr steel containing 0.5 mass% Re. The measured creep rupture life is more than 8400 hours (30.24 Ms) under the condition of 923K, 137 MPa. This Re effect emerges in the case when the steel contains more than 3 mass% W [27], so that it is considered that Re cooperates with W to increase the creep resistance of steels.

According to transmission electron microscopy (TEM) and X-ray diffraction experiments [27], there are three types of precipitates in these steels. They are $M_{23}C_6$, Laves phase, and MX carbonitrides of (V,Nb)(C,N) [27]. EDA analysis was performed on the precipitates in the 0.2 mass% Re containing B2 steel, aged for 12,465 hours (44.874 Ms) at 923K. It was found that a few % Re is partitioned into the Laves phase with the $(Fe,Cr)_2W$-type. The micrographs of the Laves phase in the B2 steel are shown in Fig. 4-20C. Before aging, the fine Laves phase is dispersed in the martensite lath as shown in the left micrograph. After aging for 12,465 hours (44.874 Ms) at 923K, the Laves phase still exists in the lath, although its size becomes large. This result is in contrast to the observation in the Re-free A2 steel aged for 11,033 hours (39.7188 Ms) at 923K. Because the Laves phase is no longer present in the lath, but the large Laves phase exists along the lath boundary together with $M_{23}C_6$ and MX in the A2 steel, the existence of Re in the Laves phase works to suppress the coarsening of the Laves phase. Murata et al. [28–30] suggested that the atomic diffusion of W is suppressed significantly in the presence of Re, which probably retards the coarsening of the Laves phase, and retains the Laves phase in the martensite lath. In other words, the Laves phase dispersed in the martensite lath contributes to the strengthening of the steel, and works to maintain the high creep rupture strength even for a long period of time at 923K. Thus, the addition of a small amount of Re is beneficial to the enhancement of the creep rupture strength of ferritic steels.

4.6 Tracing of Evolutionary Process of Ferritic Steels on the *Bo-Md* Diagram

Table 4-5 lists chemical compositions for typical 9%–12% Cr ferritic steels for turbine applications [24]. Masuyama [24] illustrates the evolutionary process of these steels as shown in Fig. 4-21. In this figure, for example, the evolution of rotor steels from GE to TMK1 is interpreted as due mainly to the compositional change expressed as (−C, + Mo). This means that TMK1 has the lower C content but higher Mo content than GE (see Table 4-5). Following the evolutionary process shown in Fig. 4-21, the steels for turbine rotor have been developed in the sequence, H46→ GE→TMK1→TMK2→HR1200 (or New 12Cr). The value of the 10^5 h creep rupture strength at 600°C (873K) is given in the parentheses above each alloy. It increases from 60 MPa for H46 to 190 MPa for HR1200 (or New 12Cr).

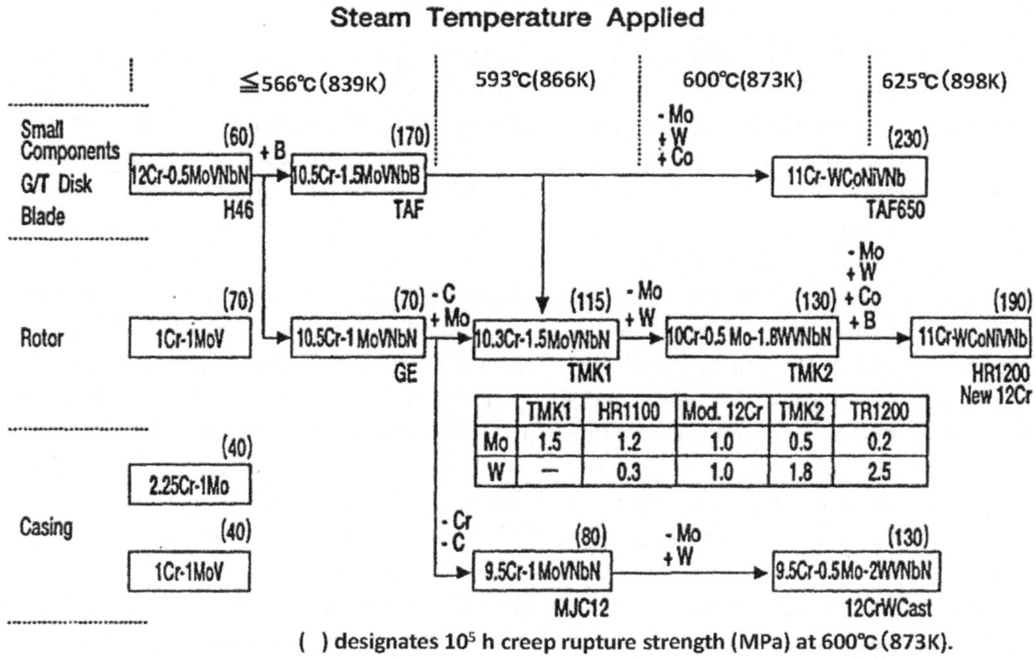

FIGURE 4-21 Evolutionary process of 9%–12% Cr steels for turbine applications. *By courtesy of Dr. F. Masuyama, History of power plants and progress in heat resistant steels, ISIJ International 41 (2001) 612–625.*

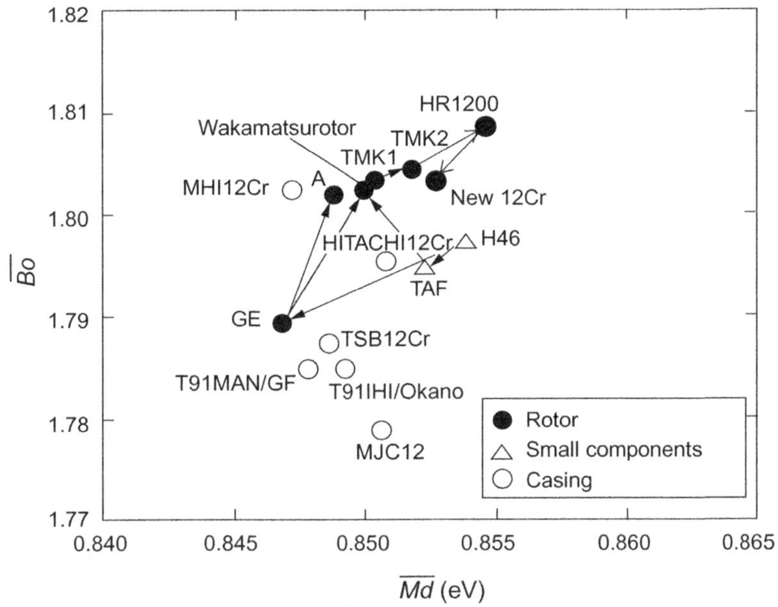

FIGURE 4-22 Tracing of evolutionary process of the ferritic steels for turbine applications using the $\overline{Bo}$-$\overline{Md}$ diagram.

This evolutionary process can be understood using the $\overline{Bo}$-$\overline{Md}$ diagram as shown in Fig. 4-22. For example, the modification from H46 for small components to GE for a rotor is the change toward the lower $\overline{Md}$ and lower $\overline{Bo}$, and hence this is a moderate modification so as to diminish the solute segregation in a large-scale rotor. However, the evolution from GE→TMK1→TMK2→HR1200 (or New 12Cr) is definitely the change toward the higher $\overline{Md}$ and the higher $\overline{Bo}$. An important difference in the alloy compositions among these steels is in the Co content, since Co is absent in GE, TMK1, and TMK2, but present in HR1200 (or New 12Cr). As explained earlier, the presence of 3 mass% Co in the steel gives room for the further addition of Mo and W and increases the creep resistance without forming δ-ferrite in the steel. The $\overline{Md}$ value is 0.852 (eV) for TMK2, being close to the critical $\overline{Md}$ value for the δ-ferrite formation, but the $\overline{Md}$ value is about 0.855 (eV) for HR1200 owing to the presence of 3 mass% Co in it as explained in Fig. 4-19. New 12 Cr is located near the position of $\overline{Md} = 0.853$(eV) and $\overline{Bo} = 1.803$, so it approaches TMK2 on the $\overline{Bo}$-$\overline{Md}$ diagram. As compared with these rotor steels, MJC12 for the casing has lower $\overline{Bo}$ and lower $\overline{Md}$ values as shown in Fig. 4-22. Besides these turbine steels, the evolutionary process of the boiler steels is similarly traceable on the $\overline{Bo}$-$\overline{Md}$ diagram [26].

4.7 Design of Ferritic Steels for Turbine Rotor

Eleven experimental steels of A3 to K3 are designed and their compositions are listed in Table 4-6 together with their $\overline{Md}$ and $\overline{Bo}$ values [31–33]. They are located on the $\overline{Bo}$-$\overline{Md}$

Table 4-6 Chemical Compositions (Mass%) of Experimental Ferritic Steels

Steel	C	Si	Mn	Ni	Cr	Mo	V	Nb	W	Co	B	N	$\overline{Md}$	$\overline{Bo}$
A3	0.12	0.02	0.01	$<$0.01	10.09	0.40	0.20	0.08	1.89	3.01	0.008	0.016	0.852	1.804
B3	0.12	0.03	0.01	$<$0.01	11.38	0.40	0.20	0.08	1.91	3.03	0.008	0.015	0.855	1.811
C3	0.10	0.03	0.01	0.20	8.56	0.11	0.15	0.07	3.45	2.99	0.009	0.020	0.852	1.800
D3	0.09	0.03	0.01	0.21	9.04	0.11	0.15	0.06	3.55	2.98	0.009	0.020	0.853	1.803
E3	0.10	0.03	0.01	0.21	9.51	0.11	0.15	0.06	3.58	2.99	0.009	0.020	0.854	1.805
F3	0.09	0.03	0.01	0.21	10.06	0.11	0.15	0.06	3.55	3.02	0.011	0.020	0.856	1.808
G3	0.10	0.03	0.01	0.20	10.50	0.10	0.15	0.06	3.54	3.00	0.010	0.021	0.856	1.810
H3	0.10	0.03	0.01	0.21	11.59	0.11	0.15	0.07	3.60	3.04	0.010	0.020	0.860	1.815
I3	0.09	0.03	0.01	0.21	9.98	0.11	0.15	0.06	3.15	2.99	0.010	0.020	0.854	1.807
J3	0.09	0.02	0.01	0.20	10.06	0.11	0.15	0.06	4.09	3.01	0.011	0.020	0.858	1.810
K3	0.09	0.03	0.01	0.20	9.98	0.11	0.15	0.07	4.51	2.98	0.011	0.021	0.859	1.810

diagram in the range of $1.800 \leqq \overline{Bo} \leqq 1.815$ and $0.852(\text{eV}) \leqq \overline{Md} \leqq 0.860(\text{eV})$. The $\overline{Md}$ value is higher than 0.852 (eV), because every steel contains 3 mass% Co. These steels are heat treated at 1343K for 5 hours (18 ks) and quenched, followed by two-step tempered treatments at 843K for 20 hours (72 ks) and then at 953K for 20 hours (72 ks). These tempered treatments are performed to remove the retained austenite completely from the steel. The tempered martensite is obtained by this heat treatment.

These 11 steels are characterized by the compositions of Cr, Mo, and W as follows. Hereafter, unless otherwise noted, the compositions are expressed in mass% units.

1. (A3, B3): The Cr content is about 10% in A3, but about 11.5% in B3. About 0.4% Mo and 1.9% W contents are the same between them.
2. (C3 ~ H3): The Cr content is about 8.5% in C3, 9.0% in D3, 9.5% in E3, 10.0% in F3, 10.5% in G3, and about 11.5% in H3. About 0.1% Mo and 3.5% W contents are the same among them.
3. (I3 ~ K3, F3): The W content is about 3% in I3, 3.5% in F3, 4.0% in J3, and 4.5% in K3. About 10% Cr and 0.1% Mo contents are the same among them.

Besides these differences, the Ni content is 0% in A3 and B3, but about 0.2% in the other steels. Also, about 3% Co content is the same among all the steels. The C content is 0.12% in A3 and B3, and about 0.1% in the other steels. A3 resembles the New 12Cr partially in the composition (see Table 4-5).

In Fig. 4-23, creep rupture time vs. applied stress curves at 923K is shown for the steels of C3 to H3, where the Cr contents are varied from 8.5% to 11.5%, while keeping the W content constant, 3.5%. It is apparent from this figure that the creep strength for a short term until about 300 hours (1.08 Ms) increases with increasing Cr content, owing to the increase in the tensile strength with the Cr content in the steels. However, the creep strength for a long term tends to increase with decreasing Cr content. As the result, for example, under the low stress of 98 MPa, the 9.0% Cr steel (i.e., D3) has the longest creep rupture life, about 25,000 hours (2.9 years, 90 Ms), but the 11.5% Cr steel (i.e., H3) has a shortest creep rupture life, about

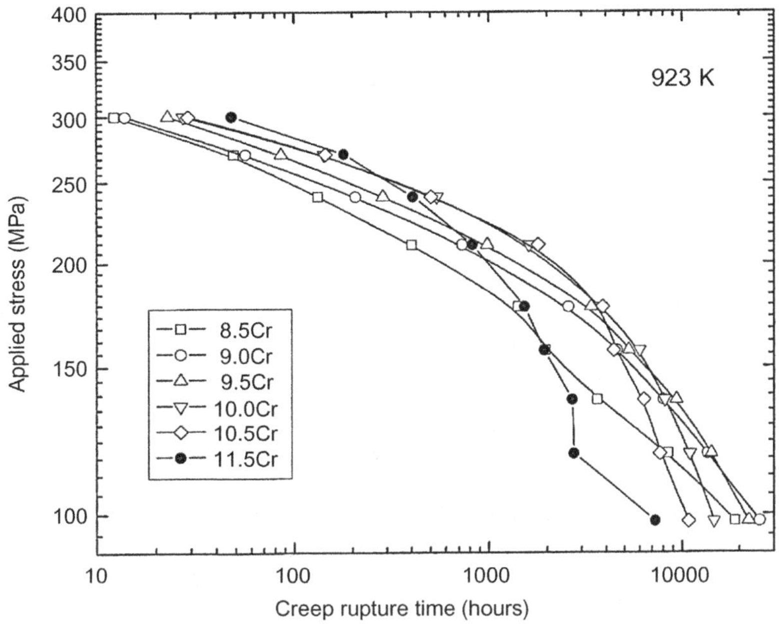

FIGURE 4-23 Creep rupture time vs. applied stress curves at 923K for steels containing 3.5 mass% W and various Cr contents of 8.5 mass% Cr to 11.5 mass% Cr.

7000 hours (0.8 years, 25.2 Ms). Thus, the creep rupture life becomes reversed simply due to the Cr content in the steels. This is also the case between A3 (10% Cr) and B3 (11.5%Cr), both containing about 1.9% W. Thus, high Cr content reduces the creep strength for a long term, regardless of the W content in the steels.

To see this trend more clearly, the Cr compositonal dependence of the creep rupture life is shown in Fig. 4-24 [31]. The creep rupture lives are plotted against the Cr contents of the steels measured at 923K under various applied stresses of 98 to 300 MPa. In case of the short term creep under the high stress of 300 MPa, the rupture life increases with increasing Cr content in the steels. However, as the rupture life increases under the lower stress condition, the peak position of the rupture life shifts gradually to the lower Cr content. Finally, in case of the lowest stress of 98 MPa, the peak appears at 9% Cr.

As shown in Fig. 4-23, there is a remarkable decrease in the long term creep strength when the Cr content is high. For example, this is seen in the creep curves around 10,000 hours for the steels containing 10.0–11.5% Cr. Such a remarkable decrease in the creep strength is one of important issues in advanced high Cr ferritic steels. However, the reason still remains unknown although several models have been proposed [27,34]. Cr is a principal alloying element to strengthen steels and also to enhance the oxidation and corrosion resistance of steels. However, its excess addition promotes the microstructural change at high temperatures, and weakens the long term creep strength under the low stress condition as shown in Fig. 4-23.

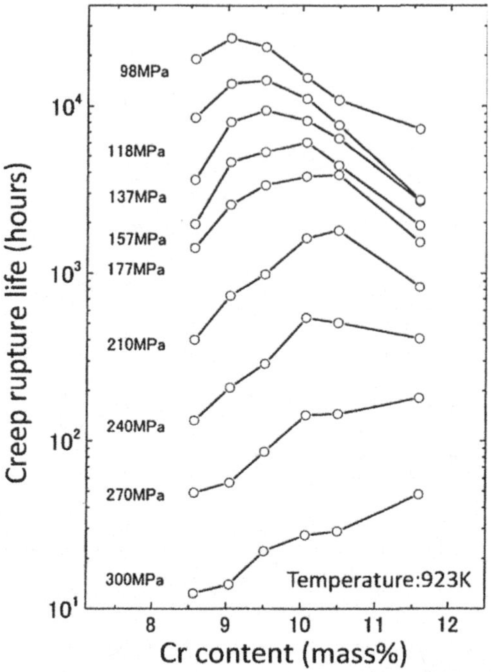

FIGURE 4-24 Cr compositional dependence of creep rupture life at 923K tested under various stress conditions.

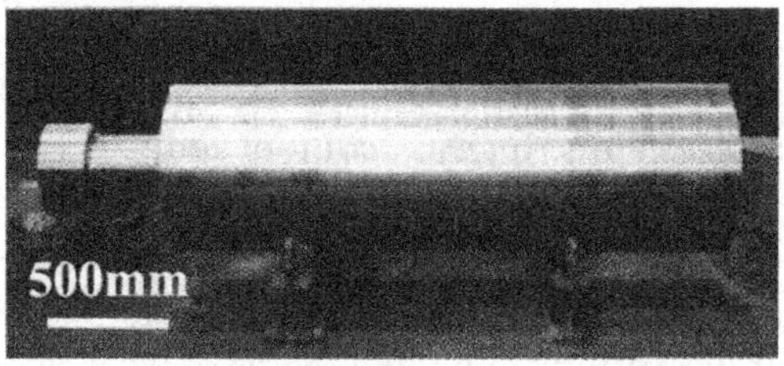

FIGURE 4-25 A 10-ton model steam turbine rotor made of designed steel.

On the basis of a series of experiments, a 10-ton model steam turbine rotor (diameter: 640 mm, length: 2270 mm) is made as shown in Fig. 4-25 [35]. The creep rupture life is about 4000 hours (14.4 Ms) under the creep condition of the temperature, 923K, and the applied stress, 157 MPa, which is about several times longer than the currently used rotor steel.

References

[1] S. Wakoh, J. Yamashita, Band structure of ferromagnetic iron self-consistent procedure, J. Phys. Soc. Jpn 21 (1966) 1712–1726.

[2] J. Hubbard, The Approximate calculation of electronic band structures III, J. Phys. C: Solid State Phys. 2 (1969) 1222–1229.

[3] J. Callaway, S. Wang, Energy bands in ferromagnetic iron, Phys. Rev. B 16 (1977) 2095–2105.

[4] K.J. Duff, T.P. Das, Electron states in ferromagnetic iron, I. band properties, Phys. Rev. B 3 (1971) 192–208.

[5] R.J. Weiss, G. Mazzone, The electronic structure of the 3*d* transitionmetals, J. Appl. Crystallogr 14 (1981) 401–416.

[6] M. Morinaga, N. Yukawa, H. Adachi, Alloying effect on the electronic structure of bcc Fe, J. Phys. F: Met. Phys. 15 (1985) 1071–1084.

[7] J. Friedel, C.M. Sayers, Correlation effects in bulk modulus and equilibrium lattice spacing of transition-metals, J. Physique Lett. 38 (1977) L263–L265.

[8] M. Morinaga, N. Yukawa, H. Adachi, Alloying effect on the electronic structure of Ni_3Al (γ'), J. Phys. Soc. Jpn 53 (1984) 653–663.

[9] H. Ezaki, M. Morinaga, K. Kusunoki, Y. Tsuchida, Electronic approach to the prediction of phase stability in Cr-Mo ferritic steels, Tetsu-to-Hagane 78 (1992) 1377–1382.

[10] R.S. Mulliken, Electronic population analysis on LCAO–MO molecular wave functions. I, II, III, IV, J. Chem. Phys 23 (1955) 1833–1840, 1841–1846, 2338–2342, 2343–2346.

[11] L. Pauling, The Nature of the Chemical Bond, 3rd edn, Cornell University, Ithaca, 1960.

[12] C.R. Austin, C.R. St. John, R.W. Lindsay, Creep properties of some binary solid solutions of ferrite, Trans. AIME 162 (1945) 84–105.

[13] M. Morinaga, Y. Murata, H. Ezaki, Design and development of structural alloys for nuclear applications, Proc. Inter. Symposium on Materials Chemistry in Nuclear Environment, Tsukuba, Japan, March 12-13 (1992) pp.241–252.

[14] M. Murata, K. Koyama, Y. Matsumoto, M. Morinaga, N. Yukawa, Equilibrium phase diagram of Fe-Cr-Mn ternary system, ISIJ International 30 (1990) 927–936.

[15] M. Morinaga, N. Yukawa, H. Adachi, T. Mura, Electronic state of interstitial atoms (C, N, O) in FCC Fe, J. Phys. F: Met. Phys. 18 (1988) 923–934.

[16] S. Nagakura, S. Oketani, Structure of transition metal carbides, Trans. ISIJ 8 (1968) 265.

[17] S. Nagakura, Electronic structure of iron nitrides studied by electron diffraction. I. γ'-Fe_4N, J. Phys. Soc. Jpn 25 (1968) 488–498.

[18] T. Bell, W.S. Owen, Martensite in iron-nitrogen alloys, J. Iron Steel Inst. 205 (1967) 428.

[19] M. Morinaga, N. Yukawa, H. Adachi, T. Mura, Electronic stability effect on local strain in martensite, J. Phys. F: Met. Phys 17 (1987) 2147–2162.

[20] C. Demangeat, M.A. Khan, J.C. Parlebas, Valency effect of light interstitials (B, C, N, O) in α-iron, J. Magn. Magn. Mater 15-18 (1980) 885–886.

[21] H. Okamoto, Phase Diagrams for Binary Alloys, Second Edition, ASM International, Materials Park, Ohio, 2010.

[22] M. Akai, H. Akai, J. Kanamori, Electronic structure of impurities in ferromagnetic iron. III. light interstitials, J. Phys. Soc. Jpn 56 (1987) 1064–1077.

[23] W. Schlachter, G.H. Gessinger, Innovation in power engineering: Role of materials, in: E. Bachelet, et al. (Eds.), High Temperature Materials for Power Engineering, 1990, pp. 1–24.

[24] F. Masuyama, History of power plants and progress in heat resistant steels, ISIJ International 41 (2001) 612–625.

[25] M. Morinaga, Alloy design based on molecular orbital method, Materials Transactions 57 (2016) 213–226.

[26] M. Morinaga, R. Hashizume, Y. Murata, An electronic approach to the design of ferritic steels for power plants, Materials for Advanced Power Engineering 1994 (Part I) (1994) 319–328. Liège, Belgium, October 3–6.

[27] R. Hashizume, Development of Ni-based superalloys and heat-resistant high Cr ferritic steels, Ph.D. Thesis, Nagoya University, Japan, November 2009.

[28] Y. Murata, K. Takami, M. Kamiya, M. Morinaga, R. Hashizume, K. Miki, et al., Coherent precipitate of the laves phase in heat-resistant 10Cr ferritic steel containing high W content, Tetsu-to-Hagane 88 (2002) 46–53.

[29] T. Kunieda, K. Yamashita, Y. Murata, T. Koyama, M. Morinaga, Effect of rhenium addition on tungsten diffusivity in iron-chromium alloys, Materials Transactions 47 (2006) 2106–2108.

[30] T. Kunieda, Y. Murata, M. Morinaga, T. Koyama, Effect of refractory elements (X: Mo, W, Re) on the microstructure evolution of Fe-8Cr-0.1C-X martensitic steels during tempering, Materials Transactions 45 (2004) 392–395.

[31] R. Hashizume, K. Miki, T. Azuma, T. Ishiguro, Y. Murata, M. Morinaga, Optimization of Cr content in high Cr heat resistant steels on the basis of long term creep strength, Tetsu-to-Hagane 88 (2002) 793–800.

[32] K. Miki, T. Azuma, T. Ishiguro, R. Hashizume, Y. Murata, M. Morinaga, Optimization of Cr content for long-term creep strength in high Cr heat resistant steel, in: Proc. 4th Int. Conf. on Advances in Materials Technology for Fossil Power Plants, South Carolina, USA, 2004, pp. 1115–1123.

[33] K. Miki, T. Azuma, T. Ishiguro, R. Hashizume, Y. Murata, M. Morinaga, Effect of Cr content on the creep strength and microstructural change in high Cr heat resistant steel, in: Proc. 7th Liège Conference on Materials for Advanced Power Engineering, Liège, Belgium, 2002, pp. 1497–1504.

[34] A. Iseda, H. Teranishi, F. Masuyama, Effects of chemical compositions and heat treatments on creep rupture strength of 12 wt%Cr heat resistant steels for boiler, Tetsu-to-Hagane 76 (1990) 1076–1083.

[35] Patent registration numbers: JP 3315702, US 6174385, DE 778356, JP 3759776, JP 3798456, JP 4222705, JP 4262414.

5 Titanium Alloys

5.1 Molecular Orbital Calculation and Alloying Parameters

5.1.1 Cluster Models for bcc Ti and hcp Ti

An allotropic transformation takes place between hcp Ti (α-Ti) and bcc Ti (β-Ti) around 1155K in pure Ti. Both bcc and hcp cluster models are employed for the electronic structure calculation [1]. For example, the bcc cluster model is shown in Fig. 5-1A. It consists of a central Ti atom, 8 $Ti^{(1)}$ atoms in the first nearest neighbor, and 6 $Ti^{(2)}$ atoms in the second nearest neighbor. An alloying element, M, is substituted for a central Ti atom. So, the bcc cluster model is expressed as MTi_{14}. The lattice parameter used is 0.3320 nm.

On the other hand, the hcp cluster model is shown in Fig. 5-1B. It consists of a central Ti atom, 12 $Ti^{(1)}$ atoms in the first nearest neighbor, and 6 $Ti^{(2)}$ atoms in the second nearest neighbor. An alloying element, M, is substituted for a central Ti atom. So, the hcp cluster model is expressed as MTi_{18}. The lattice parameters used are $a = 0.2950$ nm and $c = 0.4683$ nm.

By using these cluster models, a series of calculations is carried out for various alloying elements, M, to determine alloying parameters relevant to titanium.

5.1.2 Alloying Parameters

In Fig. 5-2 the energy level structure is given of 3*d* alloying elements in bcc Ti [1,2]. For example, the result of pure bcc Ti calculated by using the Ti_{15} cluster model (i.e., M = Ti) is shown in the left side of the figure, where the position of the Fermi energy level, E_f, is indicated by an arrow (←). This E_f energy level is set at zero and used as a reference. The levels of $18t_{1u}$ to $5e_u$ are derived mainly from Ti 3*d* orbitals, and form the Ti 3*d* band where the E_f lies. The lower energy levels of $17t_{1u}$ to $15a_{1g}$ and the higher energy levels of $13e_g$ to $8t_{1g}$ are composed of Ti 3*d*, 4*s*, and 4*p*. Two energy levels, $16t_{2g}$ and $13e_g$, existing around 2.5 eV, are the *d*-levels of a central Ti atom in the cluster model shown in Fig. 5-1A. The fractions of the *d* component are about 40% in $16t_{2g}$ and 59% in $13e_g$.

In alloyed bcc Ti with 3*d* transition elements, the energy levels originated mainly from the M-*d* orbitals appear above the E_f. These M-*d* orbital levels are the e_g and t_{2g} levels indicated by dotted lines in Fig. 5-2. The height of these levels decreases monotonously with the atomic number of M in the 3*d* series. The average value of these two levels is defined as *Md* for bcc Ti alloys.

Another alloying parameter is the bond order (*Bo*). The bond order due to the *d-d* covalent bond between atoms is calculated following the Mulliken population analysis [3]. The results are shown in Fig. 5-3A for bcc Ti and in Fig. 5-4A for hcp Ti. In each figure, M-Ti and

A Quantum Approach to Alloy Design. DOI: https://doi.org/10.1016/B978-0-12-814706-1.00005-4

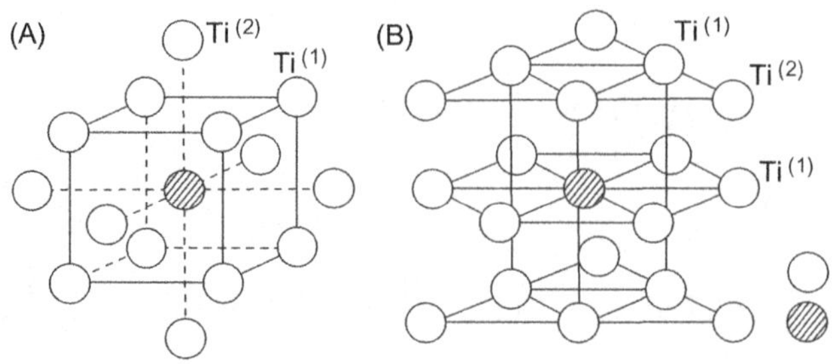

FIGURE 5-1 Cluster models used for calculations, (A) bcc cluster and (B) hcp cluster.

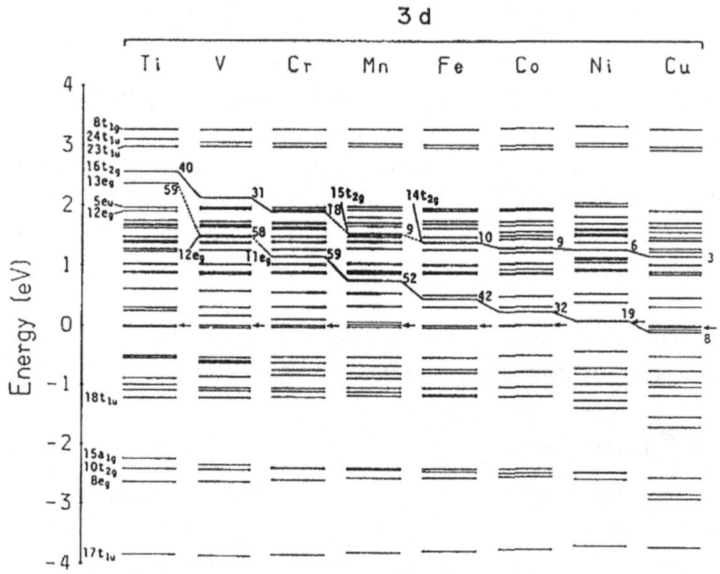

FIGURE 5-2 Energy level structure of bcc Ti containing 3*d* transition elements.

Ti-Ti are the bond order between M-*d* and Ti-3*d* electrons and between Ti-3*d* and Ti-3*d* electrons in a cluster, respectively. Total is the sum of them. The M-Ti bond order varies depending largely on the alloying element, M, in both bcc Ti and hcp Ti. There is a small peak near V, and then it decreases with the order of elements, M. Total bond order varies in a similar manner as does the M-Ti bond order. The magnitude of the total bond order is smaller in 3*d* elements than in 4*d* and 5*d* elements. The trend of the bond order change with M resembles between bcc Ti and hcp Ti.

As explained in the Ni alloys and Fe alloys in Chapters 3 and 4, the high bond order elements are principal alloying elements in most of the alloys used for structural applications. This is also the case in the titanium alloys, since the principal alloying elements such as V, Cr, Zr, Nb, and Mo have the high bond order. To allow for the density and the specific

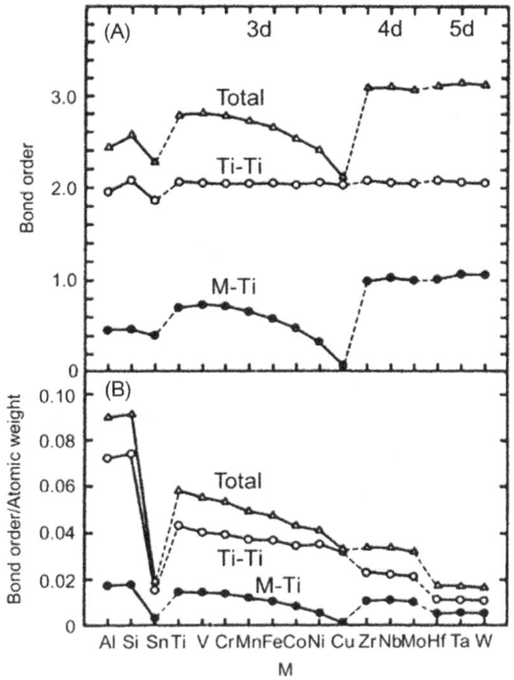

FIGURE 5-3 (A) Bond order and (B) ratio of bond order to atomic weight for M in bcc Ti.

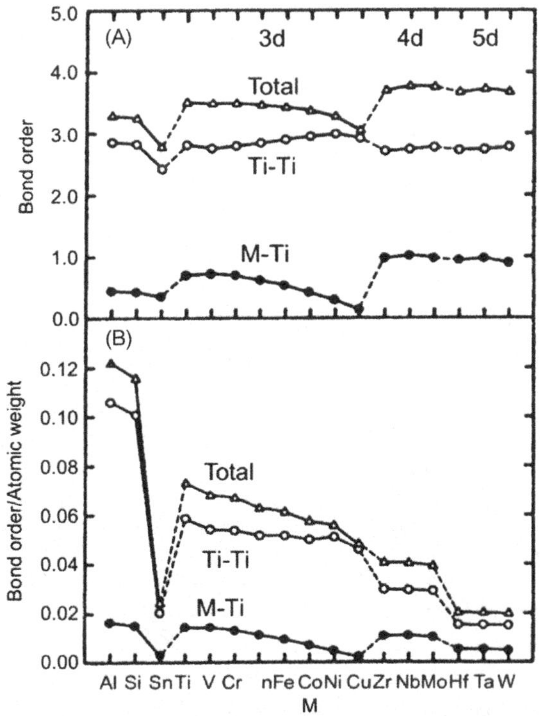

FIGURE 5-4 (A) Bond order and (B) ratio of bond order to atomic weight for M in hcp Ti.

strength of alloys, the bond order of each element is divided by its atomic weight. The results are given in Fig. 5-3B for bcc Ti and in Fig. 5-4B for hcp Ti. It is apparent that Fig. 5-3B is very similar to Fig. 5-4B. The ratio of bond order to atomic weight is large for Al, (Ti), V. It may be said from this result that the selection of alloying elements in a Ti-6Al-4V system is very reasonable.

In Table 5-1 the *Bo* and *Md* parameters are listed for various alloying elements, M, in bcc Ti. To see a general trend, the *Md* values for bcc Ti are compared with those for bcc Fe in Fig. 5-5. The relation between them is expressed as, $y = 1.13x^{0.914}$, where y is the *Md* value for bcc Ti and x is the *Md* value for bcc Fe, and the correlation coefficient is $R = 0.984$. But this nonlinear relation is different from a linear relation observed between fcc Fe and Ni_3Al (see Fig.4-9B). This is probably due to the electronic structure difference between Ti and Fe,

Table 5-1 List of *Bo* and *Md* Values for Alloying Elements in bcc Ti

3d	*Bo*	*Md* (eV)	4d	*Bo*	*Md* (eV)	5d	*Bo*	*Md* (eV)	Other	*Bo*	*Md* (eV)
Ti	2.790	2.447	Zr	3.086	2.934	Hf	3.110	2.975	Al	2.426	2.200
V	2.805	1.872	Nb	3.099	2.424	Ta	3.144	2.531	Si	2.561	2.200
Cr	2.779	1.478	Mo	3.063	1.961	W	3.125	2.072	Sn	2.283	2.100
Mn	2.723	1.194	Tc	3.026	1.294	Re	3.061	1.490			
Fe	2.651	0.969	Ru	2.704	0.859	Os	2.980	1.018			
Co	2.529	0.807	Rh	2.736	0.561	Ir	3.168	0.677			
Ni	2.412	0.724	Pd	2.208	0.347	Pt	2.252	0.146			
Cu	2.114	0.567	Ag	2.094	0.196	Au	1.953	0.258			

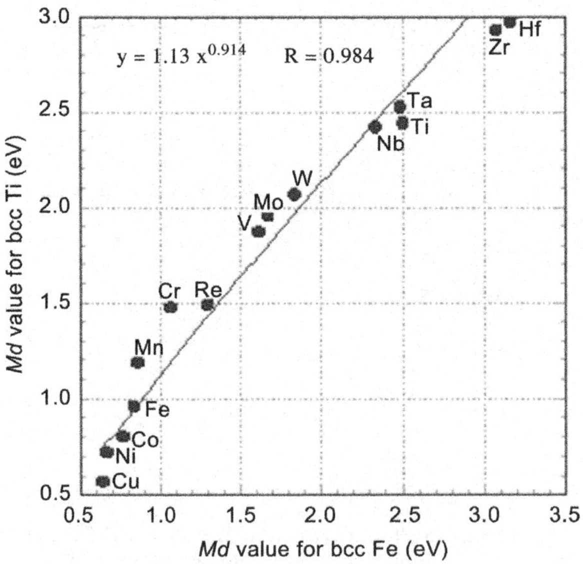

FIGURE 5-5 Comparison of *Md* value for bcc Ti with that for bcc Fe.

since Ti is far away from Fe in the periodic table. However, the whole trend of the *Md* parameter change with M still resembles between them.

Hereafter, the parameters obtained for bcc Ti are employed for the following analysis of alloy properties, since these parameters appear to be relatively insensitive to the crystal structure [1,2].

5.2 Correlation of Alloying Parameters With Alloy Properties

5.2.1 Classification of Binary Phase Diagrams

Three typical phase diagrams of binary Ti-M alloys are illustrated in Fig. 5-6A–C, where M's are transition metals. Here, (A) is the α and β perfect solid solution type, (B) is the β-isomorphous type, and (C) is the β-eutectoid type. As shown in Fig. 5-6, these three types of phase diagrams can be distinguished clearly on the *Bo* vs. *Md* diagram. The Ti-W phase

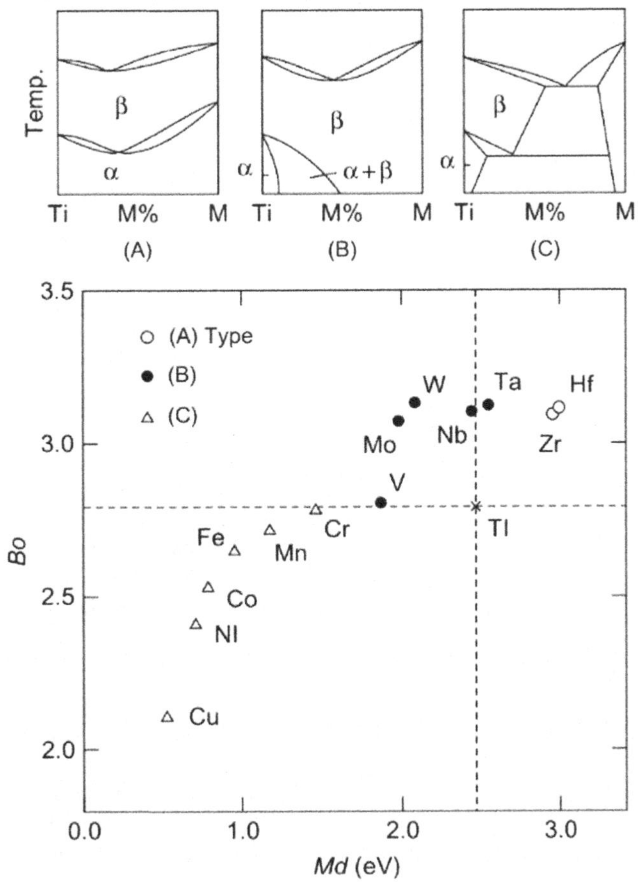

FIGURE 5-6 Representation of binary Ti-M alloys on the *Bo* vs. *Md* diagram.

diagram was previously grouped into (C), but according to a recent study it was revised and grouped into (B) [4], in agreement with the present classification.

5.2.2 Classification of Practically Used Alloys Into α, $\alpha+\beta$, and β-Types

We usually classify titanium alloys into the α, $\alpha+\beta$, and β types from the phases existing in the alloy. About 40 practically used alloys are presented on the $\overline{Bo}$-$\overline{Md}$ diagram shown in Fig. 5-7, where the $\overline{Bo}$ and $\overline{Md}$ values for each alloy are calculated from the alloy composition using Eqs. (3.1) and (3.2) in Chapter 3 [1,2]. Here, the alloy compositions are denoted in mass%. In binary Ti-M alloys the high-temperature β-phase is retained metastably even at room temperature when the alloy composition exceeds a certain critical value. The critical value changes with M, and such data, (a)–(g), for various M are plotted on this diagram.

It is evident from Fig. 5-7 that three types of alloys are located separately on this diagram. For instance, Ti-6Al-4V alloy (no. 9) exists in the $\alpha+\beta$ type field. Also, Ti-8Mn alloy (no. 6) exists in the β field despite the $\alpha+\beta$ type alloy. However, this is not a contradiction. Because Ti-8Mn is really a β-type alloy, since the 8% Mn content in the alloy exceeds a critical value of 6% Mn, as shown in Fig. 5-7. But the alloy is heat-treated in the ($\alpha+\beta$) temperature range to improve the mechanical property, and incidentally grouped into the $\alpha+\beta$ type alloy. Even if the alloy type is uncertain, it is predictable by calculating the $\overline{Bo}$ and $\overline{Md}$ values from the composition and plotting the alloy location on the $\overline{Bo}$-$\overline{Md}$ diagram shown in Fig. 5-7. No experiments are needed for this prediction.

Furthermore, the alloying vector for binary Ti-M alloys is presented in Fig. 5-8 to understand the alloying behavior of respective elements on the $\overline{Bo}$-$\overline{Md}$ diagram. Alloying vector

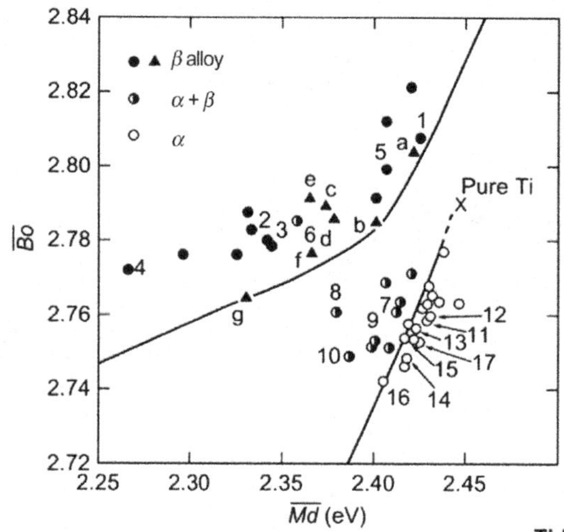

1. Ti-4.5 Sn-11.5 Mo-6 Zr (Beta III)
2. Ti-3 Al-8 V-4 Zr-4 Mo-6 Cr (Beta C)
3. Ti-3 Al-8 V-8 Mo-2 Fe (8-8-2-3)
4. Ti-3 Al-13 V-11 Cr (13-11-3)
5. Ti-15 Mo-5 Zr-3 Al
6. Ti-8 Mn (8 Mn)
7. Ti-6 Al-6 Mo-2 Sn-4 Zr (6-2-4-6)
8. Ti-5 Al-2 Sn-2 Zr-4 Mo-4 Cr (Ti-17)
9. Ti-6 Al-4 V (6-4)
10. Ti-6 Al-6 V-2 Sn (6-6-2)
11. Ti-2.25 Al-11 Sn-5 Zr-1 Mo-0.2 Si (IMI-679)
12. Ti-6 Al-0.5 Mo-5 Zr-0.2 Si (IMI-685)
13. Ti-6 Al-2 Mo-2 Sn-4 Zr (6-2-4-2)
14. Ti-5 Al-6 Sn-2 Zr-1 Mo-0.2 Si (5621 S)
15. Ti-5 Al-2.5 Sn (A-110)
16. Ti-8 Al-1 V-1 Mo (8-1-1)
17. Ti-5.5 Al-3.5 Sn-3 Zr-0.3 Mo-1 Nb-0.3 Si (IMI-829)

Ti-M(mass%) (a) M=Mo(10%), (b) Fe(4%), (c) Cr(8%), (d) Mn(6%), (e) V(15%), (f) Co(6%), (g) Ni(8%)

FIGURE 5-7 Grouping of commercial titanium alloys into the three types of α, $\alpha+\beta$, and β alloys on the $\overline{Bo}$-$\overline{Md}$ diagram.

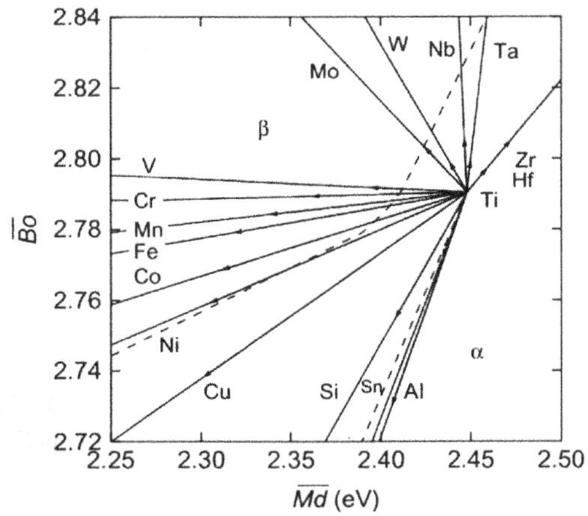

FIGURE 5-8 Alloying vectors of the elements in Ti. Each vector represents the location of Ti-10 mass% M alloy.

starts at the position of pure Ti and ends at the position of Ti-10 mass% M alloy. By comparing Fig. 5-7 with Fig. 5-8, we may see that for example the alloying vector for the Ti-Al alloy gets into the α-phase field as the Al content increases, indicating that Al is an α-stabilizing element. On the other hand, for example, V, Nb, and Ta are the β-stabilizing elements, because their vectors are directed toward the β-phase field. These results are in agreement with the well-known alloying behavior of elements in titanium.

5.2.3 Boundary Between Slip and Twin Deformation

In the course of the deformation of β-type alloys either the slip or the twin mechanism is operating, depending on the stability of the β-phase [5]. The slip mechanism is dominant when the alloy has high β-phase stability, whereas the twin mechanism is dominant when the alloy has relatively low β-phase stability. The boundary compositions between the slip and twin mechanism are determined by examining deformation bands that appear around a Vickers indentation. As shown in Fig. 5-9, wavy slip bands appear when the slip mechanism is dominant, whereas straight twin bands appear when the twin mechanism is dominant. This observation is carried out for a variety of binary alloys such as Ti-V, Ti-Cr, Ti-Mn, Ti-Fe, Ti-Co, Ti-Nb, Ti-Mo, Ti-W, and also for multiple-component Ti-Al-V-Cr-Mo-Zr alloys with various compositions. The results are summarized in the $\overline{Bo}$-$\overline{Md}$ diagram shown in Fig. 5-9. It is noticed that most of commercially available alloys exist along the slip/twin boundary. For instance, Ti-10%V-2%Fe-3%Al (10-2-3) is in the region of the twin mechanism, and Ti-15%V-3%Cr-3%Sn-3%Al (15-3-3-3) is in the region of the slip mechanism near the slip/twin boundary.

In addition to the slip or twin deformation region, the region for the appearance of martensite phase in the alloy is specified on the $\overline{Bo}$-$\overline{Md}$ diagram shown in Fig. 5-9 [2,5].

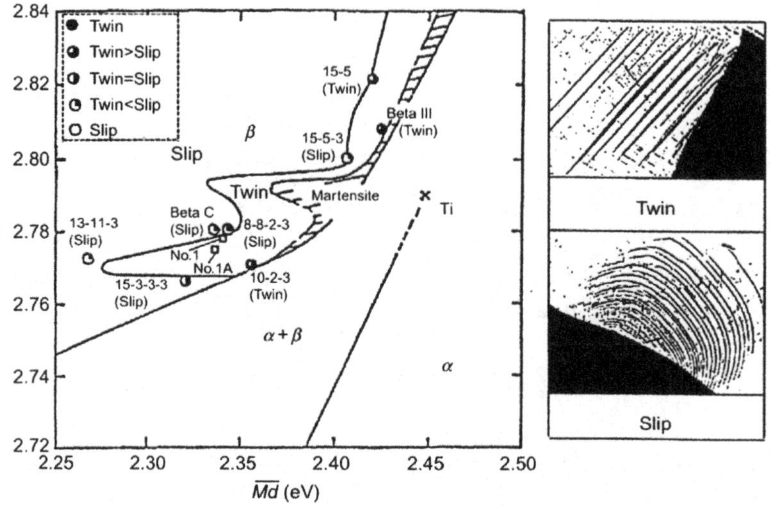

FIGURE 5-9 Slip/twin boundary and martensite region on the $\overline{Bo}$-$\overline{Md}$ diagram.

5.2.4 Corrosion Resistance

High corrosion resistance is one of the characteristics of titanium, but it is still corroded in a reducing atmosphere such as hydrochloric acid and sulfuric acid. But the corrosion resistance of titanium is improved by alloying. To understand the alloying behavior of elements on the corrosion resistance, a series of polarization curves is measured at 343K in both 10 mass% H_2SO_4 and 10 mass% HCl solutions with a variety of binary Ti-M alloys (M = Al, V, Cr, Fe, Co, Ni, Zr, Nb, Mo, Hf, and Ta) [6]. It is found that titanium alloys with higher $\overline{Bo}$ values show lower critical anodic current density in the polarization curve and hence the higher corrosion resistance even in a 10 mass% H_2SO_4 solution at 343K, as shown in Fig. 5-10A–B. A similar result is also obtained in a 10 mass% HCl solution at 343K [6]. Thus, the $\overline{Bo}$ parameter is convenient to predict the corrosion resistance of Ti alloys in the acid environments such as H_2SO_4 and HCl solution.

On the other hand, it is well known that passivation causes excellent corrosion resistance of titanium alloys, even though the alloying effect on the passivation is still unknown. Morishita et al. [7] simulated local electronic states of the cathode using the DV-Xα cluster method, and proposed a local electrochemical cell model shown in Fig. 5-11. The alloying elements discussed here are M = Re, Os, Ir, Pt, Au, Tc, Ru, Rh, Pd, and Ag. Every element has a lower *Md* value than Ti as shown in Table 5-1, so that charge transfer takes place in the direction from Ti to M. Therefore, in the cell model shown in Fig. 5-11, the titanium matrix region behaves like the anode and the alloying element containing region behaves like the cathode.

The anodic reaction in aqueous corrosion is expressed as, $Ti \rightarrow Ti^{2+} + 2e^-$, and the cathodic reaction is expressed as, $2H^+ + 2e^- \rightarrow H_2$. This cathodic reaction is supposed to be enhanced when conduction electrons are localized in the cathode region and then discharged smoothly to H^+, so as to promote the hydrogen evolution on the cathode surface.

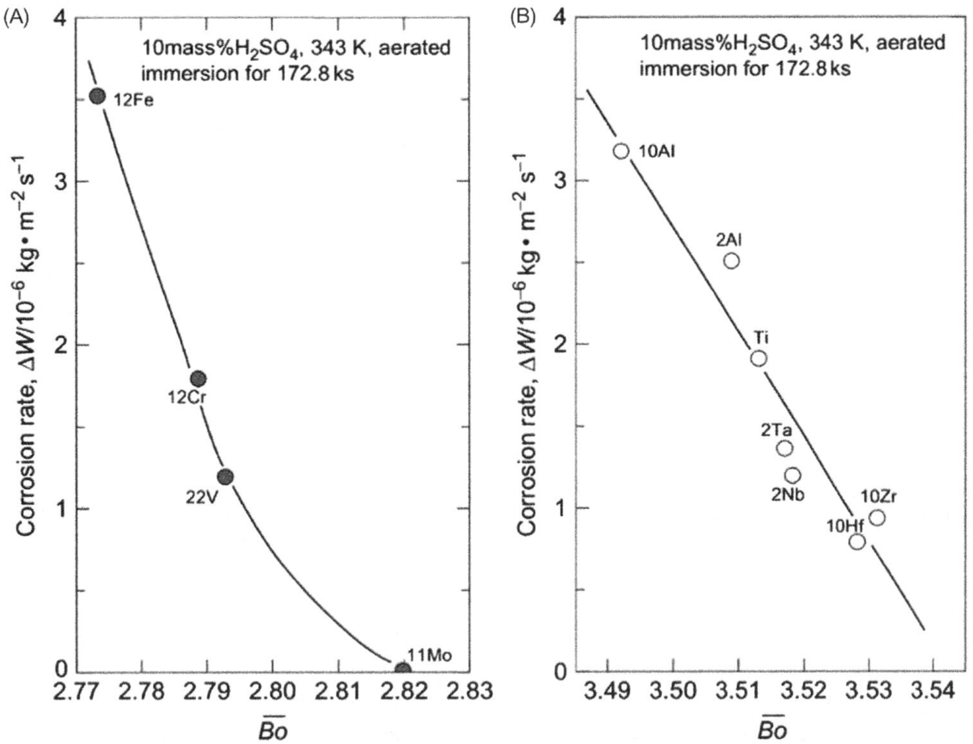

FIGURE 5-10 Correlation of corrosion rate, ΔW, with $\overline{Bo}$ for (A) bcc and (B) hcp Ti-M binary alloys in 10% H_2SO_4 at 343K.

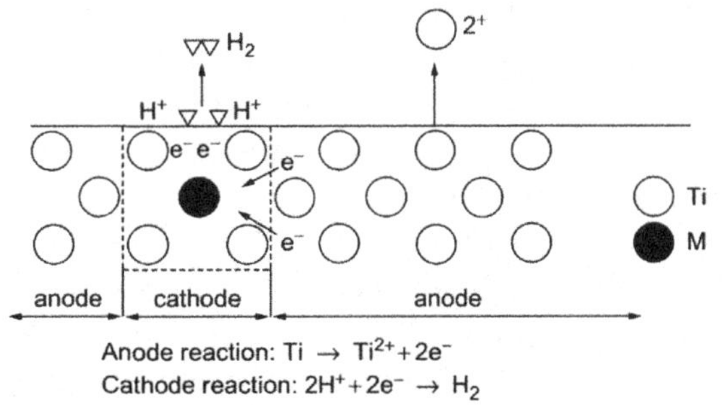

FIGURE 5-11 Schematic model of the electrochemical cell.

Following this proposed model, the tendency to the electron localization is estimated from the density of states near the Fermi level, as it is a measure to show the capacity of the electrons in the cathode region that are transferred from the anode region. Also, the tendency to discharge electrons to H^+ is related to the height of the Fermi level. This is because the high Fermi level gives a circumstance for the electrons to be loosely bound to the cathode region and hence to be discharged easily from the cathode surface.

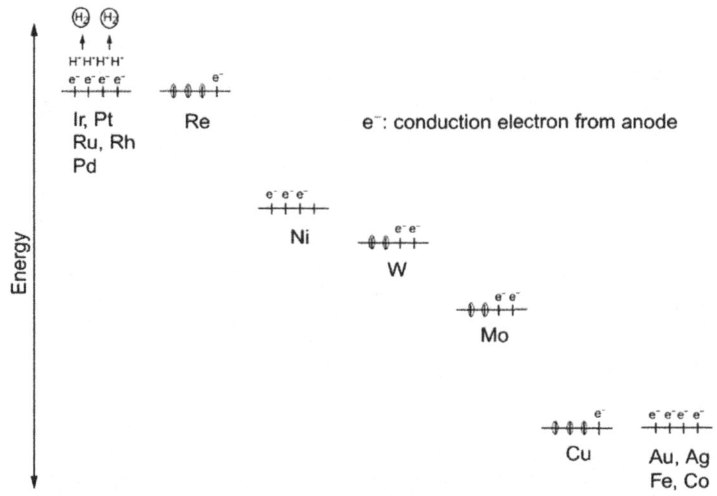

FIGURE 5-12 Electronic model of cathodic reaction.

This proposed model is confirmed experimentally from the measurement of the polarization curves of a variety of Ti-0.1mol%M alloys. For instance, the cathodes containing Ir, Pt, Ru, Rh, and Pd meet the previously mentioned two conditions to activate the cathodic reaction. This is because as shown in Fig. 5-12, for all these elements the density of states near the Fermi level is high enough to promote the electron localization and also the Fermi level itself is high enough to promote the smooth electron discharging. It is also found experimentally that for any alloy containing these elements, the measured hydrogen overpotential is small and the cathodic polarization curve shifts to the noble potential side. These experimental results clearly indicate that the cathodic reaction is actually enhanced, which is consistent with the theoretical prediction. Thus, the proposed model based on the molecular orbital calculation is reasonably good for representing the local electronic states of the cathode in aqueous corrosion. The hydrogen overpotential of transition metal–based alloys and compounds will be presented in Chapter 7.

Using the present approach, Ti-0.5%Ni-2%Ta-0.05%Ir alloy has been developed by Morishita et al. This alloy exhibits high aqueous corrosion resistance even in any acid atmospheres including nitric acid. Also, new $\alpha+\beta$ type alloys of higher corrosion resistance in acid atmosphere than Ti-6Al-4V (ELI) alloy have been developed by Okazaki et al. [8]. The alloy compositions for medical implants are Ti-15%Zr-4%Nb-2%Ta-0.2%Pd and Ti-15%Sn-4% Nb-2%Ta-0.2%Pd (mass %), which are free from the toxic elements V and Al.

5.3 Alloy Design of Titanium Alloys

5.3.1 High Strength β-Type Alloys

First, one example of alloy design is explained in a concrete manner. Once a specific alloy system and a specific $\overline{Bo}$-$\overline{Md}$ region are set on the diagram shown in Fig. 5-7, the alloy composition is obtained by applying the rule of the vector sum to the alloying vectors shown in Fig. 5-8. It is no longer needed to optimize the alloy composition from trial-and-error experiments.

As described earlier, Ti-15-3-3-3 deforms by the slip mechanism, but Ti-10-2-3 deforms by the twin mechanism. In general, the lower yield stress and the longer fracture elongation are expected in the twin-deformed alloy as compared with the slip-deformed alloy [5]. This means that the twin deformation enhances the cold workability, whereas the slip deformation yields the high strength of Ti alloys. To meet this opposite nature between cold workability and high strength, it is proposed to design a new alloy in which the twin mechanism works in the solution-treated state, but the slip mechanism works in the aged state [5]. Namely, it is supposed to be convenient for the Ti alloy to keep a large elongation while it is cold-rolled in the solution-treated state, but to have a high strength in the final aged state to guarantee many years of service.

According to the $\overline{Bo}$-$\overline{Md}$ diagram, if the alloy is located inside the twin region in Fig. 5-9, the twin deformation will take place in the solution-treated state. However, once the α-phase is decomposed in the β-type alloy by the subsequent aging, the alloy composition of the remaining matrix shifts to the more β-stabilized region in the diagram, so that inevitably the slip deformation works in the aged state.

From this point of view, high strength β-type alloys Ti-13%V-3%Cr-2%Nb-(2-3)%Al (mass %) have been designed [5]. Hereafter, the 2% Al—containing alloy is named no. 1 alloy, and the 3% Al—containing alloy is named no.1A alloy. Their locations are indicated in Fig. 5-9.

In these designed alloys, V and Al are chosen because these elements have high values of the (bond order/atomic weight) ratio, as shown in Figs. 5-3B and 5-4B, and hence the specific strength may be high. Needless to say, Al is a strengthening element of the α-phase, which precipitates in the β matrix by proper aging treatments. Cr is chosen to improve the ductility of alloys. Nb is chosen to distribute the alloy position in a proper region on the $\overline{Bo}$-$\overline{Md}$ diagram. For this Ti-V-Al-Cr-Nb system, the alloy location is selected inside the twin region but close to the slip/twin boundary as shown in Fig. 5-9. The no. 1 and no. 1A alloys are positioned near the region between Ti-10-2-3 and Ti-15-3-3-3 on the $\overline{Bo}$-$\overline{Md}$ diagram.

The results of the tensile test are listed in Table 5-2 for the alloys in the solution-treated state and the aged state as well. It is found that the twin mechanism works in both the designed alloys (no. 1 and no. 1A) when they are solution-treated at 1093K. The yield strength (YS) is about 700 MPa and the ultimate tensile strength (UTS) is about 800 MPa with the elongation of about 30%. However, the slip mechanism works when aged at 733K for 18 ks in the no. 1 alloy and for 28.8 ks in the no. 1A alloy. The size of the α-phase dispersed in the no. 1 and no. 1A alloys is about 60 nm, being finer than the size of the needle-like α phase in Ti-15-3-3-3, about 0.35 μm in length and 0.05 μm in width. As listed in Table 5-2, the UTS for no. 1 and no. 1A alloys is as high as 1400 ~ 1500 MPa, and much stronger than about 1200 MPa for Ti-15-3-3-3. The measured alloy density is 4.684 (g/cm^3) for no. 1 alloy and 4.660 (g/cm^3) for no. 1A alloy, being lower than 4.718 (g/cm^3) for Ti-15-3-3-3. Also, the designed alloys are superior in the fracture toughness to Ti-15-3-3-3 [5]. Thus, following a theoretical alloy design method, new β-type alloys, Ti-13%V-3%Cr-2%Nb-(2-3)% Al (mass%) have been developed without repeating any trial-and-error experiments.

Table 5-2 Experimental Results of the Tensile Test for the Designed Alloys (no. 1 and no. 1A) and Ti-15-3-3-3

Alloy no.	Chemical Composition (Mass %)	Heat Treatment	Yield Strength (MPa)	Ultimate Tensile Strength (MPa)	Elongation (%)
1	Ti-13V-2Al-3Cr-2Nb	S.T.	685	799	29.5
		S.T.A.	1464	1507	7.0
1A	Ti-13V-3Al-3Cr-2Nb	S.T.	719	800	30.0
		S.T.A.	1321	1446	6.0
Ti-15-3-3-3	Ti-15V-3Al-3Cr-3Sn	S.T.	804	813	15.0
		S.T.A.	1161	1242	11.0

S.T., Solution treatment; S.T.A., aging after solution treatment.

5.3.2 β-Type Alloys for Biomedical Applications

Recently, great attention has been directed toward the β-type alloys for biomedical applications, because they possess superior properties including shape memory effect, superelasticity, and low Young's modulus [9]. For example, to replace human bone with some alloy, we have invested a lot of time and effort in the development of those β-type alloys that exhibit low Young's modulus and high strength. The present molecular orbital approach has been applied to the design of such bioimplant Ti-based alloys.

For instance, Niinomi et al. [10] developed a new β-type alloy with the composition of Ti-29%Nb-13%Ta-4.6%Zr (TNTZ) (mass%), which is composed only of nontoxic elements such as Nb, Ta, and Zr. The Young's modulus is about 50 GPa, which is much lower than that of Ti-6Al-4V(ELI). Saito et al. [11] also reported that gum metals appear in the neighborhood of the TNTZ alloy. Namely, their positions are close to $\overline{Bo} = 2.87$ and $\overline{Md} = 2.45$(eV) on the $\overline{Bo}$-$\overline{Md}$ diagram. Another criterion for the appearance of gum metals is the *e/a* value of 4.24. The alloy compositions are, for instance, Ti-12%Ta-9%Nb-3%V-6%Zr-1.5%O and Ti-23%Nb-0.7%Ta-2%Zr-1.2%O (mol%). When gum metals undergo 90% cold working, the superelasticity and low elastic modulus emerge besides other superior properties. The gum metals are going to be used for the tooth-straightening wire.

5.3.3 Extension of $\overline{Bo}$-$\overline{Md}$ Diagram Over the Higher $\overline{Bo}$ Region

Recently, the $\overline{Bo}$-$\overline{Md}$ diagram shown in Fig. 5-7 has been extended to the region where the $\overline{Bo}$ value is higher than 2.84, because superior β-type alloys such as TNTZ alloys and gum metals appear in such a high $\overline{Bo}$ region. The extended diagram is shown in Fig. 5-13 [12].

The β-phase stability depends on the alloy composition, or on the temperature even at the same composition. The martensite phase and ω (omega) phase appear in some compositional ranges of the β-type alloys. Related phase constitution is illustrated on the extended $\overline{Bo}$-$\overline{Md}$ diagram in Fig. 5-13. As to the martensitic transformation, it starts at the *Ms* temperature and finishes at the *Mf* temperature, both of which vary largely depending on the alloy

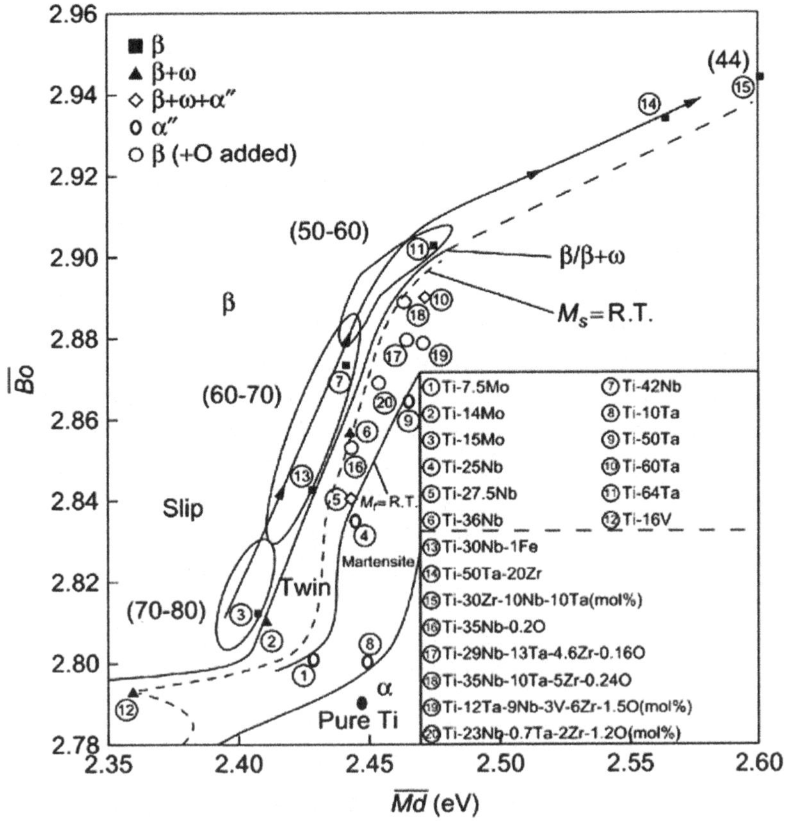

FIGURE 5-13 Extended $\overline{Bo}$-$\overline{Md}$ diagram to the region where $\overline{Bo}$ is higher than 2.84. The $\beta/\beta + \omega$ phase boundary is shown together with the martensite boundaries for Ms = RT and for Mf = RT. The Young's modulus (GPa) is also given in parentheses for typical alloys.

composition. So, the iso-Ms temperature curves of Ms = RT (room temperature) and Mf = RT are drawn in the $\overline{Bo}$-$\overline{Md}$ diagram.

In particular, a least stable β-phase alloy is newly defined here as the alloy containing the least amount of the β-stabilizing elements to get a β single phase. A least stable β-phase alloy lies along the narrow region of the left or upper side of the $\beta/\beta + \omega$ phase boundary indicated by a solid line in Fig. 5-13. As explained in the next section, the lowest Young's modulus is observed in such a least stable β-phase alloy.

5.3.4 Correlation of Phase Stability With Alloy Properties

The phase stability is related closely to the alloy properties such as the Young's modulus, the superelasticity, and the shape memory effect. So, this is examined here from a viewpoint of alloy design, using the extended $\overline{Bo}$-$\overline{Md}$ diagram [12].

5.3.4.1 Young's Modulus

It is known that the α" martensite phase has a lower Young's modulus than the α' martensite phase. The ω phase has the highest modulus, so its precipitation should be suppressed in the alloy to keep the Young's modulus low [13]. As compared with these phases, the β-phase has the lowest modulus in most titanium alloys [14,15]. Also, the β-phase stability increases with the content of the β-stabilizing elements, M, in various binary Ti-M alloys. For instance, when M is either V, or Nb, or Ta, the measured Young's modulus increases monotonously with the M content in the β-phase alloy. In other words, the least stable β-phase alloy has the lowest Young's modulus among the β-phase alloys, as mentioned above. For instance, the lowest Young's modulus is about 50 GPa for Ti-64Ta, 60 GPa for Ti-40Nb, and 72 GPa for Ti-26V (mass%), so that the Young's modulus changes in the order, Ti-Ta < Ti-Nb < Ti-V. Also, there is a trend that the Young's modulus decreases with increasing $\overline{Bo}$ value, since it is $\overline{Bo} = 2.794$ for Ti-26V, $\overline{Bo} = 2.869$ for Ti-40Nb, and $\overline{Bo} = 2.903$ for Ti-64Ta.

This is also seen from Fig. 5-13, where there is a region of the least stable single β-phase along the β/β + ω phase boundary. Among the least stable β-phase alloys lying along the phase boundary, the Young's modulus decreases with the $\overline{Bo}$ value, as is illustrated in Fig. 5-13 using an arrow direction. For example, the Young's modulus decreases from 70–80 GPa in ③ Ti-15Mo (mass%) to 44 GPa in ⑮ Ti-30Zr-10Nb-10Ta (mol%), as the $\overline{Bo}$ value increases. Therefore, the alloying of high *Bo* elements such as Zr, Nb, Mo, Hf, Ta, and W is preferable for the design of β-phase alloys with low Young's modulus.

5.3.4.2 Change in β-phase Stability With the Addition of O, Al, Sn, and Zr

The addition of a small amount of the α-stabilizing elements such as O, Sn, and Al to β-phase alloys modifies the phase boundary and influences the mechanical properties [11,13,16–19]. For example, it is reported that the O addition lowers the modulus [16]. In the β-type alloys with the high $\overline{Bo}$ values, the elements, O, Sn, and Al, are no longer the α-stabilizing elements, but behave like the β-stabilizing elements. For instance, in the Ti-35Nb-7Zr- 5Ta- (0.06-0.6) O alloys (mass%) [18], the ω-phase formation is suppressed by the addition of O, so that O behaves like a β-stabilizing element in the single β-phase. This is probably interpreted as due to the pinning of the linear defects by oxygen, which hinders the collapse of $\{111\}_\beta$ atomic planes in the β-phase, and hence suppresses the ω-phase formation in the O-containing alloys. This may be one of the reasons why a small amount of O is added to most of the recently developed Ti alloys [11,17,18,20,21]. The oxygen addition suppresses even the martensite formation [22]. However, oxygen still behaves like a α-stabilizing element in the aging process, because the amount of the α-phase precipitated during aging increases as the oxygen content increases in the alloy [18].

Here, it is noted that any β-type alloys of ⑯ to ⑳ shown in Fig. 5-13, contain a small amount of oxygen. However, the oxygen contents in these alloys are not included in the calculation, because the *Bo* and *Md* parameters are unknown for oxygen. As described previously, oxygen does not follow the general feature of the α-stabilizer, but instead it behaves like the β-stabilizer. Then, with the addition of oxygen the single β-phase boundary shifts to the region of the lesser amount of β-stabilizing elements, as is indicated by the symbol ● in Fig. 5-13. For instance, for Ti-Nb-Ta-Zr-V and Ti-Nb alloys, the O addition appears to make a single β-phase region wider as is presented by the alloy numbers ⑯ to ⑳ in Fig. 5-13. In other words, they are located in the $\beta + \omega$ (or martensite) region lying below the $\beta/\beta + \omega$ phase boundary, despite that they are actually the single β-phase alloys. This is indeed a limitation of the present approach.

Also, zirconium (Zr) has been considered to be a neutral element with respect to the phase stability. However, it behaves in the nearly same way as does the oxygen in the β-phase alloys. Namely, Zr behaves like a weak β-stabilizing element in Ti-Nb-Zr alloy [23] and Ti-Cr-Zr alloy [19]. Its β-stabilizing effect is, however, weaker as compared with Al and Sn.

One example is shown in Fig. 5-14 for Ti-5Zr-*p*Nb-10Ta-0.23O (mass%) alloys ($p = 20$, 25, 30, 35, and 40) and Ti-5Zr-30Nb-*r*Ta-0.23O (mass%) alloys ($r = 0$, 5, 10, 15, and 20) [12,20]. It is apparent from Fig. 5-14 that the β-phase becomes more stable as Nb or Ta content

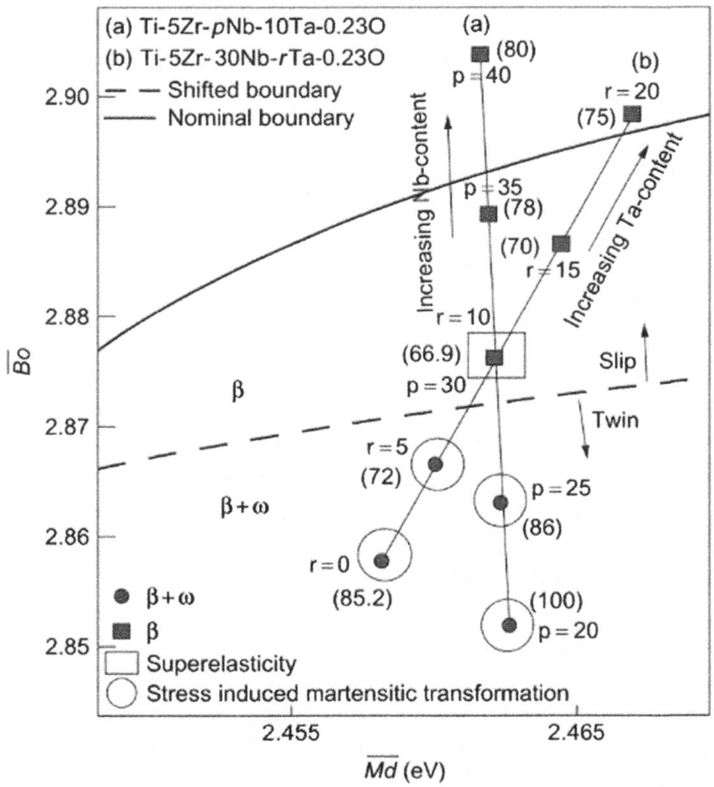

FIGURE 5-14 Phase stability changes with Nb and Ta contents in Ti-5Zr-Nb-Ta-0.23O alloys. The Young's modulus (GPa) is given in parentheses. The coalloying of O and Zr shifts the $\beta/\beta + \omega$ phase boundary from a solid curve to a dotted curve.

increases, so that the ω-phase formation is suppressed in them. Also, the coalloying of O and Zr shifts the β/β + ω phase boundary to the further lower compositional region of β-stabilizing elements (i.e., Nb and Ta), as indicated by a dashed curve in Fig. 5-14. The nominal boundary for the O- and Zr-free alloys is also indicated by a solid curve in Fig. 5-14. When these two curves are compared with each other, it is evident that the phase boundary shifts largely on the $\overline{Bo}$-$\overline{Md}$ diagram by the coalloying of O and Zr into the β-phase alloys.

The alloying effect of newly considered β-stabilizing elements on the β-phase stability seems to change in the order, O > Al > Sn > Zr. Special alloying behavior emerges by the addition of these elements into the β-phase alloys [12].

5.3.4.3 Superelasticity and Shape Memory Effect

For Ti-Nb binary alloys, superelasticity appears in the least stable β-phase alloy (i.e., Ti-[40-42] Nb [mass%]) [24], in which the reversible transformation takes place between the β-phase and the martensite at RT. This is because the energy difference between the β-phase and the martensite α″ phase is very small in such a least stable β-phase alloy. Also, the shape memory effect emerges in the lower Nb-content alloy (i.e., Ti-[35-40] Nb) [24], in which the *Ms* temperature is close to RT. However, neither the superelasticity nor the shape memory effect appears in the more stable β-phase alloys (>42Nb), in which the *Ms* temperature is well below RT. For a lower Nb-content alloy (i.e., Ti-30Nb) that is located in the martensite zone, neither the superelasticity nor the shape memory effect emerges. However, as indicated by alloy number ⑬ in Fig. 5-13, the addition to 1% Fe to this Ti-30Nb alloy moves the alloy location to the least stable β-phase zone, so it exhibits the superelasticity.

A compositional range of the alloys where either the superelasticity or the shape memory effect emerges, is modified by the addition of O, Al, Sn, and Zr. For instance, as mentioned above, the shape memory effect emerges in a Ti-38.5Nb alloy, but the further addition of 1.5% Al to this alloy yields a 4.3% superelastic strain [25]. Such a variation in the alloy property is also seen in Ti-36Nb by changing the O content in it. Namely, the shape memory effect still emerges in Ti-36Nb-(0-0.12) O alloys, but the further addition of O causes the superelasticity in the Ti-36Nb-(0.22−0.43) O alloys, in which the least stable single β-phase exists at RT [17]. Thus, the emergence of the superelasticity is further promoted in the O-containing Ti-Nb alloys than in the O-free Ti-Nb alloys [17].

This is also true in the multiple-components Ti-5Zr-*p*Nb-10Ta-0.23 O alloys ($p = 20$, 25, 30, 35, and 40) [20]. Neither phenomenon is observed in the stable β-phase alloys (i.e., 35 Nb). However, as shown in Fig. 5-14, the superelasticity emerges in the least stable β-phase alloys (i.e., 30 Nb). The stress-induced martensitic transformation, which is associated with the shape memory effect, is observed in the lower Nb content alloys (i.e., 20 Nb and 25 Nb) [21]. It is noted here that the least stable β-phase alloy (i.e., 30 Nb) has the lowest Young's modulus, as explained previously.

Thus, the β-phase stability of the alloys correlates well with the alloy properties such as the low Young's modulus, the superelasticity, and the shape memory effect, and such a correlation is traceable on the $\overline{Bo}$-$\overline{Md}$ diagram. The special alloying of O, Al, Sn, and Zr induces a remarkable phase boundary shift, which will give us plenty of scope for the modification of the phase stability and hence the alloy properties in a reasonable way.

Thus, the present alloy design theory is constructed not only for the fundamental understanding of alloys, but also for the practical alloy design without repeating trial-and-error experiments [26,27]. In particular, this theory is useful for the design of titanium alloys, since the $\overline{Bo}$-$\overline{Md}$ diagram represents well the phase constitution in the alloys, and such knowledge of the phase stability presented on the $\overline{Bo}$-$\overline{Md}$ diagram is important in predicting various alloy properties such as the mechanical strength, corrosion resistance, Young's modulus, superelasticity, and shape memory effect. Recalling that all these properties are essential to the biomedical titanium alloys, it is greatly expected that the present theory works well in practice in the course of the alloy design for biomedical applications.

References

[1] M. Morinaga, N. Yukawa, H. Adachi, Electronic structure and phase stability of itanium alloys, Tetsu-to Hagane 72 (1986) 555–562.

[2] M. Morinaga, N. Yukawa, T. Maya, K. Sone, H. Adachi, Theoretical design of titanium alloys, Proc. Sixth World Conf. on Titanium, Cannes, France, June 6–9, 1988, Société Française de Métallurgie, Paris, (1988), pp. 1601–1606.

[3] R.S. Mulliken, Electronic population analysis on LCAO-MO molecular wave functions.1.,2.,3.,4, J. Chem. Phys 23 (1955) 1833–1840, 1841–1846, 2338–2342, 2343–2346.

[4] H. Okamoto, Phase Diagrams for Binary Alloys, Second Edition, ASM International, Metals Park, Ohio, 2010.

[5] M. Morinaga, M. Kato, T. Kimura, M. Fukumoto, I. Harada, K. Kubo, Theoretical design of β-type titanium alloys, Proc. 7th World Conference on Titanium, San Diego, California, June 29-July 2, 1992, TMS, Warrendale, Pennsylvania (1992), pp. 217–224.

[6] M. Morishita, Y. Ashida, M. Chikuda, M. Morinaga, N. Yukawa, H. Adachi, Active corrosion rate for Ti-based alloys in aqueous corrosion and its correlation with the bond order obtained by electron theory, ISIJ Int. 31 (1991) 890–896.

[7] M. Morishita, M. Chikuda, Y. Ashida, M. Morinaga, N. Yukawa, H. Adachi, Electronic states of the cathodes of titanium-based alloys in aqueous corrosion, Mater. Trans. JIM 32 (1991) 264–271.

[8] Y. Okazaki, A. Ito, T. Tateishi, Y. Ito, Effect of alloying elements on anodic polarization properties of titanium alloys in acidic solutions, Mater. Trans. JIM 35 (1994) 58–66.

[9] M. Semlitsch, F. Stab, H. Webber, Titanium-aluminium-niobium alloy, development for biocompatible, high strength surgical implants, Biomed. Tech 30 (1985) 334–339.

[10] D. Kuroda, M. Niinomi, M. Morinaga, Y. Kato, T. Yashiro, Design and mechanical properties of new β type titanium alloys for implant materials, Mater. Sci. Eng. A 243 (1998) 244–249.

[11] T. Saito, T. Furuta, J.H. Hwang, S. Kuramoto, K. Nishino, N. Suzuki, et al., Multifunctional alloys obtained via a dislocation-free plastic deformation mechanism, Science 300 (2003) 464–467.

[12] M. Abdel-Hady, K. Hinoshita, M. Morinaga, General approach to phase stability and elastic properties of β-type Ti-alloys using electronic parameters, Scripta Materialia 55 (2006) 477–480.

[13] H. Ikehata, N. Nagasako, T. Furura, A. Fukumoto, K. Miwa, T. Saito, First-principles calculations for development of low elastic modulus Ti alloys, Phys. Rev. B 70 (2004) 174113.

[14] L.A. Matlakhova, A.N. Matlakhova, S.N. Monteiro, S.G. Fedotov, B.A. Goncharenko, Properties and structural characteristics of Ti-Nb-Al alloys, Mater. Sci. Eng. A 393 (2005) 320–326.

[15] Y.L. Zhou, M. Niinomi, T. Akahori, Effects of Ta content on young's modulus and ensile properties of binary Ti-Ta alloys for biomedical applications, Mater. Sci. Eng. A 371 (2004) 283–290.

[16] T. Ozaki, H. Matsumoto, S. Watanabe, S. Hanada, Beta Ti alloys with low young's modulus, Mater. Trans 45 (2004) 2776–2779.

[17] J.I. Kim, H.Y. Kim, H. Hosoda, S. Miyazaki, Shape memory behavior of Ti-22Nb-(0.5-2.0)O (at%) biomedical alloys, Mater. Trans. 46 (2005) 852–857.

[18] J.I. Qazi, B. Marquardt, H.J. Rack, High–strength metastable beta-titanium alloys for biomedical applications, JOM 56 (2004) 49–51.

[19] S. Ishiyama, S. Hanada, O. Izumi, Effect of Zr, Sn and Al additions of deformation mode and beta phase-stability of metastable beta Ti alloys, ISIJ Int 31 (1991) 807–813.

[20] N. Sakaguchi, M. Niinomi, T. Akahori, T. Saito, T. Furuta, Effects of alloying elements on elastic modulus of Ti-Nb-Ta-Zr system alloy for biomedical applications, Mater. Sci. Forum 449-4 (2004) 1269–1272.

[21] N. Sakaguchi, M. Niinomi, T. Akahori, Tensile deformation behavior of Ti-Nb-Ta-Zr biomedical alloys, Mater. Trans. 45 (2004) 1113–1119.

[22] M. Tahara, H.Y. Kim, T. Imamura, H. Hosoda, S. Miyazaki, Lattice modulation and superelaticity in oxgen-added beta-Ti alloys, Acta Mater 59 (2011) 6208–6218.

[23] J.I. Kim, H.Y. Kim, T. Inamura, H. Hosoda, S. Miyazaki, Shape memory characteristics of Ti-22Nb-(2-8) Zr(at%) biomedical alloys, Mater. Sci. Eng. A 403 (2005) 334–339.

[24] H.Y. Kim, H. Satoru, J.I. Kim, H. Hosoda, S. Mitazaki, Mechanical properties and shape memory behavior of Ti-Nb alloys, Mater. Trans 45 (2004) 2443–2448.

[25] Y. Fukui, T. Inamura, H. Hosoda, K. Walashima, S. Miyazaki, Mechanical properties of a Ti-Nb-Al shape memory alloy, Mater. Trans. 45 (2004) 1077–1082.

[26] M. Morinaga, Y. Murata, H. Yukawa, Molecular orbital approach to alloy design, in: G. Bozzolo, et al. (Eds.), Applied Computational Materials Modeling-Theory, Simulation and Experiment, Springer, New York, 2007, pp. 255–306.

[27] M. Morinaga, Alloy design based on molecular orbital method, Mater. Trans 57 (2016) 213–226.

6 Aluminum Alloys and Magnesium Alloys

6.1 Parameters Related to Mechanical Properties

Mechanical properties of metals and alloys have been understood by dislocation theory. For instance, for aluminum alloys, the precipitation hardening is interpreted as due to the dislocation bowing and subsequent dislocation loop formation around a precipitate (called Orowan's mechanism) or due to the dislocation cutting a precipitate [1–5]. High densities of dislocations introduced to the alloy by cold working cause the strain hardening [5,6]. In addition, the solid solution hardening arises from the disturbing effect of solute atoms on the dislocation motion [7]. However, despite great efforts there is still a large barrier to the quantitative prediction of alloy strength with the aid of dislocation theory. This is probably due to the complexities in the interactions between dislocations and solute atoms, and also the poor understanding of dislocation core structures [5–7].

6.1.1 Al Alloys

To get an alternative method for the prediction of mechanical properties, it is strongly required to find out the parameters that are related closely to the mechanical properties of Al alloys [1,2]. Among various parameters, the electronegativity (Φ) and the atomic radius (R) are chosen [8,9], since these are classical parameters to represent the nature of chemical bond between atoms in solids [10]. For an alloy, the average difference values of these parameters are defined simply by taking the compositional average. $\Delta\overline{\Phi}$ and $\Delta\overline{R}$ are expressed as,

$$\Delta\overline{\Phi} = \Sigma\, X_i|\Phi_i - \Phi_{Al}|, \tag{6.1}$$

$$\Delta\overline{R} = \Sigma\, X_i|R_i - R_{Al}|. \tag{6.2}$$

Here, X_i is the mole fraction of component i in the alloy, Φ_i and R_i are the electronegativity and the atomic radius of component i, respectively, and the summation extends over the components, $i = 1, 2, \ldots, n$.

The atomic radius difference, $|R_i - R_{Al}|$, is an indicator to show the magnitude of local lattice strain around the alloying element i in aluminum, and hence $\Delta\overline{R}$ means the overall average value of the lattice strain in the alloy. On the other hand, the electronegativity difference, $|\Phi_i - \Phi_{Al}|$, is an indicator to show the amount of the transferred charges between the i atom and the surrounding Al atom, and hence $\Delta\overline{\Phi}$ means the overall average value of

A Quantum Approach to Alloy Design. DOI: https://doi.org/10.1016/B978-0-12-814706-1.00006-6

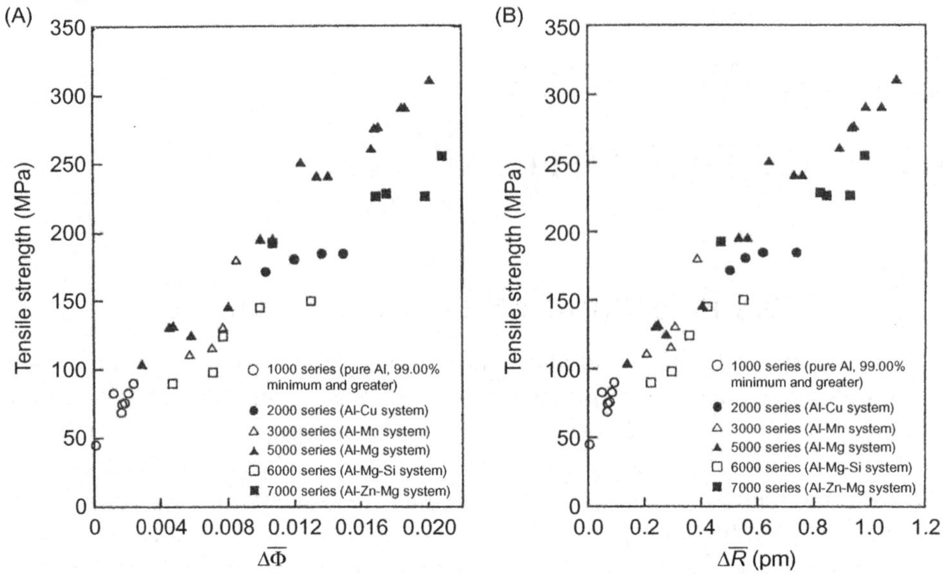

FIGURE 6-1 Correlation of tensile strength with (A) $\Delta\overline{\Phi}$ and (B) $\Delta\overline{R}$ for the O temper Al alloys.

the transferred charges between alloying element and aluminum atom in the alloy. Such a charge transfer yields some changes in the chemical bond strength between constituent atoms in the alloy. As the amount of the transferred charges increases, the chemical bond strength increases in the alloy [9].

The tensile strengths of commercially available wrought aluminum and aluminum alloys are shown in Fig. 6-1A for $\Delta\overline{\Phi}$ and Fig. 6-1B for $\Delta\overline{R}$. The standard strength data are taken from the Metals Handbook [11] and the Aluminium Handbook [12]. Unless otherwise noted, the strength data used here are measured ones at room temperature. In the figure, every alloy is identified by the four digits, following the Aluminum Association Designations [13]. It is apparent that the tensile strength increases with increasing $\Delta\overline{\Phi}$ and $\Delta\overline{R}$, even though the data are somewhat scattered, in particular for $\Delta\overline{\Phi}$. This scattering may arise partially from the reason that both these parameters are assigned to pure elements and inevitably any alloying effect is not taken into account in them. To reduce the scattering, it is desirable to introduce a new alloying parameter that reflects explicitly the alloying behavior of elements in aluminum, while still retaining a correlation with both the electronegativity and the atomic radius of elements. Such an alloying parameter is obtained from the molecular orbital calculation using alloy cluster models, as previously explained in Chapters 3, 4, and 5.

6.1.2 Magnesium Alloys

Ninomiya et al. [3] measured the Vickers hardness at room temperature for a variety of magnesium alloys to get a fundamental data set of mechanical properties. The alloy compositions are given in Ref. [3]. In Fig. 6-2, the measured hardness data are plotted against

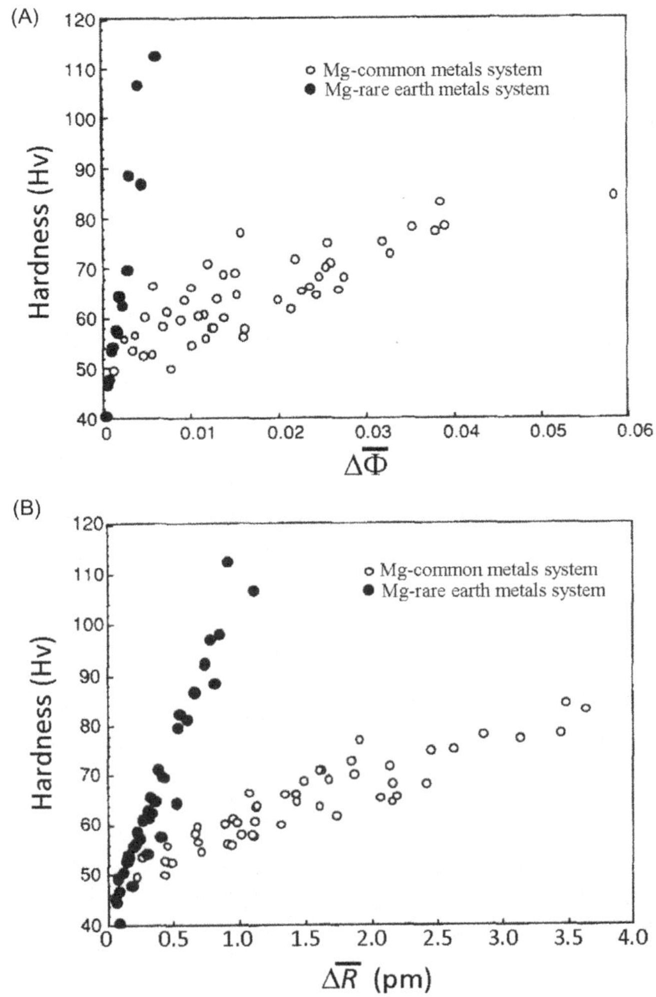

FIGURE 6-2 Correlation of the Vickers hardness with (A) $\Delta\overline{\Phi}$ and (B) $\Delta\overline{R}$ for Mg alloys.

Fig. 6-2A the electronegativity difference, $\Delta\overline{\Phi}$ and Fig. 6-2B the atomic radius difference, $\Delta\overline{R}$ [3]. Similarly to Eqs. (6.1) and (6.2) for Al, $\Delta\overline{\Phi}$ is defined as, $\Delta\overline{\Phi} = \Sigma X_i|\Phi_i - \Phi_{Mg}|$, and $\Delta\overline{R}$ is defined as, $\Delta\overline{R} = \Sigma X_i|R_i - R_{Mg}|$. It is apparent from these figures that the hardness increases with increasing $\Delta\overline{\Phi}$ and $\Delta\overline{R}$. The correlation seems better in $\Delta\overline{R}$ than in $\Delta\overline{\Phi}$. But there is a separation in the plots between the Mg–common metals system and Mg–rare earth metal system.

In either Al alloys or Mg alloys, it is required to first determine a new alloying parameter that correlates with the electronegativity and atomic radius of elements. For this purpose, a series of molecular orbital calculations is performed for Al and Mg.

6.2 Molecular Orbital Calculation and Alloying Parameter *Mk* for Al

6.2.1 Alloying Parameter *Mk* for Al

The electronic structure calculations are performed using a MAl_{18} cluster model shown in Fig. 6-3A [1]. This is the same cluster model as used for fcc Fe (see Fig. 4-7). It consists of alloying element, M, at a center and the surrounding 12 $Al^{(1)}$ atoms in the first nearest neighbor and 6 $Al^{(2)}$ atoms in the second nearest neighbor in fcc Al. The DV-Xα molecular orbital calculation is carried out with this MAl_{18} cluster model for various alloying elements, M.

The calculated energy level structures of alloyed Al with nontransition metals are shown in Fig. 6-4 together with the result of pure Al (i.e., M = Al) [1]. In this figure the Fermi energy

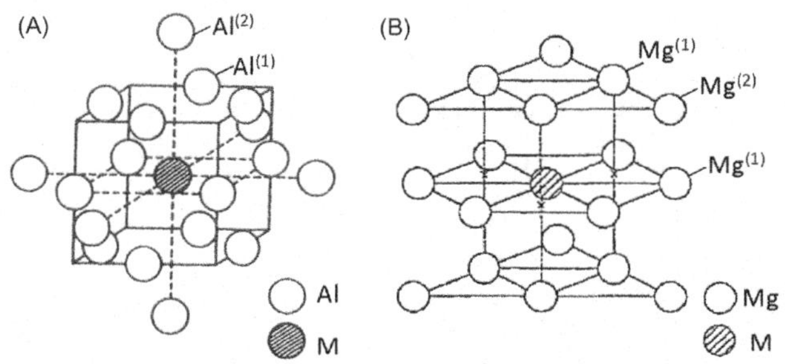

FIGURE 6-3 (A) Cluster model, MAl_{18}, for fcc Al, and (B) cluster model, MMg_{18} for hcp Mg.

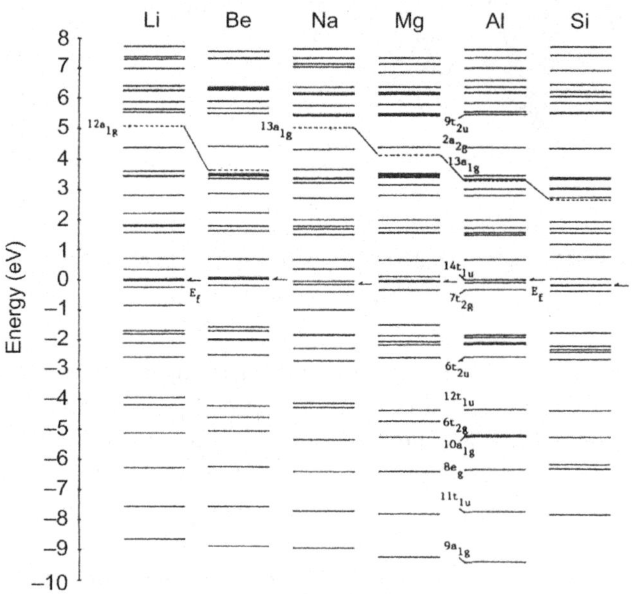

FIGURE 6-4 Energy level structures of pure and alloyed Al with nontransition metals. The Fermi level of pure and alloyed Al is indicated by an arrow (←).

level, E_f, of pure Al is set at zero and used as a reference. The position of the E_f level is indicated by an arrow (←) in the figure. In pure Al, the levels of $9a_{1g}$ to $14t_{1u}$ are originated mainly from the Al 3*s* and 3*p* orbitals, and form an s, p conduction band. The Fermi energy level lies on the $14t_{1u}$ level, which consists of the 16.6% 3*s*, 80.1% 3*p*, and 3.3% 3*d* components. The fraction of the *d* component increases in higher energy levels than the E_f level. For example, the $2a_{2g}$ level existing near 4.5 eV consists of 0.0% 3*s*, 43.1% 3*p*, and 56.9% 3*d* components. Similarly, the $9t_{2u}$ level existing near 5.5 eV consists of 2.4% 3*s*, 27.8% 3*p*, and 69.8% 3*d* components, since the level position further approaches the atomic Al 3*d* orbital energy position located in the higher energy region.

The *s* component extends to a wide energy range, and its fraction of the occupancy in any level is generally small, as explained in the $2a_{2g}$ and $9t_{2u}$ levels. However, in the $13a_{1g}$ level existing near 3.3 eV, there is a relatively high fraction of the *s* component (16.7%) that originates from the Al atom at the center of a cluster. Although the major components are still the *p* (55.3%) and the *d* (28.0%) orbitals, the height of this energy level changes with alloying elements as indicated by the dotted line in Fig. 6-4. For example, it decreases monotonously with increasing atomic number from Na to Si. For Li and Be, the $12a_{1g}$ level corresponds to this level. For transition metals such a level also appears and changes with the order of elements in the periodic table [1]. Thus, this a_{1g} level is considered to be a dominant *s*-component level depending on the alloying elements, M, in Al. For simplicity, this level is hereafter referred to as the *Mk* level. The values of *Mk* are listed in Table 6-1 for a variety of alloying elements, M.

The *p*-orbital energy level may be usable instead of the *s*-orbital energy level. But a spherically symmetrical *s*-orbital is probably better than a directional *p*-orbital for the purpose of investigating the mechanical properties of aluminum alloys.

The *Mk* level correlates with the electronegativity of M, as shown in Fig. 6-5A. For nontransition metals denoted by open circles, *Mk* levels change linearly with the electronegativity. For transition metals denoted by solid circles, they are located well above a dotted line of

Table 6-1 List of *Mk* Values for Alloying Elements in Al

Element	*Mk* (eV)	Element	*Mk* (eV)
Li	5.096	Mn	4.443
Be	3.650	Fe	4.328
Na	5.036	Co	4.314
Mg	4.136	Ni	4.248
Al	3.344	Cu	4.037
Si	2.680	Zn	3.290
K	6.196	Ga	3.013
Ca	5.550	Ge	2.614
Sc	5.200	Zr	5.433
Ti	5.009	Nb	5.227
V	4.782	Mo	5.079
Cr	4.601		

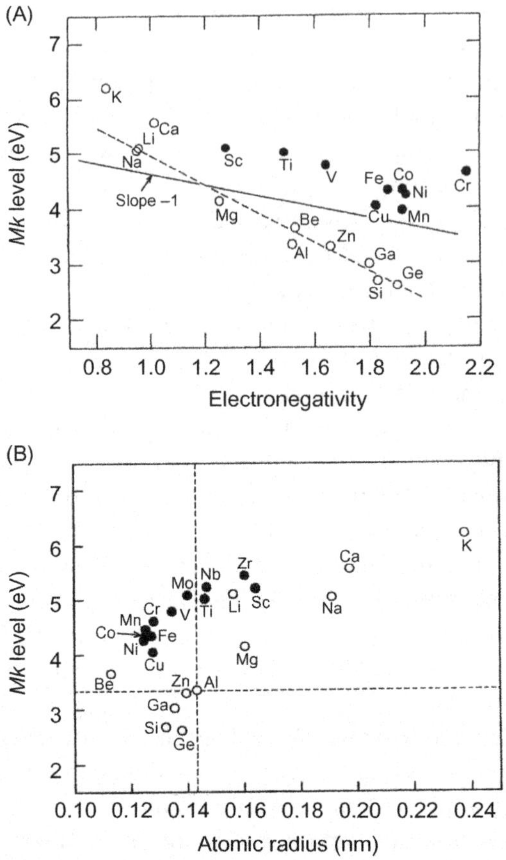

FIGURE 6-5 Correlation of the *Mk* level with (A) electronegativity and (B) atomic radius of elements; ●, transition metals and ○, nontransition metals.

nontransition metals. This indicates that the existence of 3*d* electrons affects the energy of 4*s* electrons. It is known that the energy eigenvalue obtained by the DV-Xα calculation represents the electronegativity [14], although the interaction between neighboring atoms in a cluster necessitates some modification of this idea for a free atom. In Fig. 6-5A, a line of slope of -1 is drawn as a reference, since the *M-s* level (eV) may be expressed using the electronegativity Φ (eV) as, *M*-s level $= -\Phi + c$, where c is an arbitrary constant. This line is located between the transition metals and nontransition metals.

Also, the *Mk* level is associated with the atomic radius of M as shown in Fig. 6-5B. The *Mk* level increases with increasing atomic radius of elements, although there are some differences in the correlation between transition metals and nontransition metals. Thus, the *s*-orbital energy level, *Mk*, is obtained for the nontransition metal, Al, that is different from the *d*-orbital energy level, *Md*, for the transition metals such as Ni, Fe, and Ti. However, both the *Mk* and the *Md* parameters correlate with the electronegativity and the atomic radius of elements. These parameters reflect the nature of chemical bond between atoms in the alloys. The mechanical properties of Al alloys are predictable using the *Mk* parameter as explained later.

6.2.2 Charge Transfer Between Atoms

The calculation of the bond order has no physical meaning in Al when the alloying elements M are nontransition metals, since the *s*, *p* electrons of both Al and M atoms spread over the crystal lattice without forming local covalent bond with M. Instead the contribution of charge transfer to the total cohesive energy is large in a nearly-free-electron metal like aluminum. This contribution increases with increasing transferred charges between atoms [9,15]. To estimate the transferred charges, the ionicity difference, $(I_M - I_{Al}{}^{(1)})$, between M and its first nearest neighbor $Al^{(1)}$ atoms is calculated for various M. Here, I_M and $I_{Al}{}^{(1)}$ are respective ionicity values of M and $Al^{(1)}$ atoms evaluated from the Mulliken population analysis [14]. The results are shown in Fig. 6-6, together with the electronegativity difference, $(\Phi_M - \Phi_{Al}{}^{(1)})$ [9]. In this figure, the result of Al corresponds to the ionicity difference between a central Al atom and the surrounding $Al^{(1)}$ atoms in a pure Al cluster.

As might be expected, the ionicity difference is related to the electronegativity difference. But there is a deep drop at Cr in the curve of the electronegativity difference, despite the absence of such a drop in the curve of the ionicity difference. This is probably due to the estimation methods of the electronegativity of the elements. The electronegativity for Cr reported by Watson and Bennett [10] is intermediate between V and Mn, in coincidence with the present trend for the ionicity difference. However, for a series of K, Ca, and Sc, the ionicity difference does not follow the electronegativity difference. A similar irregularity also

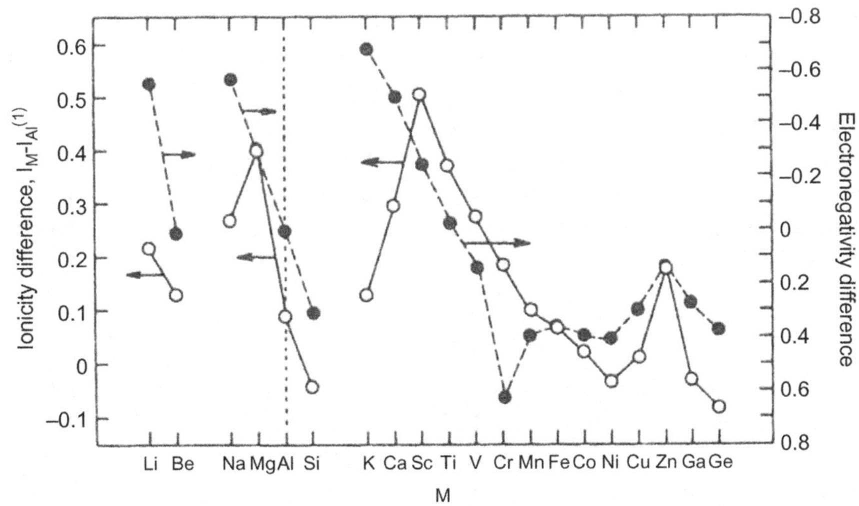

FIGURE 6-6 Ionicity difference between M and its first nearest neighbor aluminum atom, $Al^{(1)}$. The dashed line shows the electronegativity difference between M and $Al^{(1)}$ atoms.

occurs in Na and Mg. This is probably due to the difference in the number of valence electrons among elements. For example, in case Na and K are completely ionized, Na^{+1} and K^{+1} are formed and the ionicity is +1. On the other hand, for Ca and Mg, the ionicity is +2, which is larger than the ionicity of +1 for Na and K. This trend probably appears in the calculated values of the ionicity shown in Fig. 6-6. From these results it is expected that non-transition elements showing the strong ionic bonding with Al are Li, Mg, and Zn for each row of the periodic table. These elements are principal alloying elements in commercially available Al-Zn-Mg alloys and Al-Li based alloys. Besides these elements, Sc shows a large ioniocity difference, but it is not soluble in the Al solid solution.

6.3 Prediction of Mechanical Properties of Al Alloys

6.3.1 *Mk* Approach to Mechanical Properties

In Table 6-1, the *Mk* parameters are listed for a variety of alloying elements in aluminum. Most of the elements have a higher *Mk* value than Al except for Si, Zn, Ga, and Ge. For an alloy, two kinds of average values are defined as follows [2],

$$\overline{Mk} = \Sigma X_i Mk_i, \tag{6.3}$$

$$\Delta\overline{Mk} = \Sigma X_i |Mk_i - Mk_{Al}|. \tag{6.4}$$

Here, Mk_i is the *Mk* value for component *i* listed in Table 6-1. When all the elements in the alloy have a higher (or lower) *Mk* value than Al, these two averages differs only by a constant bias of Mk_{Al}, 3.344 eV, and hence there is no essential difference between them. However, when the elements of higher and lower *Mk* values are mixed in the alloy, the two averages have different meaning. For instance, for an Al-Zn-Mg alloy the larger *Mk* for Mg is somewhat compensated by the smaller *Mk* for Zn in the calculation of $\overline{Mk}$, whereas both Mg and Zn elements act in an additive way in the calculation of $\Delta\overline{Mk}$. Thus, a certain interaction between Mg and Zn atoms is counted in $\overline{Mk}$, but not in $\Delta\overline{Mk}$. However, it is still unknown which of the average is more suitable for the present problem. $\overline{Mk}$ may overestimate the interaction between Mg and Zn atoms, but $\Delta\overline{Mk}$ may underestimate it. So, these two limiting averages are assumed here and used for the estimation of the mechanical properties of Al alloys using the *Mk* parameter.

6.3.2 Non-Heat-Treatable Alloys

To verify the *Mk* approach a variety of data compiled in the Metals Handbook [11] and the Aluminium Handbook [12] are examined. In Fig. 6-7, the tensile strengths are plotted against Fig. 6-7A $\overline{Mk}$ and Fig. 6-7B $\Delta\overline{Mk}$ for a variety of wrought pure Al and Al alloys that are annealed enough to obtain the lowest strength. This treatment is called the O temper according to the Temper Designation System [11]. The $\overline{Mk}$ parameter shows a better correlation with the tensile strength as shown in Fig. 6-7A, as compared with the $\Delta\overline{Mk}$ parameter shown

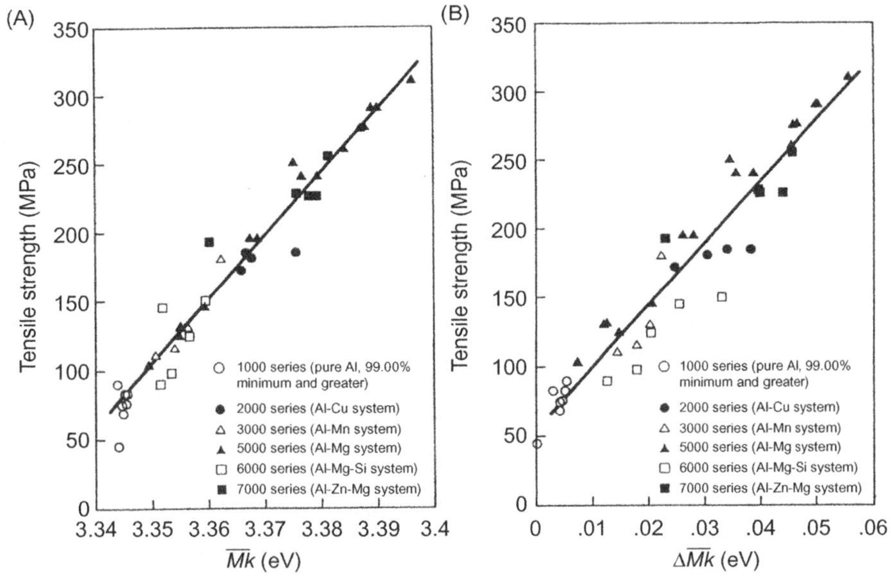

FIGURE 6-7 Correlation of tensile strength with (A) $\overline{Mk}$ and (B) $\Delta\overline{Mk}$ for the O temper Al alloys.

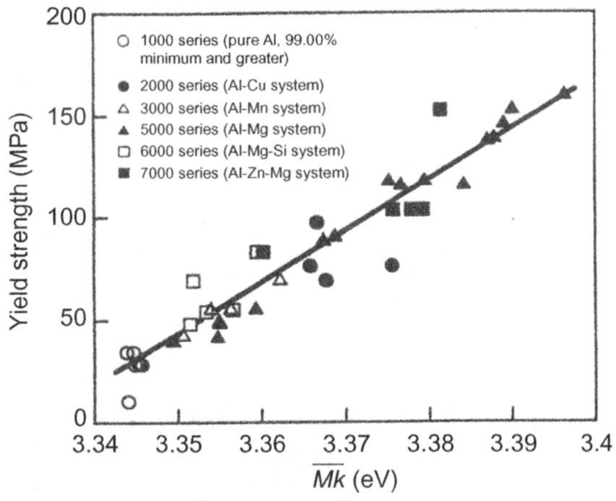

FIGURE 6-8 Correlation of yield strength with $\overline{Mk}$ for the O temper Al alloys.

in Fig. 6-7B. Also, the scattering of the data is significantly reduced in the plots of $\overline{Mk}$, as compared with the plots of (A) $\Delta\overline{\Phi}$ and (B) $\Delta\overline{R}$ shown in Fig. 6-1. It is interesting that the tensile strength varies linearly with the $\overline{Mk}$, irrespective of the alloy types of the 1000 to the 7000 series in the specification. This indicates that the tensile strength is not so sensitive to the respective elements in the alloys, but dependent largely on the average compositions weighted by the *Mk* parameter. The yield strengths of these alloys also vary linearly with $\overline{Mk}$ as shown in Fig. 6-8.

6.3.3 Strain Hardening of Non-Heat-Treatable Alloys

Cold working increases the strength of wrought Al alloys and this is called strain hardening [2,4]. The yield strength and the tensile strength are plotted against $\overline{Mk}$ in Fig. 6-9 and in Fig. 6-10, respectively. Here, the Hn8 temper means cold working to make about a 75% area reduction of the specimen, and the Hn4 temper means cold working to guarantee a tensile strength intermediate between the O and the Hn8 temper specimens [11]. Both the yield strength and the tensile strength of the Hn4 and Hn8 temper alloys vary approximately linearly with $\overline{Mk}$. In each figure, the strength difference between the O temper and the Hn4

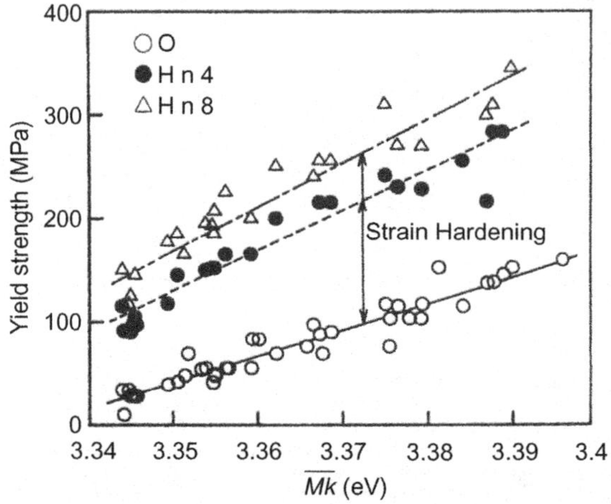

FIGURE 6-9 Correlation of yield strength with $\overline{Mk}$ for the O, Hn4, and Hn8 temper Al alloys.

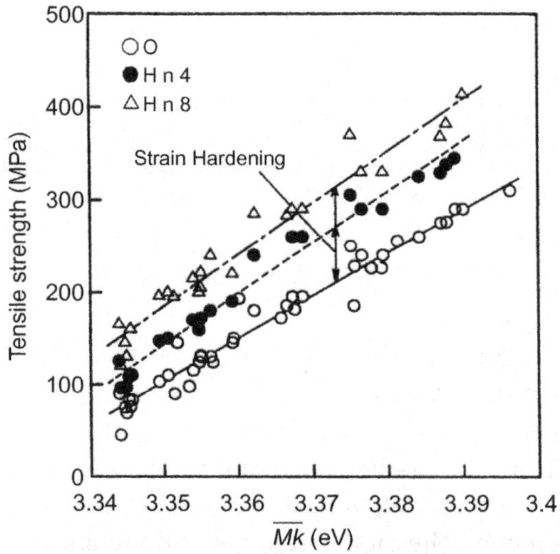

FIGURE 6-10 Correlation of tensile strength with $\overline{Mk}$ for the O, Hn4, and Hn8 temper Al alloys.

Table 6-2 Approximate Equations for the Mechanical Properties of Non-Heat-Treatable Al Alloys

Mechanical Properties	Temper Designation	$MP = a + b\overline{Mk}$		Correlation Coefficient
		a	*b*	
Yield strength	O	−8410	2520	0.962
(MPa)	Hn4	−12,790	3860	0.962
room temperature	Hn8	−13,960	4220	0.957
Yield strength	O	−8860	2660	0.984
(MPa)	Hn4	−14,860	4490	0.994
77K	Hn8	−16,330	4940	0.971
Tensile strength	O	−15,400	4630	0.979
(MPa)	Hn4	−18,340	5520	0.987
room temperature	Hn8	−18,630	5620	0.977
Tensile strength	O	−16,170	4890	0.986
(MPa)	Hn4	−24,520	7390	0.989
77 K	Hn8	−24,860	7500	0.966

(or Hn8) temper alloys is attributable to the increase in the strength due to the strain hardening of alloys.

Assuming that each mechanical property (MP) of non-heat-treatable alloys can be expressed as $MP = a + b\overline{Mk}$, the coefficients a and b are determined using the least-squares method. The results are summarized in Table 6-2.

6.3.4 Heat-Treatable Alloys

The Al alloys are strengthened by the precipitates dispersed in the matrix and this is called precipitation hardening. The tensile strengths of the T6 temper alloys are plotted in Fig. 6-11A for $\overline{Mk}$ and Fig. 6-11B for $\Delta\overline{Mk}$ [2]. Here, the T6 temper means that the alloy is solution treated at high temperatures and then artificially aged at a certain temperature (for example, 420–480K) to yield the precipitation hardening of alloys. In contrast to the O temper alloys shown in Fig. 6-7, the scattering of the data is smaller in the plots of (B) $\Delta\overline{Mk}$ than (A) $\overline{Mk}$ in Fig. 6-11. The difference between $\Delta\overline{Mk}$ and $\overline{Mk}$ will be discussed later.

The tensile strength varies almost linearly with $\Delta\overline{Mk}$ in Fig. 6-11B, and it is less dependent on the alloy types of Al-Cu (2000 series), Al-Mg-Si (6000 series), and Al-Zn-Mg (7000 series), despite the substantial difference in the type of precipitates among them. It is surprising that the tensile strength depends only on $\Delta\overline{Mk}$ even for these two-phase alloys. The difference in the strength between the T6 and the O temper specimens is attributable to the increase in the strength due to the precipitation hardening of alloys. Not only the tensile strength, but also the yield strength changes linearly with $\Delta\overline{Mk}$, as shown in Fig. 6-12.

Another example is shown in Fig. 6-13A–B for the T4 temper alloys. T4 means that the alloy is solution treated at high temperatures and then aged at room temperature. Both the

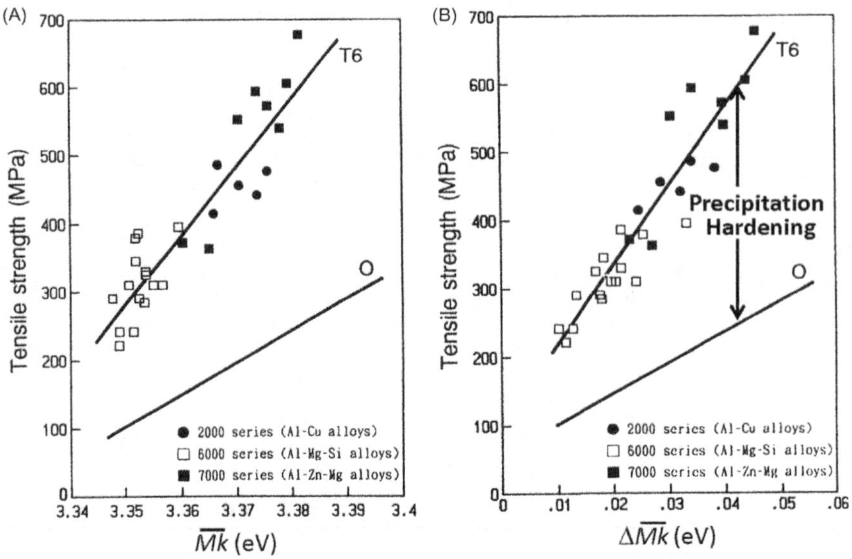

FIGURE 6-11 Correlation of tensile strength with (A) $\overline{Mk}$ and (B) $\Delta\overline{Mk}$ for the T6 temper Al alloys.

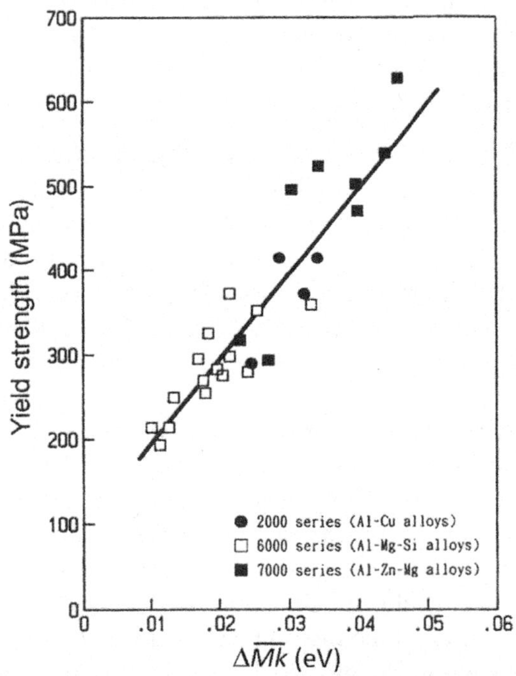

FIGURE 6-12 Correlation of yield strength with $\Delta\overline{Mk}$ for the T6 temper Al alloys.

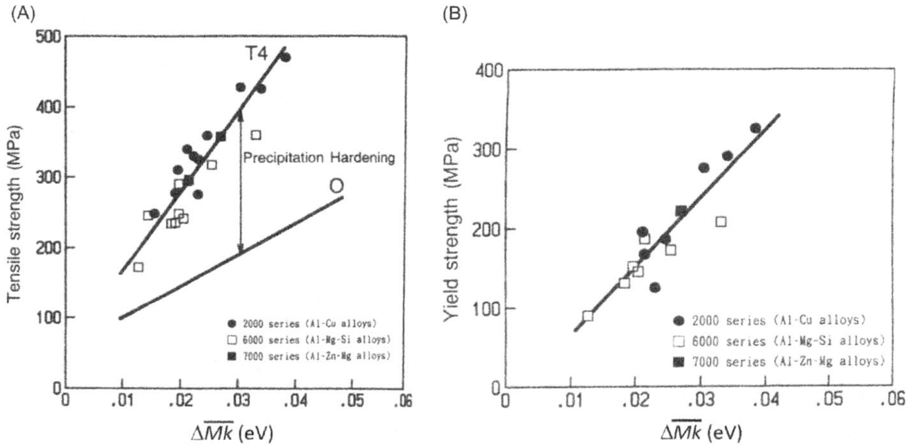

FIGURE 6-13 Correlation of (A) tensile strength and (B) yield strength with $\Delta\overline{Mk}$ for the T4 temper Al alloys.

Table 6-3 Approximate Equations for the Mechanical Properties of Heat-Treatable Al Alloys

Mechanical Properties	Temper Designation	$MP = a + b\Delta\overline{Mk}$		Correlation Coefficient
		a	*b*	
Yield strength (MPa) room temperature	T4	−22.7	8650	0.908
	T6	93.4	10,040	0.907
Yield strength (MPa) 77K	T6	74.3	12,080	0.957
Tensile strength (MPa) room temperature	T4	44.4	11,190	0.929
	T6	102.8	11,480	0.934
Tensile strength (MPa) 77K	T6	161.6	12,360	0.977

tensile strength and the yield strength change almost linearly with $\Delta\overline{Mk}$, regardless of the alloy types.

Assuming that each mechanical property (MP) of heat-treatable alloys can be expressed as $MP = a + b\overline{Mk}$, the coefficients *a* and *b* are determined using the least-squares method. The results are summarized in Table 6-3.

6.3.5 Strength Map for Alloy Design

As shown in Fig. 6-14, a strength map is constructed for the Al-Mg based alloys (the 5000 series alloys) to show the utility of the *Mk* approach to practical alloy design [2]. Here, the alloying elements other than Mg are counted using the Mn equivalent. For instance, the Mn equivalent (mol%) for an Al-Mg-Mn-X-Y alloy is defined as,

$$\text{Mn equivalent (mol\%)} = [\text{Mn}] + (1/Mk_{Mn}) \cdot \{Mk_{X}[\text{X}] + Mk_{Y}[\text{Y}]\} \tag{6.5}$$

Here, [Mn], [X], and [Y] are the respective compositions (mol%) of Mn, X, and Y in the alloy. A similar equation will be obtained readily even for the alloy containing more elements. If necessary, this Mn equivalent in the mol% units is converted easily to the mass % units using a standard conversion method. In fact, in Fig. 6-14, both axes are drawn in the mass% units for the sake of convenience to represent commercially available alloys in the figure. The number 5XXX below each closed circle is the alloy name and the number above the circle is the tensile strength (MPa). Also, the tensile elongation (%) is indicated in parentheses. For example, the 5182 alloy located near 4.5 mass% Mg and 0.8 mass% Mn equivalent has a tensile strength of 276 MPa and tensile elongation of 25%. The calculated isotensile strength lines are also drawn in the figure. It is apparent that the strength of the commercially available aluminum alloys is well represented on this map.

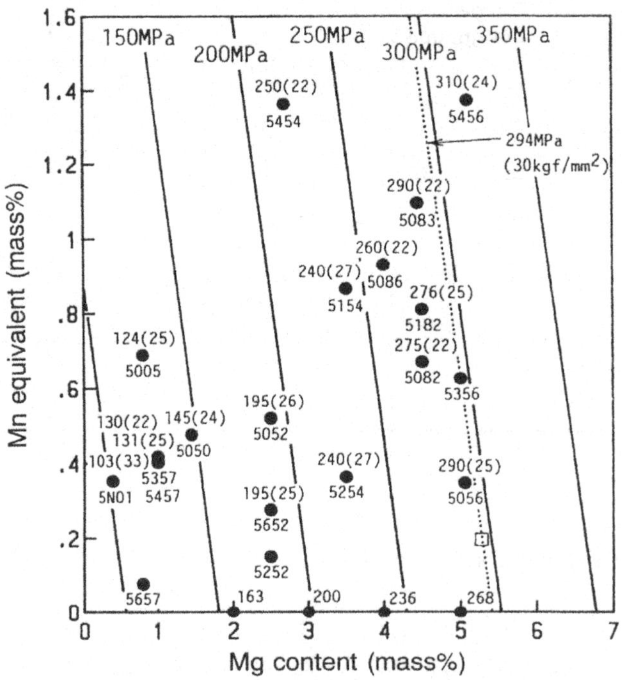

FIGURE 6-14 Strength map for Al-Mg based alloys (the 5000 series alloys).

Alloy design is carried out with this *Mk* strength map by setting a target of 294 MPa (30 kgf/mm^2) for the tensile strength and 30% for the tensile elongation for use of the body of automobiles. The tensile elongation may not be treated by the present *Mk* approach alone, because it may vary even in the presence of minor impurities, which scarcely affect the compositionally averaged $\overline{Mk}$ value. So, in the present design the Mn equivalent is set empirically to be 0.2 mass%, as this is probably the optimum amount to maximize the tensile elongation of alloys [2]. A designed alloy is denoted by a small square on the dotted line showing 294 MPa in the figure and it is Al-5.3 mass% Mg - 0.2 mass% Mn equivalent alloy. This alloy is prepared in practice and tested experimentally in a conventional way. As the result, it is confirmed that the target values of 294 MPa and 30% are indeed satisfied for this designed alloy.

6.4 Energy Approach to Mechanical Properties of Al Alloys

6.4.1 Energy Range of $\overline{Mk}$ and $\Delta\overline{Mk}$ Parameters

The use of both $\overline{Mk}$ and $\Delta\overline{Mk}$ parameters opens a door to the energy approach to the mechanical properties. The values of $\overline{Mk}$ and $\Delta\overline{Mk}$ are very different, but as shown in Figs. 6-7 and 6-11, their values change in the energy range, $3.34\text{ eV} < \overline{Mk} < 3.4\text{ eV}$ with the interval of 0.06 eV and $0\text{ eV} < \Delta\overline{Mk} < 0.06\text{ eV}$. The interval is the same between $\overline{Mk}$ and $\Delta\overline{Mk}$. In this section, to make the units clear, eV/atom is used instead of eV unless otherwise noted. So, the value of 0.06 eV is expressed here as 0.06 eV/atom.

A bias of 3.34(4) eV in $\overline{Mk}$ is originated from the *Mk* value of Al. In fact, since $Mk_{Al} = \Sigma X_i Mk_{Al} = 3.344$, $(\overline{Mk} - 3.344)$ is derived from Eq. (6.3) as,

$$\overline{Mk} - 3.344 = \Sigma X_i(Mk_i - Mk_{Al}). \tag{6.6}$$

This expression is comparable in style to the definition of $\Delta\overline{Mk}$ given in Eq. (6.4). Only a difference lies between $(Mk_i - Mk_{Al})$ in Eq. (6.6) and $|Mk_i - Mk_{Al}|$ in Eq. (6.4). $\Delta\overline{Mk}$ is always larger than $(\overline{Mk} - 3.344)$, as long as the alloying elements of higher and lower *Mk* values than 3.344 (eV) are mixed in the alloy.

It is evident from Figs. 6-7 and 6-11 that all the mechanical properties of Al alloys change within the energy range of 0.06 eV/atom in both $\overline{Mk}$ and $\Delta\overline{Mk}$. This energy of 0.06 eV/atom is very small, that is, only about 1.8% of the cohesive energy of pure Al, 3.39 eV/atom [16]. It is rather comparable to the thermal energy, $k_B T = 8.62 \times 10^{-5}\,T$ (eV/atom), where k_B is the Boltzmann constant and T is the absolute temperature. Using this equation, the value of 0.06 eV/atom is converted into the temperature, T = 696K (423°C). Therefore, it is only natural that mechanical properties change largely with temperatures. One example is shown in Fig. 6-15. The tensile strengths of heat-treatable Al alloys are weak at 589K (316°C), being less than 60 MPa [11]. It is much weaker as compared with the tensile strengths at room temperature shown in Fig. 6-11. The high temperature data so far reported may be up to about 698K (425°C), although the data is very limited and incomplete. For example, the

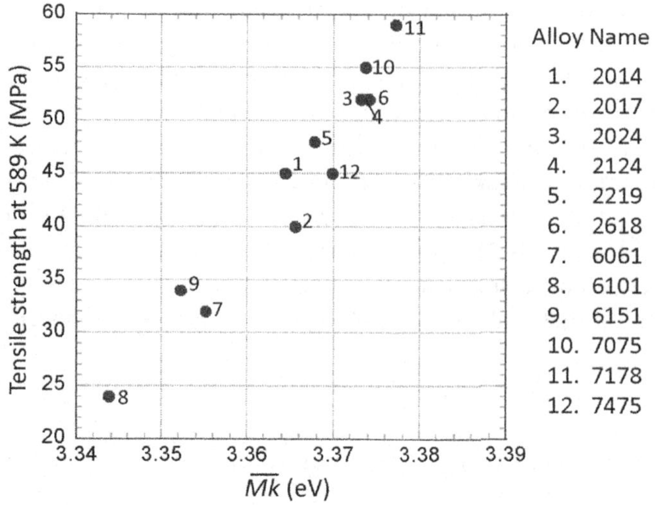

FIGURE 6-15 Correlation of tensile strength at 589K with $\overline{Mk}$.

tensile strength of alloy 2124 is 30 MPa at 698K [11]. No strengthening mechanism is operating at this temperature.

Thus, the measured temperature range of the tensile strength is consistent with the one expected from the thermal energy of 0.06 eV/atom. Therefore, 0.06 eV/atom is indeed a reasonable value to treat the mechanical properties of Al alloys. As discussed later, the strain energy to drive the solid solution hardening or the precipitation hardening will be the order of magnitude 0.06 eV/atom.

6.4.2 Solid Solution Hardening and Precipitation Hardening

6.4.2.1 Solid Solution Hardening

Following dislocation theory, the solid solution hardening will arise from the hindering effects of solute atoms on the dislocation motion [17,18]. This may be due either to the pinning of dislocations by solute atom in the core or to the friction between moving dislocations and solute atoms [7]. For example, the solute–dislocation interaction may be attributable to the elastic interaction (the so-called Cottrell atmosphere) [17,18]. Such an elastic interaction with dislocations is primarily caused by the local lattice strain introduced by the mismatch in size between solute and solvent atoms.

As explained in detail in Chapter 11, such local lattice strains are calculated for various alloying element in metals. The energy gained by the local lattice strain is also obtained (see Chapter 11). Such an energy gained by the local lattice strain is hereafter called strain energy. In case of Al, for convenience, calculated results of the strain energy are shown in Fig. 6-16. For example, the strain energy is about 0.01 eV/atom for M = Fe, but it is less than 0.002 eV/atom for soluble elements (e.g., M = Li, Mg, Cu, Zn) in the Al solid solution. These

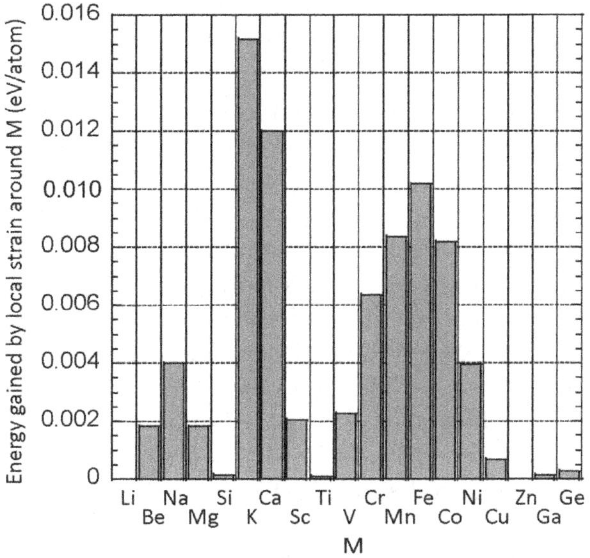

FIGURE 6-16 Energy gained by local strain around an isolated alloying element, M in fcc Al. Local strain is introduced into the Al matrix by the mismatch in size between M and Al atoms.

values of the strain energy are small, but still comparable to the energy range of (0–0.06) eV/atom. The strain energy is calculated from the local lattice strain around an isolated alloying element in Al. Being different from such a dilute alloy system, a variety of elements are alloyed into commercial available Al alloys, and the alloy compositions are relatively high ($1 \sim 7$mol%). Therefore, the strain energies of these alloys will approach 0.06 eV/atom as the alloy composition increases. Thus, the solid solution hardening is understood in view of the strain energy. As ($\overline{Mk} - 3.344$) is probably related to the strain energy of the alloy, the tensile strength increases monotonously with the $\overline{Mk}$ as shown in Fig. 6-7 for the O temper alloy.

6.4.2.2 Precipitation Hardening

Precipitation hardening is associated with the formation of coherent or semicoherent precipitates in Al alloys. For example, the G.P. zones are a coherent solute rich cluster and the intermediate phase (e.g., η' phase in Al-Zn-Mg) is a semicoherent precipitate in the Al alloys. An important factor to determine which precipitate is either coherent or semicoherent is the closeness of match in the lattice spacing between the matrix and the precipitate. It is generally considered that the precipitates, and even more importantly, the strain fields in the matrix surrounding the precipitates retard the movement of dislocations, depending on the closeness of the mismatch [19].

Thus, the precipitation hardening is related closely to the strain fields in the matrix surrounding the precipitates such as the coherent G.P. zones or semicoherent intermediate phase. The coherent or semicoherent strain fields in the matrix surrounding the precipitates are large enough to make a high barrier to the movement of dislocations in the alloy, so that the precipitation hardening takes place in the T6 or T4 temper alloys as shown in Figs. 6-11 to 6-13.

In view of the strain energy, large coherent or semicoherent strain fields in the matrix surrounding the precipitates are accompanied with the high strain energy. However, the strain energy must be low in the remaining matrix distant from the precipitates. This is because the appearance of solute-rich precipitates reduces the solute concentration in the remaining matrix. It is stressed here that not only the solute distributions, but also the strain distributions, are modified largely in the matrix in the course of the precipitation.

The tensile strengths of either the T6 temper alloys or the O temper alloys are treated in the same energy range of (0–0.06) eV/atom, as shown in Figs. 6-11 and 6-7. In general, any alloys of the high $\Delta\overline{Mk}$ values have the high $(\overline{Mk} - 3.344)$ values, too. Therefore, the total strain energy stored in the whole alloy is supposed to be less changed before and after the precipitation. The change in the total strain energy is probably at most 0.012 eV/atom, judging from the maximum difference between $\Delta\overline{Mk}$ and $(\overline{Mk} - 3.344)$ values of the alloys, as explained later.

From the slope of the line drawn in Fig. 6-11 or Fig. 6-7, the increment in the tensile strength due to the strain energy is estimated to be about 115 MPa per 0.01 eV/atom for the T6 temper alloys and about 46 MPa per 0.01 eV/atom for the O temper alloys. Thus, the increment in the tensile strength is about 2.5 times larger in the T6 temper alloys as compared with the O temper alloys. Reminding that the strain energy is distributed rather uniformly in the matrix of the O temper alloys, the large strain energy localized in the matrix surrounding the solute rich precipitates causes this strength difference. This is the precipitation hardening of Al alloys.

Furthermore, it is found from Fig. 6-11 that the tensile strength increases linearly with $\Delta\overline{Mk}$, irrespective of the alloy types and hence the kinds of precipitates dispersed in the matrix. This result clearly indicates that the strain fields in the matrix surrounding the precipitates are more crucial to the mechanical properties of alloys than the kinds of precipitates. However, the strain fields depend on the closeness of match in the lattice spacing between the matrix and the precipitate. The amount of the precipitate should be as large as possible to increase the extent of the matrix neighboring the precipitate. In this sense the precipitate is also very important to increase the alloy strength, as discussed later.

6.4.3 Utility Difference Between $\overline{Mk}$ and $\Delta\overline{Mk}$ Parameters

A remaining question is the reason why the $\overline{Mk}$ parameter is used for the prediction of the mechanical properties for the O temper (annealed) alloys (see Fig. 6-7), and why the $\Delta\overline{Mk}$ is used for the T4, T6 temper alloys (see Figs. 6-11 to 6-13). It is difficult to answer this question, because information on the strain energy is very limited. However, let's assume here that both $(\overline{Mk} - 3.344)$ and $\Delta\overline{Mk}$ are parameters to show the strain energies of alloys, and that the strain energies of alloys increase with increasing these parameters. As explained in

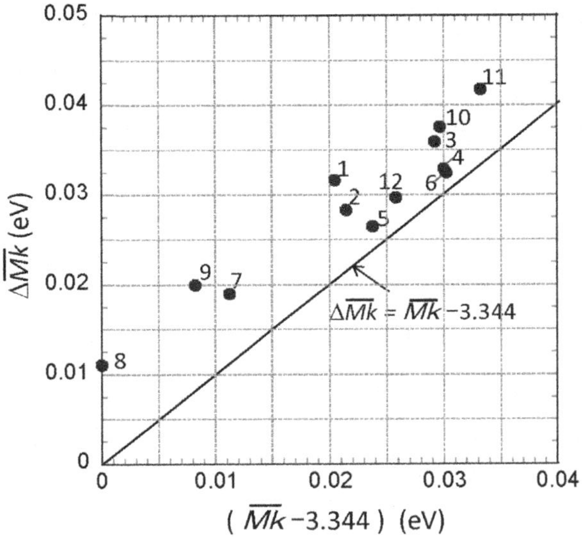

FIGURE 6-17 Comparison between $\Delta\overline{Mk}$ and ($\overline{Mk}$ − 3.344) for heat-treatable alloys.

Eq. (6.6), a constant bias of 3.344 (eV) that is the *Mk* value of Al is subtracted in the formula of ($\overline{Mk}$ − 3.344), since the strain energy is zero for pure Al.

In Fig. 6-17, the $\Delta\overline{Mk}$ values are compared with the ($\overline{Mk}$ − 3.344) values for various heat-treatable alloys. The alloy numbers used in Fig. 6-17 are the same as in Fig. 6-15. For comparison, a line of showing an equation, $\Delta\overline{Mk} = \overline{Mk} - 3.344$, is drawn in Fig. 6-17. It is apparent from this figure that $\Delta\overline{Mk}$ is larger by 0.002 ~ 0.012 eV/atom than ($\overline{Mk}$ − 3.344). In other words, the strain energy of $\Delta\overline{Mk}$ is probably higher by 0.002–0.012 eV/atom than that of ($\overline{Mk}$ − 3.344). The higher strain energy yields the higher strength. Therefore, it seems reasonable that the $\Delta\overline{Mk}$ parameter is used for the T4 or T6 temper alloys with stronger tensile strength and that the $\overline{Mk}$ parameter is used for the O temper alloys with weaker tensile strength.

However, even for T4 or T6 temper alloys the $\overline{Mk}$ parameter should be used at high temperatures, since local strain energies are released a great deal from the alloy by the thermal energy at high temperatures. This example is shown in Fig. 6-15. The tensile strength increases with increasing $\overline{Mk}$. If room temperature, 297K (24°C), is used as a reference of the tensile strength, the change in the thermal energy between 297K and 589K is estimated to be $8.62 \times 10^{-5} \times (589\text{-}297) = 0.025$ eV/atom. This value is larger than the difference value, 0.002 ~ 0.012 eV/atom between $\Delta\overline{Mk}$ and ($\overline{Mk}$ − 3.344). In other words, the thermal energy overcomes the energy difference between $\Delta\overline{Mk}$ and ($\overline{Mk}$ − 3.344). Also, as most of the precipitates are resolved into the matrix at 589K, the alloys are nearly in the O temper state, resulting in the weak strength at 589K. Therefore, it is more reasonable to use the $\overline{Mk}$ parameter at high temperature instead of the $\Delta\overline{Mk}$ parameter even for heat-treatable alloys as shown in Fig. 6-15. Such examples will also be presented in Mg alloys.

6.5 Approach to Design of High Strength Al Alloys

6.5.1 Maximum Tensile Strength Change With Precipitates

The next topic is an approach to the design of high strength Al alloys. The highest strength Al alloys produced by a conventional method is about 700 MPa as shown in Fig. 6-18. As explained before, high strength Al alloys are strengthened by the precipitates dispersed in the matrix. Namely, the alloy is solution heat treated at high temperatures and quenched. Subsequently, it is aged at a certain temperature for a proper time to form the precipitates in the matrix. This is the so-called aging treatment and the increase in the alloy strength attendant on this treatment is called age-hardening.

For example, besides the G.P. zones, the precipitate-hardening phases are θ″, θ′, and θ (Al_2Cu) in Al-Cu (2000 series); β″, β′, and β (Mg_2Si) in Al-Mg-Si (6000 series); and η′ and η ($MgZn_2$) in Al-Zn-Mg (7000 series) [19]. As shown in Fig. 6-18, there is a certain upper limit in the tensile strength depending on the type of alloys. The limit strength is about 400 MPa for the 6000 series, about 500 MPa for the 2000 series, and about 700 MPa for the 7000 series. This limit strength is dependent mainly on the amount of the precipitate in the matrix. In fact, as shown in Fig. 6-18, the upper limit of the tensile strength (hereafter called maximum tensile strength) increases linearly with the maximum solubility (mol%) of the precipitate in the Al solid solution, although the solubility changes with temperature [19]. The amount of the precipitate in each alloy increases with increasing maximum solubility of the precipitate in Al. As the result, a great number of largely distorted regions appear in the matrix surrounding the precipitates and disturb the dislocation motion effectively, resulting in the

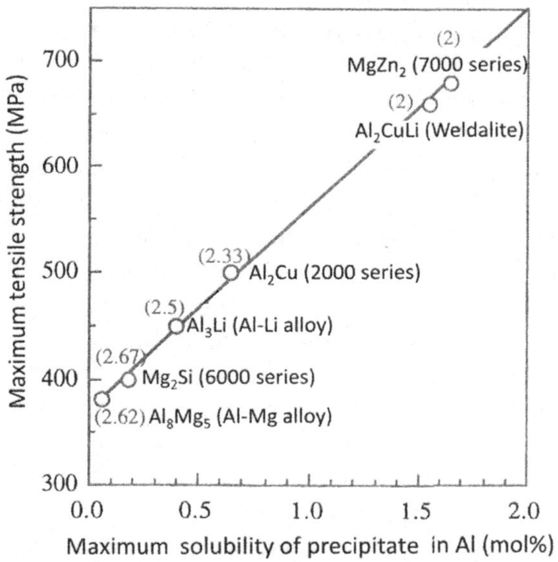

FIGURE 6-18 Relationship between maximum tensile strength and maximum solubility of precipitate in Al. The *e/a* value for each precipitate is given in parentheses.

high strength of the alloy. Therefore, for the design of high strength alloys, it is required to find out a new type of precipitate that forms in abundance in the alloy.

6.5.2 Average *e/a* Values for Precipitates

To continue the present discussion, the average electrons-per-atom ratio, *e/a*, is used here as an indicator to show the nature of the chemical bond between atoms in each precipitate. The average *e/a* value is denoted in the parentheses for each precipitate in Fig. 6-18. For example, in case of Al_2Cu, $(e/a)_{Al} = 3$ and $(e/a)_{Cu} = 1$, so that the average *e/a* value is $(3 \times 2 + 1 \times 1)/3 = 2.33$. As shown in Fig. 6-18 there is a clear trend that the amount of precipitate in the alloy increases with decreasing average *e/a* value. In fact, as shown in Fig. 6-19, the maximum tensile strength appears to change linearly with the average *e/a* value of the precipitate. This is not surprising since highly soluble elements in Al of $(e/a)_{Al} = 3$ are Li, Ag, Cu of $(e/a)_{Li,\ Ag,\ Cu} = 1$ and Zn, Mg of $(e/a)_{Zn,\ Mg} = 2$, all of which have the lower *e/a* values than Al. As a result, the precipitate has the low average *e/a* value, since it consists of the high soluble elements with low *e/a* values.

For the further discussion, let's consider the cohesive energy, E_{coh} (X), as a measure of the chemical bond strength between atoms in pure metal, X. The cohesive energy is the energy gained by arranging the atoms in a crystalline solid, as compared with the gas state. For example, Na, Mg, and Al are neighboring elements in the same row of the periodic table. For these three metals a relative ratio of the cohesive energy is approximately expressed as, E_{coh} (Na): E_{coh} (Mg) : E_{coh} (Al) = 1.0 : 1.4 : 3.0 [16], so that the chemical bond strength changes in the order, Na $<$ Mg $<$ Al. This E_{coh} (X) ratio is comparable to the *e/a* ratio of X, $(e/a)_{Na} : (e/a)_{Mg} : (e/a)_{Al} = 1 : 2 : 3$. Thus, the cohesive energy increases with increasing

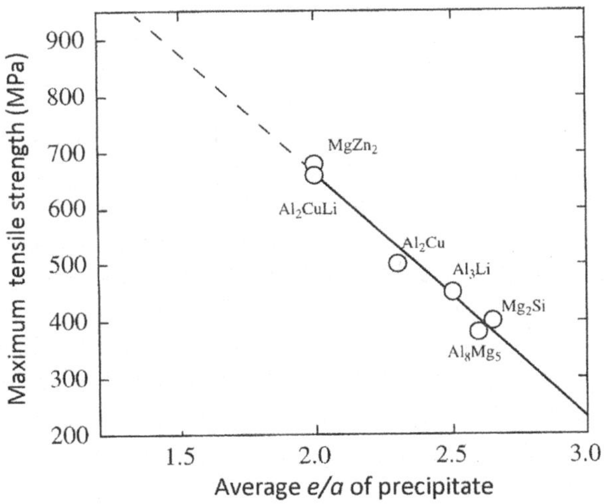

FIGURE 6-19 Relationship between maximum tensile strength and average electrons-per-atom ratio, *e/a*, of precipitate.

e/a value of metal, since the *e/a* value is the number of electrons to contribute to the chemical bonding between atoms in simple metals.

6.5.3 Age-Precipitation Hardening

As far as the precipitate consists only of simple metals, the average *e/a* value is a good measure of the chemical bond strength of the precipitate. The lower average *e/a* value of the precipitate may be a sign of the weaker chemical bond operating between atoms in the precipitate. As described earlier, the precipitate in the age-hardening alloys has a lower average *e/a* value. A role of the precipitate in strengthening alloys is to introduce the high strain fields into the matrix surrounding the precipitate. The formation of the energetically stable precipitate is not necessarily required, but it is rather preferable for the precipitate to have a weak chemical bond between atoms. This is because the driving force for the precipitate formation is supposed to be weak. As a result, the precipitate reaction progresses gradually following the step sequence, G.P. zones → intermediate precipitate → equilibrium precipitate. Fine precipitates appear in each step and contribute to the mechanical strength of the alloy by introducing the coherent or semicoherent strain around the precipitate. Such a precipitation sequence occurs not only in Al alloys, but also in Mg alloys, as might be expected from the previous discussion. For example, the precipitation sequence in Mg-Zn alloys is, G.P. zones → β_1', β_2' ($MgZn_2$) → $\beta(Mg_2Zn_3)$, where the average *e/a* value is 2 for the $MgZn_2$ precipitate. Recently the ordered G.P. zones have been observed in Mg-Al-Ca alloys [20]. Thus, the strengthening mechanism by the precipitate with the low average *e/a* value matches the well-known concept of "age-precipitation hardening." Strictly speaking, the formation energy of the precipitate should be used instead of the cohesive energy. But the age-precipitation phenomenon of Al alloys is probably interpreted as due to the formation of the *s, p* compound (precipitate) with the average *e/a* value of less than 3 in a sea filled with nearly free electrons of Al with $(e/a)_{Al} = 3$.

6.5.4 A Perspective on Design of High Strength Al Alloys

As is evident from Fig. 6-19, it is probably key to find a new precipitate with the average *e/a* value of less than 2 to develop high strength Al alloys. For instance, Li of the lowest $e/a = 1$ is the most suitable alloying element to fit this end. However, the average *e/a* value is not less than 2 for any precipitates so far observed in Al-Li based alloys. For example, the average *e/a* value is 2 for AlLi and Al_2CuLi, 2.11 for Al_5CuLi_3, 2.2 for Al_6CuLi_3, and 2.25 for Al_2LiMg [19].

Here, we recall that the average *e/a* value is 2 for $MgZn_2$ since $e/a = 2$ for Mg and Zn as shown in Fig. 6-19. To the author's limited knowledge, it may be interesting to add Li into the Al-Zn-Mg alloys, because the average *e/a* value is expected to become less than 2 for any precipitate consisting of Mg, Zn, and Li. In fact, according to the ternary phase diagram of Li-Mg-Zn system at 573K [21], a number of Mg $(Li_xZn_{1-x})_2$ are present along the constant compositional line of 33.3mol% Mg. Their structures are either the Laves type or Laves polytypes depending on the Li content x, and they are very similar to the binary $MgZn_2$ Laves phase.

6.6 Molecular Orbital Calculation and Alloying Parameter *Mk* for Mg

The hcp cluster model used for the calculation is shown in Fig. 6-3B [3]. This is the same cluster model as used for hcp Ti as shown in Fig. 5-1B. It is a MMg_{18} cluster model in which an alloying element, M, is located at the center and surrounded by the 12 first nearest neighbor Mg atoms, $Mg^{(1)}$ and the 6 second nearest neighbor Mg atoms, $Mg^{(2)}$. Simple metals, transition metals, and rare earth metals are chosen for M.

The energy level structures of Mg alloyed with various M are calculated [3]. As is similar to the result of Al, the M-*s* component is extended to a wide energy range, and its fraction of the occupancy in each level is generally small. However, a'_1 levels existing above the Fermi energy level have a relatively high fraction of the *s* component of the central alloying element, M. The *s*-orbital energy, *Mk*, is then calculated by taking a weighted average of the three a'_1 levels existing above the Fermi energy level [3]. This is because the symmetry of the point group is lower in the D_{3h} of the hcp Mg cluster model as compared with the O_h of the fcc Al cluster model, and then the energy level structure for Mg is relatively more complex than that for Al shown in Fig. 6-4. The weighting factor used for averaging is the fraction of the *s*-component in each a'_1 level. The *Mk* values calculated in this way are listed in Table 6-4 for various alloying elements in magnesium.

As is found from Table 6-4, the values of *Mk* parameter are higher in rare earth metals than in transition metals than in simple metals. To confirm these results, the radial orbital

Table 6-4 List of *Mk* Values for Alloying Elements in Mg

Element	*Mk* (eV)	Element	*Mk* (eV)
Li	4.641	Y	4.957
Be	3.891	Zr	4.846
Na	4.664	Nb	4.802
Mg	4.033	Mo	4.787
Al	3.613	Ag	4.354
Si	3.959	Cd	3.999
K	5.059	In	2.737
Ca	4.608	Sn	2.463
Ti	4.341	Sb	2.184
V	4.546	La	6.432
Cr	4.498	Ce	6.400
Mn	4.405	Pr	6.391
Fe	4.393	Nd	6.391
Co	4.352	Sm	6.398
Ni	4.304	Gd	6.404
Cu	4.141	Tb	6.406
Zn	2.727	Dy	6.413
Ga	2.403		
Ge	2.106		

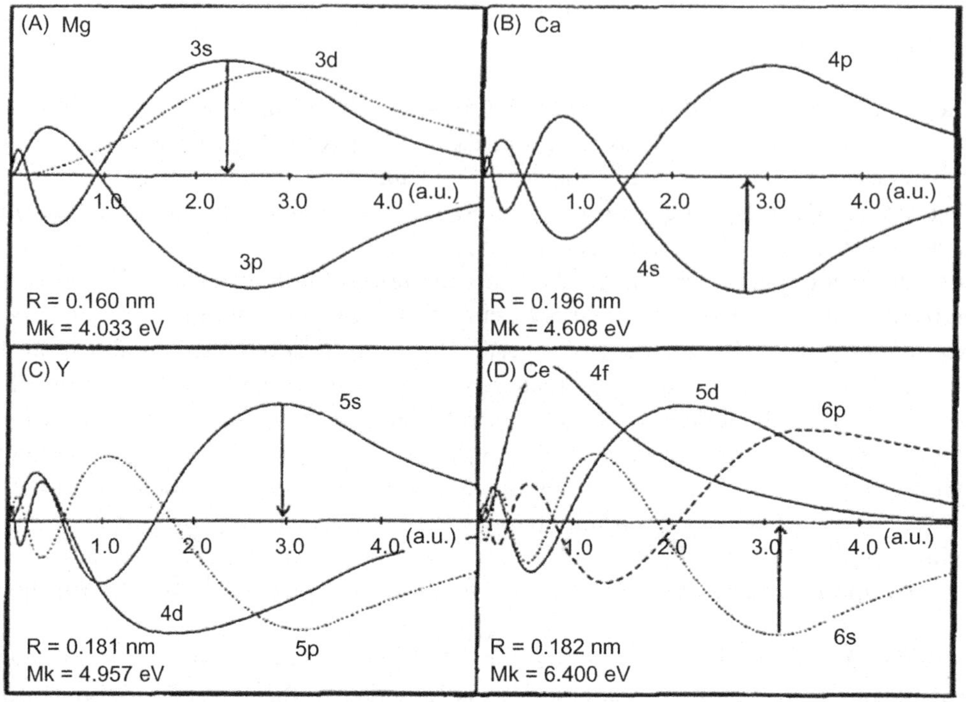

FIGURE 6-20 Radial orbital function of alloying elements in the cluster: (A) Mg, (B) Ca, (C)Y, and (D) Ce, (1 a.u. = 0.0529 nm).

functions of alloying elements in the cluster are illustrated in Fig. 6-20 for (A) Mg, (B) Ca, (C) Y, and (d) Ce. The peak position of the *s*-orbital is indicated by arrows in each figure. For comparison, the values of *Mk* parameter and the atomic radius *R* are also given in the figure. It is evident from Fig. 6-20 that the extent of the *s*-orbital is the largest for Ce among the elements, followed by Y, Ca, and Mg. This order is consistent with the order of the *Mk* parameter, but not with the order of the atomic radius assigned to pure metal.

6.7 Evaluation of Mechanical Properties for Mg Alloys

6.7.1 Vickers Hardness

Similarly to Eqs. (6.3) and (6.4), $\overline{Mk}$ and $\Delta\overline{Mk}$ are defined as, $\overline{Mk} = \Sigma X_i Mk_i$ and $\Delta\overline{Mk} = \Sigma X_i |Mk_i - Mk_{Mg}|$. In Fig. 6-21, the Vickers hardness data measured at room temperature are plotted against the $\Delta\overline{Mk}$ parameter [3], because some precipitates exist in these experimental alloys. It is apparent from this figure that there is a better correlation between the hardness and $\Delta\overline{Mk}$, as compared with the previous plots of $\Delta\overline{\Phi}$ and $\Delta\overline{R}$ shown in Fig. 6-2A–B. Also, there is no separation in the plots between the Mg–common metals system and the Mg–rare earth metal system.

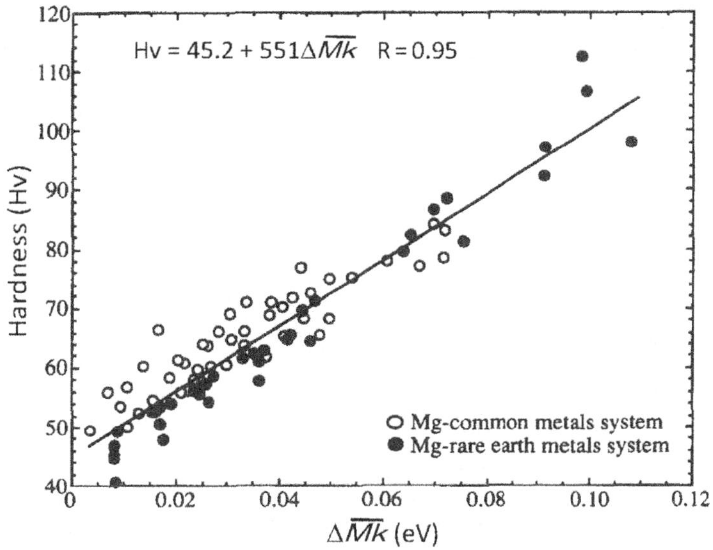

FIGURE 6-21 Correlation of the Vickers hardness with $\Delta\overline{Mk}$ for Mg alloys.

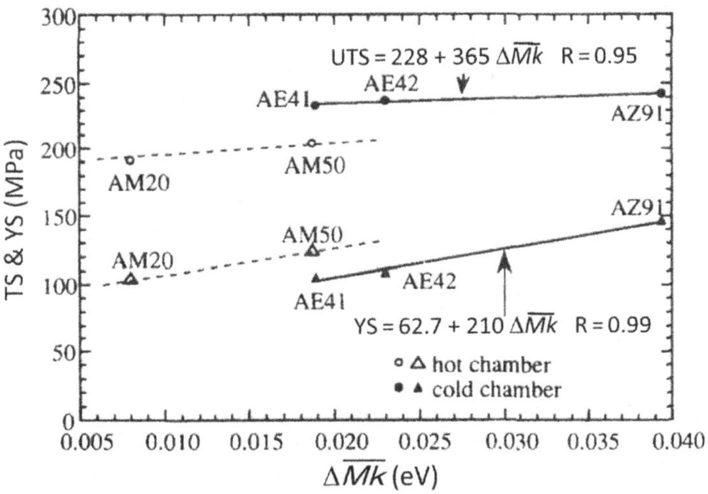

FIGURE 6-22 Correlation of tensile strength (TS) and yield stress (YS) with $\Delta\overline{Mk}$ for die-casting Mg alloys.

6.7.2 Yield Strength and Tensile Strength at Room Temperature

6.7.2.1 Casting Alloys

In Fig. 6-22 both the tensile strength and the yield strength measured at room temperature are plotted against $\Delta\overline{Mk}$ for three cold chamber die-casting alloys (AE41, AE42, and AZ91) and two hot chamber die-casting alloys (AM20 and AM50). These data are taken from Terje and Halcon [22]. In this figure the solid lines and dotted lines represent the results for cold

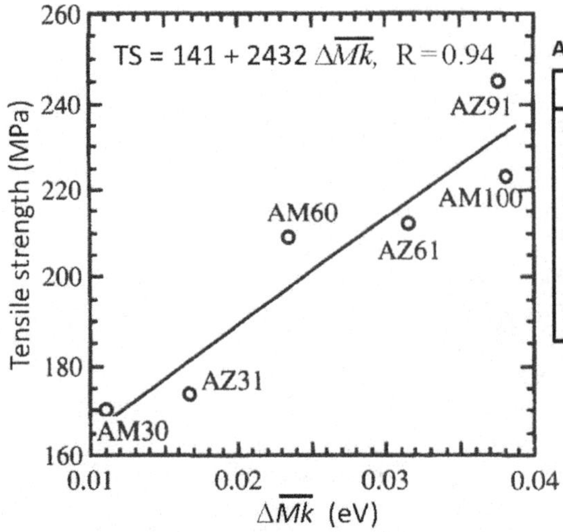

Alloy compositions (mass%)

	Al	Zn	Mn	Mg
AZ31	2.73	0.38	0.11	Bal.
AZ61	6.22	0.38	0.13	Bal.
AZ91	8.09	0.26	0.09	Bal.
AM30	2.53	0.00	0.10	Bal.
AM60	5.45	0.00	0.13	Bal.
AM100	8.93	0.00	0.14	Bal.

FIGURE 6-23 Correlation of tensile strength (TS) of typical commercially available cast Mg alloys with $\Delta\overline{Mk}$.

chamber and hot chamber die-casting alloys, respectively. For cold chamber die-casting alloys both the strengths change linearly with $\Delta\overline{Mk}$. However, for hot chamber die-casting alloys, the data are too limited to get any conclusion, but the slope of each dotted line seems similar to that of the corresponding solid line. The difference in the strengths between cold and hot chamber die-casting alloys is attributable to the difference in the casting methods between them.

Ninomiya et al. [3] fabricated six alloys of Mg-Al-Mn (AM30, AM60, AM100) and Mg-Al-Zn-Mn (AZ31, AZ61, AZ91) using a conventional casting method. The tensile strengths of as-fabricated alloys measured at room temperature are plotted against $\Delta\overline{Mk}$ in Fig. 6-23, where alloy compositions are also included. The tensile strength increases with $\Delta\overline{Mk}$. Thus, the correlation is reasonably well in case the same casting conditions are employed for the fabrication.

However, it is rather difficult to put in order the strength data of the cast alloys compiled in Metals Handbook [11] by using the $\Delta\overline{Mk}$ parameter. This is true even in case the target alloys of interest are limited within the sand and permanent mold casting alloys [11]. The reason is unknown, but the trifling difference in the casting conditions may affect the mechanical property data of cast alloys. However, as the measured temperature increases, the difference is going to disappear, as explained later.

6.7.2.2 Wrought Alloys

On the other hand, the tensile strength of wrought alloys measured at room temperature is plotted against $\Delta\overline{Mk}$ in Fig. 6-24. The data are taken from the Metals Handbook [11]. The temper designation is put together with the alloy name in Fig. 6-24. The temper designation given in the figure is as follows; F, as fabricated; H24, strain hardened and partially

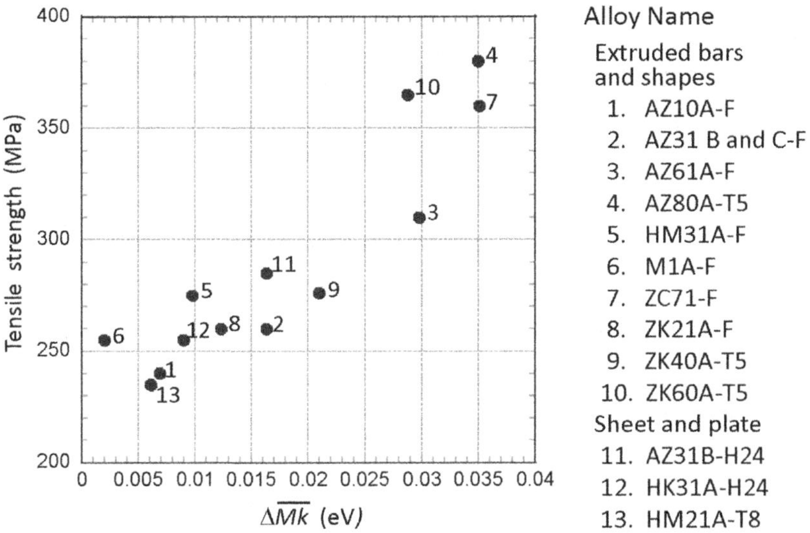

FIGURE 6-24 Correlation of tensile strength at room temperature with $\Delta\overline{Mk}$ for Mg alloys for extruded bars and shapes, and for sheet and plate.

annealed; T5, artificially aged only; T6, solution heat treated and artificially aged; T8, solution heat treated, cold worked, and artificially aged [11]. The *Mk* value for M = Th has not been calculated, so it is assumed to be 6.4 (eV), judging from the *Mk* values for other rare earth elements listed in Table 6-4.

As shown in Fig. 6-24, the tensile strength at room temperature increases with $\Delta\overline{Mk}$, irrespective of the alloys for use of extruded bars and shapes or of sheet and plate. The tensile strength appears to be less sensitive to the temper conditions, although the data are somewhat scattered. Thus, the tensile strengths of wrought alloys are predictable from the $\Delta\overline{Mk}$ parameter.

6.7.3 High Temperature Tensile Strength and Creep Stress

The tensile strengths of six die-casting Mg-Al-RE-Mn alloys (RE, rare earth element) were measured at 533K and 589K by Waltrip et al. [23]. The alloy compositions are shown in Fig. 6-25A, and the tensile strengths are plotted against (B) $\Delta\overline{Mk}$ and (C) $\overline{Mk}$. The correlation with the measured tensile strength is much better for (C) $\overline{Mk}$ than for (B) $\Delta\overline{Mk}$ at 533K and 589K.

Also, in Fig. 6-26, the creep stresses of these alloys at 394K, 450K, 533K, and 589K, are plotted against (A) $\Delta\overline{Mk}$ and (B) $\overline{Mk}$. Here, the creep stress is defined as the external stress to produce 0.1% creep strain in 1000 hours (3.6 Ms). It is apparent from Fig. 6-26 that the correlation with the creep stress is much better for (B) $\overline{Mk}$ than for (A) $\Delta\overline{Mk}$.

As another example, the tensile strengths measured at 588K are plotted against $\overline{Mk}$ in Fig. 6-27 for various Mg alloys including casting alloys as well as wrought alloys. All the data

(A)

(mass %)

Alloy	% Al	% RE	% Mn	Alloy	% Al	% RE	% Mn
AE43	3.5	3.33	0.38	AE22	1.9	1.94	0.35
AE42	3.6	2.05	0.29	AE52X1	5.4	1.70	0.30
AE41	3.7	1.30	0.34	AE52X2	5.3	2.11	0.27

% RE =% (Ce+La+Nd+Pr)

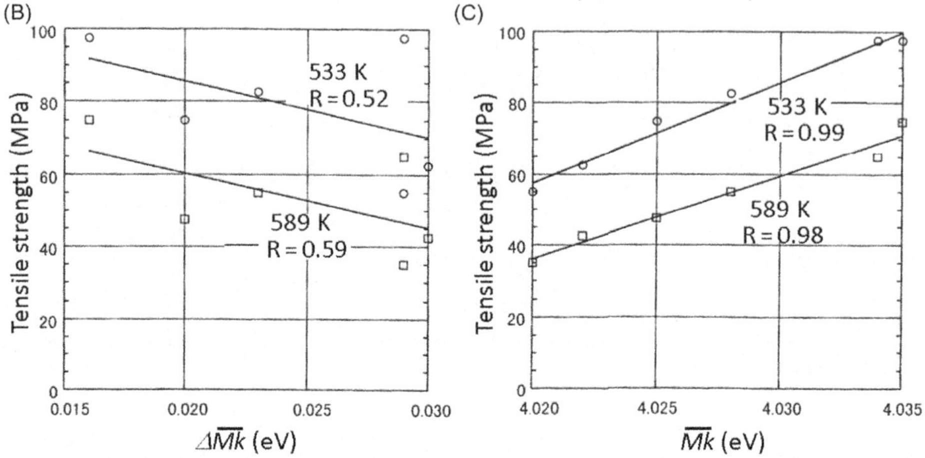

FIGURE 6-25 (A) Alloy compositions, and correlation of tensile strength at high temperatures with (B) $\Delta\overline{Mk}$ and (C) $\overline{Mk}$.

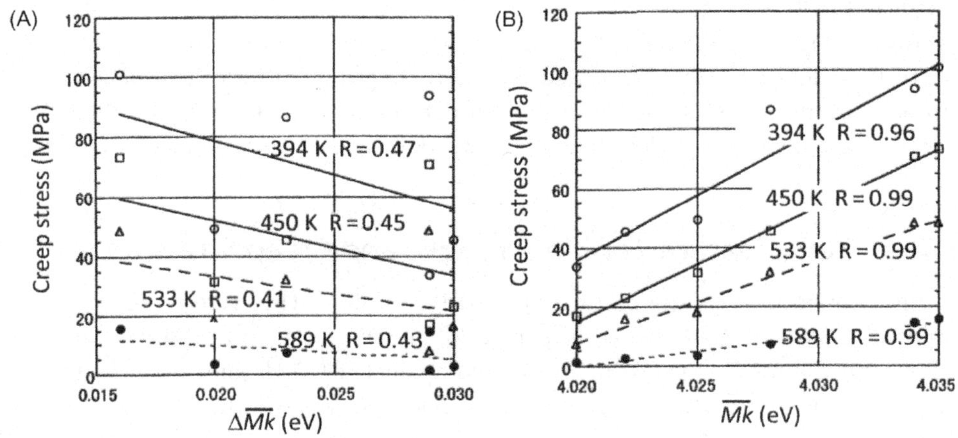

FIGURE 6-26 Correlation of creep stress with (A) $\Delta\overline{Mk}$ and (B) $\overline{Mk}$.

except for ZC63A-T6 (no. 13) are compiled in the Metals Handbook [11]. For convenience, nominal compositions of these alloys are listed in Table 6-5 [11]. The high temperature tensile strengths increase with the $\overline{Mk}$, irrespective of the casting alloys and wrought alloys, although the data are somewhat scattered. As previously explained, a similar result is also found in Al alloys as shown in Fig. 6-15 for the tensile strength at 589K.

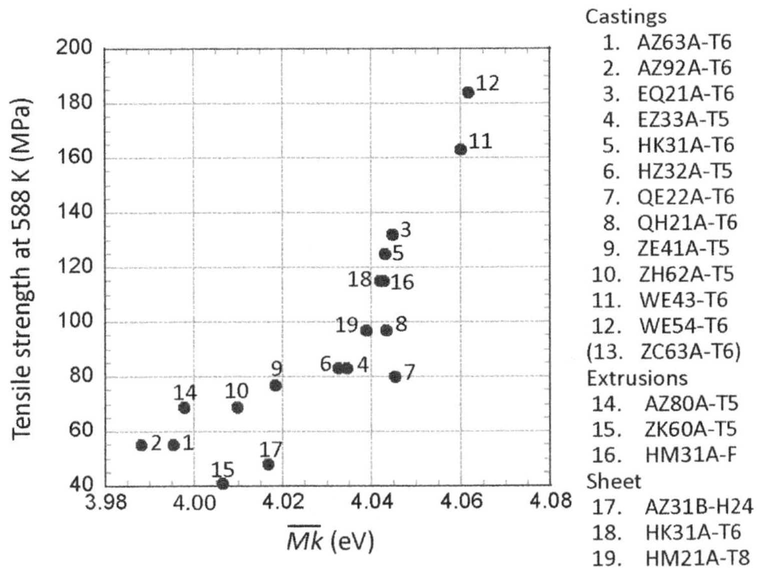

FIGURE 6-27 Correlation of tensile strength at 588K with $\overline{Mk}$.

Table 6-5 Nominal Compositions (Mass%) of Magnesium Alloys [12]

Alloy	Al	Mn	Th	Zn	Zr	Others
Castings						
AZ63A-T6	6.0	0.15		3.0		
AZ92A-T6	9.0	0.10		2.0		
EQ21A-T6					0.7	1.5Ag, 2.1Di
EZ33A-T5				2.7	0.6	3.3RE
HK31A-T6			3.3		0.7	
HZ32A-T5			3.3	2.1	0.7	
QE22A-T6					0.7	2.5Ag, 2.1Di
QH21A-T6			1.0		0.7	2.5Ag, 1.0Di
ZC63A-T6		0.25–0.75		6.0		2.7Cu
ZE41A-T5				4.2	0.7	1.2RE
ZH62A-T5			1.8	5.7	0.7	
WE43-T6					0.7	4.0Y, 3.4RE
WE54-T6					0.7	5.2Y, 3.0RE
Extrusions						
AZ80A-T5	8.5			0.5		
ZK60A-T5				4.0	0.45	
HM31A-F		1.2	3.0			
Sheet						
AZ31B-H24	3.0			1.0		
HK31A-H24			3.0		0.6	
HM21A-T8		0.6	2.0			

Di, didymium (a mixture of rare earth elements made up chiefly of Nd and Pr); *RE*, rare earth elements.

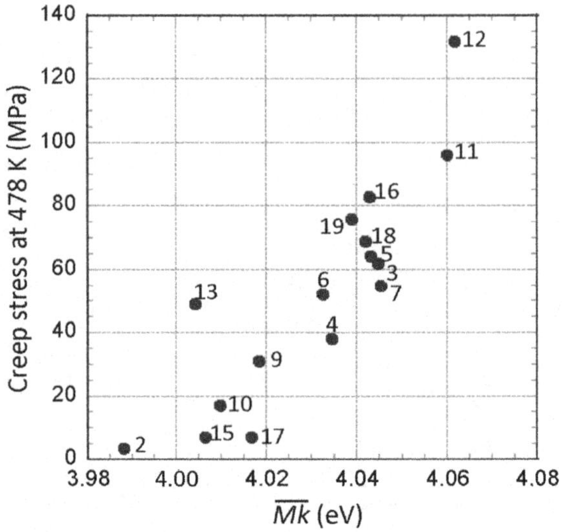

FIGURE 6-28 Correlation of creep stress at 478K with $\overline{Mk}$.

Also, the creep stresses at 478K are plotted against the $\overline{Mk}$ in Fig. 6-28. Among the alloys shown in Fig. 6-27, the creep data is available for ZC63A-T6 (no. 13) but not for AZ63A-T6 (no. 1), AZ80A-T5 (no. 14), ZK60A-T5 (no. 15), and AZ31B-H24 (no. 17) [11]. Here, the creep stress is defined as the external stress to produce 0.2% creep strain in 1000 hours (3.6 Ms) for cast alloys and in 100 hours (0.36 Ms) for wrought alloys [11]. Although the data are scattered, the creep stress tends to increase with the $\overline{Mk}$. Thus, high temperature mechanical properties are predictable using the $\overline{Mk}$ parameter.

Both tensile strength at 588K and creep stress at 478K are very high in no. 11 (WE43-T6) and no. 12 (WE54-T6), since they contain rare earth elements (RE). On the other hand, both these strengths are low in the RE-free alloys, for example, no. 2 (AZ92A-T6), no. 15 (ZK60A-T5), and no.17 (AZ31B-H24). Thus, the addition of rare earth elements is so effective in increasing heat resistant properties of Mg alloys.

6.8 Energy Approach to Mechanical Properties of Mg Alloys

6.8.1 Strain Energy of Mg Lattice Around Alloying Elements

First of all, the energy gained by local lattice strain around an isolated alloying element M in hcp Mg is calculated, and the results are shown in Fig. 6-29. As compared with the results of fcc Al shown in Fig. 6-16, the energy gained by local lattice strain (i.e., strain energy) is about the same order of magnitude. For example, it is about 0.015 eV/atom for M = Fe in Mg, which is slightly larger than the value of 0.01 eV/atom for Fe in Al. For principal alloying elements in Mg alloys (e.g., Al, Zn, Ca, Zr, Th, etc.), the strain energy is lower than

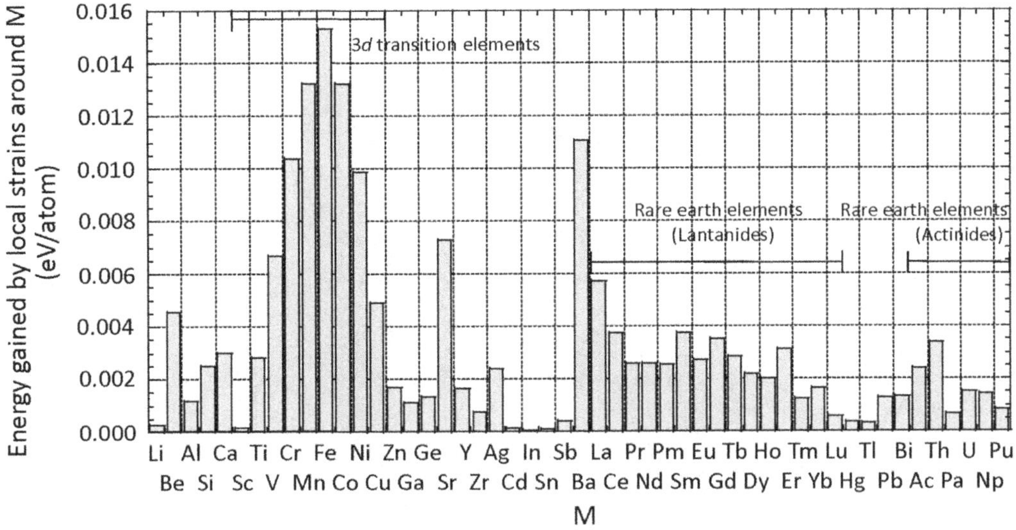

FIGURE 6-29 Energy gained by local strains around an isolated alloying element M in Mg.

0.004 eV/atom. It is interesting to note that the strain energy of Mn is as high as 0.013 eV/atom, since a small amount of Mn is often added into Mg alloys to improve mechanical properties [20]. The elastic interaction may operate between Mn and other elements in Mg alloys.

As explained before in Al alloys, such local strain energies will cause the solid solution hardening of Mg alloys. When the precipitate is formed in Mg alloys, large strains and hence high strain energies are localized greatly in the matrix neighboring the precipitate (i.e., G.P. zones and intermediate compounds). This will cause the precipitation hardening as explained in the Al alloys.

6.8.2 A Relation between $\Delta\overline{Mk}$ and $(\overline{Mk} - Mk_{Mg})$ in Mg Alloys

A relation between the $\Delta\overline{Mk}$ and $(\overline{Mk} - 4.033)$ is estimated in the same way as employed in the Al alloys. Here, 4.033 (eV) is the *Mk* value for Mg (see Table 6-4). The results are shown in Fig. 6-30 for Mg alloys. In this figure every Mg alloy shown in Figs. 6-27 and 6-28 is plotted using the same alloy number. The result of Fig. 6-30 for Mg alloys is very different from that of Fig. 6-17 for Al alloys.

Let's assume here that both $\Delta\overline{Mk}$ and $(\overline{Mk} - 4.033)$ are the parameters to show the strain energy. As shown in Fig. 6-30, the values of $(\overline{Mk} - 4.033)$ are either negative or positive, depending on the alloys. To see a general trend, two lines are drawn in the figure. One is a line showing a relation, $\Delta\overline{Mk} = \overline{Mk} - 4.033$, and the other is a line showing a relation, $\Delta\overline{Mk} = -\overline{Mk} + 4.033$. According to Eq. (6-6), the former relation is satisfied in the alloy when all the alloying elements have larger *Mk* value than Mg, whereas the latter relation is

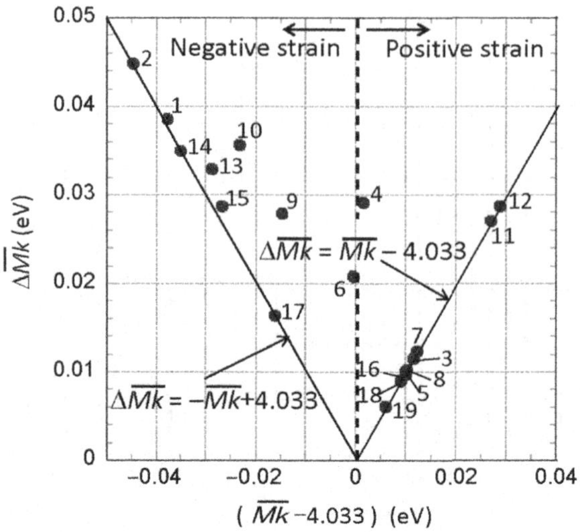

FIGURE 6-30 Comparison between $\Delta\overline{Mk}$ and $(\overline{Mk} - 4.033)$ for Mg alloys.

satisfied in the alloy when all the alloying elements have smaller *Mk* value than Mg. It is surprising that most of the alloys are located along these lines. For example, every alloy of nos. 3, 5, 7, 8, 11, 12, 16, 18, and 19 contains rare earth elements (RE) (e.g., Th) (see Table 6-5). The elements other than rare earth elements are Mn, Ag, Zr, Y, and their *Mk* values are larger than 4.033 (eV) of Mg. On the other hand, no. 14 and no. 17 alloys consist only of Al and Zn, and their *Mk* values are smaller than 4.033 (eV) of Mg. No. 1 and no. 2 alloys consist of Al, Zn, and a small amount of Mn (0.04–0.07mol%). Also, no. 13 and no. 15 alloys are RE-free alloys. Thus, RE-free alloys are located along the line, $\Delta\overline{Mk} = -\overline{Mk} + 4.033$, where $(\overline{Mk} - 4.033)$ is negative. In contrast to this, most of the RE-containing alloys are mainly located along the line, $\Delta\overline{Mk} = \overline{Mk} - 4.033$, where $(\overline{Mk} - 4.033)$ is positive. Several RE-containing alloys such as nos. 4, 6, 9, and 10 are far away from this line, where $(\overline{Mk} - 4.033)$ is either positive or negative, depending on the alloys.

To understand how such a positive or negative value of $(\overline{Mk} - 4.033)$ affects the mechanical properties, we refer to the Cottrell pinning model [17,24]. For example, let's consider the case when an element M occupies the substitutional site in pure metal X. If M atom is larger than X atom, a larger M atom yields the compressive stress field in the metal lattice neighboring M. By contrast, a smaller M atom yields the tensile stress field. Such a stress field around M interacts with the stress field around a dislocation. For example, if an edge dislocation with an extra half plane exists above the slip plane, the stress field around an edge dislocation is compressive in the region above the slip plane, but tensile in the region below the slip plane. As a result, if M can move, a large M atom accompanied with the compressive stress field goes to the region below the slip plane of the dislocation where the tensile stress field is operating, to reduce the total elastic strain energy. Finally M is pinned by the dislocation core. This is the so-called Cottrell pinning [17,24]. By analogy with this, the strain field of showing positive ($\overline{Mk}$ − 4.033) corresponds to the case of a large M atom, and hence the compressive stress field is present in the matrix neighboring M. On the other hand, the strain field of showing negative ($\overline{Mk}$ − 4.033) corresponds to the case of a small M atom, and hence the tensile stress field is present in the matrix neighboring M. If the mechanical property of the alloy is examined by a tensile test, the compressive stress field in the alloy may resist against the applied tensile stress, resulting in the increase of the tensile strength of the alloy. In other words, the tensile strength increases with increasing ($\overline{Mk}$ − 4.033) value of the alloy. In fact, as shown in Fig. 6-27, the tensile strength at 588K increases with the $\overline{Mk}$. The RE-containing alloy that is composed of large M atoms shows much stronger tensile strength than the Re-free alloy that is composed of small M atoms.

6.8.3 High Temperature Tensile Strength and Creep Stress

As shown in Fig. 6-30, the energy range of the parameter is, $0 < \Delta\overline{Mk} < 0.05$(eV), and $-0.05 < (\overline{Mk} - 4.033) < 0.04$(eV), so that it is expressed as, $0 < |\overline{Mk} - 4.033| < 0.05$(eV). Thus, the energy interval is 0.05 eV/atom for Mg alloys, and it is comparable with 0.06 eV/atom for Al alloys. The thermal energy, k_BT, is about 0.04 eV/atom at 478K (the temperature for creep stress shown in Fig. 6-28) and about 0.05 eV/atom at 588K (the temperature for tensile strength shown in Fig. 6-27). In case the tensile strength at room temperature (297K) is used as a reference, the difference in the thermal energy between 588K and 297K is about 0.025 eV/atom. Thus, the energy gain by the thermal energy is 0.025–0.04 eV/atom, which is comparable with the energy interval, 0.05 eV/atom. Therefore, both the high temperature tensile strength and creep stress are evaluated using the $\overline{Mk}$ parameter instead of the $\Delta\overline{Mk}$ parameter.

6.9 A Guiding Rule for Alloy Design

For Mg alloys consisting of two phases, the mechanical strengths may be predictable following a certain rule [25,26]. The $\Delta\overline{Mk}$ parameter is valid for the prediction of tensile strength at room temperature or lower temperatures. On the other hand, the $\overline{Mk}$ parameter is valid for the prediction of the tensile strength and the creep stress at high temperatures. In other words, the value of $\Delta\overline{Mk}$ should be large to increase the alloy strength at room temperature, whereas the value of $\overline{Mk}$ should be large to increase the alloy strength at high temperatures. For example, the WE54-T6 alloy has high strength at both room temperature and high temperatures. For example, the tensile strength is about 280 MPa at room temperature, and about 184 MPa at 588 K [11]. As shown in Table 6-5, it is composed of 3.0 mass% RE (e.g., Nd), 5.2 mass% Y, and 0.7 mass% Zr, all of which have high *Mk* values (see Table 6-4), so that both $\Delta\overline{Mk} = 0.029$(eV) and $\overline{Mk} = 4.062$(eV) are large for this WE54-T6 alloy. Similar results are also obtained by Kamado et al. [27].

6.10 Alloy Design of Mg Alloys

A heat resistant Mg alloy used at intermediate temperatures is designed following the present rule [25,26]. First, the Mg-Al alloy system is chosen, because the die-casting method is applicable to this alloy system for the production. A target of alloy design is set up toward the development of the new alloy, which exhibits a similar or even better heat resistance than the AE42 containing rare earth elements. To meet this requirement, the $\overline{Mk}$ value of the alloy should be as large as that of the AE42. So, alloying elements with the large *Mk* value are selected. The candidate elements are transition metals, rare earth metals and Ca, judging from the *Mk* values of elements listed in Table 6-4. Among them the transition metals are not soluble in the Mg solid solution and also their addition is harmful to the corrosion resistance. The rare earth metals are more expensive and have larger atomic weight than Ca. For this reason, the Mg-Al-Ca alloy is chosen as a candidate alloy. The composition of the designed alloy is shown in Table 6-6 together with the two reference alloys, AZ91 and AE42 [25,26]. The $\overline{Mk}$ value of the designed alloy is as large as that of AE42. These three alloys are produced by the die-casting method, and the creep test is performed with them. The creep temperature and applied stress conditions employed are 423K, 50 MPa and 473K, 30 MPa. The measured creep properties are summarized in Table 6-7. In either creep condition, the designed alloy has the smaller initial strain and the lower secondary creep rate than the reference alloys, AZ91 and AE42, in agreement with the prediction from the $\overline{Mk}$ parameter. Recently, the creep mechanism of the Mg-Al-Ca alloys has been investigated due to its great importance for automobile applications [20,28,29].

Finally, it is summarized that the *Mk* parameter, i.e., the *s*-orbital energy level, is first obtained from the calculation. The alloying effects are inevitably condensed in this *Mk* parameter. Despite great progress on the understanding of the mechanical properties by dislocation theory, there is still a large barrier to the quantitative prediction of alloy strength with the aid of the theory. The *Mk* approach is an alternative method to put in order the mechanical properties of the lightweight Al alloys and Mg alloys. Such an approach is also available in Zn alloys. For convenience, the *Mk* values for alloying elements in hcp Zn are listed in Table 6-8 [30]. It is important to note that the energy approach is first introduced to treat the mechanical properties of both Al and Mg alloys.

Table 6-6 Chemical Compositions of the Designed Alloy and Reference Alloys, AZ91 and AE42

	Chemical Composition (Mass%)						
	A1	**Zn**	**Mn**	**RE**	**Ca**	**Mg**	$\overline{Mk}$**(eV)**
AZ91	8.9	0.68	0.19	–	–	Bal.	3.996
AE42	4.0	–	0.22	3.4	–	Bal.	4.032
Designed alloy	3.1	–	–	–	3.2	Bal.	4.032

Table 6-7 Creep Properties of the Designed Alloy and Reference Alloys, AZ91 and AE42

423K – 50 MPa		
	Initial Strain (%)	**Secondary Creep Rate (%/Hour)**
AZ91	0.137	2.01×10^{-2}
AE42	0.117	2.28×10^{-4}
Designed alloy	0.115	1.18×10^{-4}
473K – 30 MPa		
	Initial Strain (%)	**Secondary Creep Rate (%/Hour)**
AZ91	0.089	1.10×10^{-1}
AE42	0.080	5.66×10^{-4}
Designed alloy	0.013	3.92×10^{-4}

Table 6-8 List of *Mk* Values for Alloying Elements in Zn

Element	*Mk* (eV)	Element	*Mk* (eV)
Li	6.667	Mn	4.699
Be	2.459	Fe	4.486
Na	6.925	Co	4.287
Mg	3.555	Ni	4.023
Al	1.809	Cu	3.395
Si	0.503	Zn	2.172
K	9.986	Cd	3.338
Ca	8.210	In	2.079
Sc	6.537	Sn	0.986
Ti	5.705	Sb	0.518
V	5.271		

References

[1] M. Morinaga, S. Nasu, H. Adachi, J. Saito, N. Yukawa, Alloying effect on the electronic structure of aluminium, J. Phys.: Condens. Matter. 3 (1991) 6817–6828.

[2] M. Morinaga, S. Kamado, An electronic approach to the prediction of the mechanical properties of aluminium alloys, Modelling Simul. Mater. Sci. Eng. 1 (1993) 151–164.

[3] R. Ninomiya, H. Yukawa, M. Morinaga, K. Kubota, An electronic approach to the prediction of the mechanical properties of magnesium alloys, J. Alloys. Comp. 215 (1994) 315–323.

[4] P. Haasen, Physical Metallurgy, Cambridge University Press, Cambridge, 1978.

[5] F.R.N. Nabarro, Z.S. Basinski, D.B. Holt, The plasticity of pure single crystals, Adv. Phys. 13 (1964) 193–323.

[6] W.G. Johnston, Yield points and delay times in single crystals, J. Appl. Phys. 33 (1962) 2716–2730.

[7] R. Chang, An atomistic study of fracture, Int. J. Fract. Mech. 6 (1970) 111–125.

[8] E.T. Teatum, K.A. Gshneidner Jr., J.T. Waber, Department of Commerce LA-2345 (1968).

[9] L. Pauling, Nature of the Chemical Bond, 3rd edn, Cornell University Press, Ithaca, 1960.

[10] R.E. Watson, L.H. Bennett, Transition metals: *d*-Band hybridization, electronegativities and structural stability of intermetallic compounds, Phys. Rev. B 18 (1978) 6439–6449.

[11] Metals Handbook vol.2, 10th edn, Properties and selection: Nonferrous alloys and special-purpose materials, ASM International, New York (1990).

[12] Aluminium Handbook, 4th edn, Light Metals Association of Japan, Tokyo (1990).

[13] J.E. Hatch (Ed.), Aluminium: Properties and Physical Metallurgy, American Society for Metals, Metals Park, OH, 1984.

[14] R.S. Mulliken, R.S. Mulliken, Electronic population analysis on LCAO–MO molecular wave functions. I, II, III, IV, J. Chem. Phys 23 (1955) 1833–1840, 1841–1846, 2338–2342, 2343–2346.

[15] A.R. Miedema, F.R. Deboer, P.F. Dechatel, Empirical description of the role of electronegativity in alloy formation, J. Phys. F:Metal Physics 3 (1973) 1558–1576.

[16] C. Kittel, Introduction to Solid State Physics, 5th edn, John Wiley & Sons, Inc, New York, 1976.

[17] J.P. Hirth, J. Lothe, Theory of Dislocations, McGraw-Hill, New York, 1968.

[18] H. Suzuki, Dislocations and Mechanical Properties of Crystals, John Wiley & Sons Inc, New York, 1957.

[19] G. E. Totten (VolumeEditor), ASM Handbook, Volume 4E, Heat Treating of Nonferrous Alloys, ASM International, Materials Park, OH, (2016).

[20] T. Homma, S. Nakawaki, K. Oh-ishi, K. Hono, S. Kamado, Unexpected influence of Mn addition on the creep properties of a cast Mg–2Al–2Ca (mass%) alloy, Acta Materialia 59 (2011) 7662–7672.

[21] G. Effenberg, F. Aldinger, O.I. Bodak (Eds.), Ternary Alloys, Volume 15, VCH Verlagsgesellschaft, Weiheim (Germany) and VCHPublishers, New York (USA), 1995.

[22] K.A. Terje, W. Häkon, in: B.L. Mordike, F. Hehmann (Eds.), Mechanical Properties of Pressure Die Cast Mg-Alloys, Magnesium Alloys and their Application, DGM Informationsgesellschaft, m.b.H., Verlag, 1992, pp. 221–228.

[23] J.S. Waltrip, Fresh look at some old magnesium die casting alloys for elevated temperature applications, Proceedings of the 47th Annual World Magnesium Conference, International Magnesium Association, (1990), pp.124–129.

[24] J. Weertman, J.R. Weertmsn, *Elementary Dislocation Theory*, Macmillan Company, New York, NY, 1964.

[25] M. Morinaga, R. Ninomiya, Alloy Design, in: Y. Kojima, et al. (Eds.), Handbook of Advanced Magnesium Technology, Kallos Publishing Co. Ltd, Tokyo, 2000, pp. 97–104.

[26] R. Ninomiya, T. Ojiro, K. Kubota, Improved heat resistance of Mg-Al alloys by the Ca addition, Acta Metall. Mater. 43 (1995) 669–674.

[27] S. Kamado, Y. Kojima, S. Tanike, I. Seki, S. Hama, in: B.L. Mordike, K.U. Kainer (Eds.), Magnesium Alloys and their Application, DGM Informationsgesellschaft, m.b.H., Verlag, 1998, p. 169.

[28] A.A. Luo, M.P. Balogh, B.R. Powell, Creep and microstructure of magnesium-aluminum-calcium based alloys, Metall. Mater. Trans. A 33A (2002) 567–574.

[29] Y. Terada, R. Sota, N. Ishimatsu, T. Sato, K. Ohori, A. Thousandfold Creep, Strengthening by Ca addition in die-cast AM50 magnesium alloy, Metall. Mater. Trans. A 35A (2004) 3029–3032.

[30] R. Ninomiya, K. Kubota, H. Kubo, M. Morinaga, Electronic approach to the prediction for the mechanical properties of zinc alloys, Proceedings of World Zinc '93, Australia (1993) pp. 569–573.

7

Hydrogen Overpotential of Alloys and Intermetallic Compounds

7.1 Hydrogen Overpotential

Potential greater than an equilibrium potential is required to apply to the electrode to drive a hydrogen evolution reaction on the electrode surface in an electrolytic cell. Such an extra potential deviated from the equilibrium potential is called hydrogen overpotential. The hydrogen overpotential changes largely with electrode materials. For instance, it is small for platinum and palladium, but large for mercury.

There have been numerous investigations on the catalytic activity of metals and alloys for hydrogen evolution reactions [1–17] because of the practical significance. For example, the electrolysis of water for hydrogen production requires electrode materials with low hydrogen overpotential to minimize the energy consumption in the production process. In addition to such electrocatalytic issues, it is known that the corrosion behavior of metals and alloys is dependent on the hydrogen overpotential.

Enormous attempts have been made to explain the characteristics of the hydrogen overpotential of metals, using various physical and/or electronic parameters, including atomic number [2,3], the work function [6,10], the bond strength of metals [6,11], the adsorption of hydrogen on metals [4,9,10], and the electronegativity [6,7,10]. These previous results suggest that the activity for the hydrogen evolution reaction is related closely to the electronic structures of metals, although any explicit explanation has not been given even for pure metals.

For alloys, the hydrogen overpotential changes depending on both mother metals and alloying elements, and hence the theoretical approach is more difficult for alloys than for pure metals. For this reason, the alloying effect on the hydrogen evolution reaction in transition metal–based alloys has been discussed only in a qualitative manner by using the Engel–Brewer valence bond theory [11,12]. However, the theory itself is very classical involving several assumptions, some of which are not acceptable these days, as discussed later.

In this chapter, for transition metal–based alloys and intermetallic compounds, the hydrogen overpotential is elucidated by combining experiments with calculations.

A Quantum Approach to Alloy Design. DOI: https://doi.org/10.1016/B978-0-12-814706-1.00007-8

7.2 Experiments of Hydrogen Overpotential for Transition Metals and Alloys

7.2.1 Specimen Preparation and Hydrogen Overpotential Measurement

A series of experiments on the hydrogen overpotential is carried out with the pure 3*d* transition metals, M, and also with the Ti-5 mol% M, Fe-5 mol%, and Ni-5 mol% M binary alloys, where M's are the 3*d* transition metals, M = Ti, V, Cr, Mn, Fe, Co, Ni, and Cu [18]. In addition, for Ti–Ni binary alloys the alloy compositions are varied from 1 mol% Ni to 95 mol% Ni to examine the alloying effect over a wide compositional range. All the metals and alloys are prepared from high purity raw metals [18].

The hydrogen overpotential is then evaluated by measuring cathodic polarization curves in a 1 N sulfuric acid solution at 303K. A potentiostat is used for the polarization measurements. The potential is swept from a corrosion potential at a constant rate of 20 mV/min. Before starting the polarization measurement, high purity argon (99.9999%) is passed through the electrolyte for 1 hour (3.6 ks) to remove any dissolved oxygen from it. A saturated calomel electrode (sce) and a platinum plate are used as the reference and counter electrodes, respectively. Surface preparation is carried out by abrasion, finishing with no. 600 silicon carbide waterproofed paper. Nickel leading wires are spot-welded to the specimen, and then, a part of the specimen surface is sealed with plastic resin to give an approximate surface area of 1 cm^2. Cathodic polarization measurements are repeated three times for every specimen to examine the reproducibility of the data. The hydrogen overpotential is obtained by reading the potential values at apparent current densities, 0.3, 1, 3, and 10 mA/cm^2. Attention is directed mainly toward the relative change in the hydrogen overpotential with alloying elements.

7.2.2 Cathodic Polarization Curves for Pure Ti, Fe, and Ni

Typical cathodic polarization curves are shown in Fig. 7-1 for pure Ti, Fe, and Ni. It is evident from the curves of pure Fe and Ni that following the Tafel's equation, the logarithm of the current density, log $|\, i \,|$, increases linearly with the electrode potential, as the hydrogen evolution reaction progresses on the electrode surface of these pure metals. The Tafel's equation is expressed as, $\eta = a + b \log |\, i \,|$, where η is the overpotential, i is the current density, a and b are constants. In contrast to this result, the positive current density appears in a part of the cathodic polarization curve of pure Ti, as indicated by a dotted line (see Fig. 7-1). However, the curve follows the Tafel's equation in a lower potential range than −0.6 V (vs sce). Thus, the hydrogen overpotential for these metals increases in the order, Ni < Fe < Ti. The average Tafel's slopes are about 0.145, 0.110, and 0.135 V/decade for Ti, Fe, and Ni, respectively. These values imply that the one-electron transfer step (the so-called Volmer step) may be dominant in the present hydrogen evolution reaction.

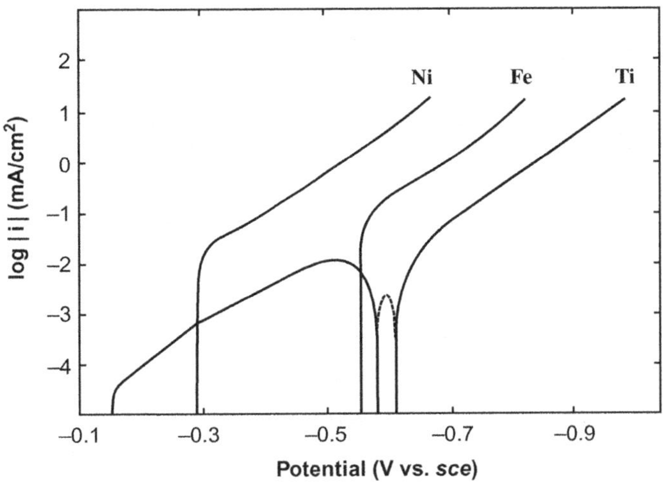

FIGURE 7-1 Cathodic polarization curves for pure Ti, Fe, and Ni in 1 N sulfuric acid at 303K.

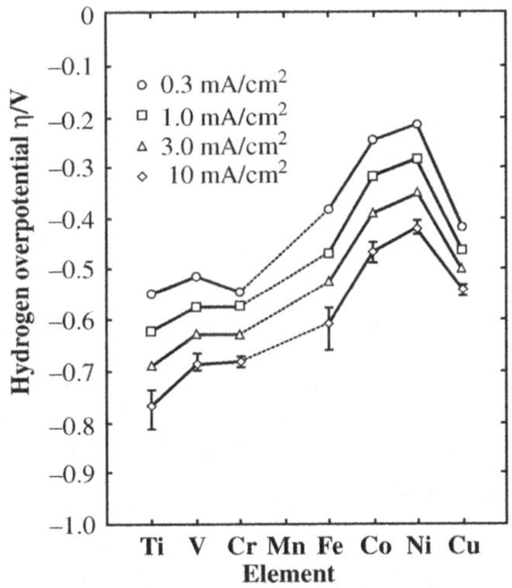

FIGURE 7-2 Hydrogen overpotential for 3*d* transition metals measured in 1 N sulfuric acid at 303K.

7.2.3 Hydrogen Overpotential for Transition Metals and Alloys

The measured hydrogen overpotential is shown in Fig. 7-2 for pure metals. The data plotted here are the average values of three measurements. In the overpotential curve at 10 mA/cm^2, the scattering range of the data among the three measurements is indicated by error bars. In this figure the data for Mn is missing, because the bulk Mn metal is too brittle

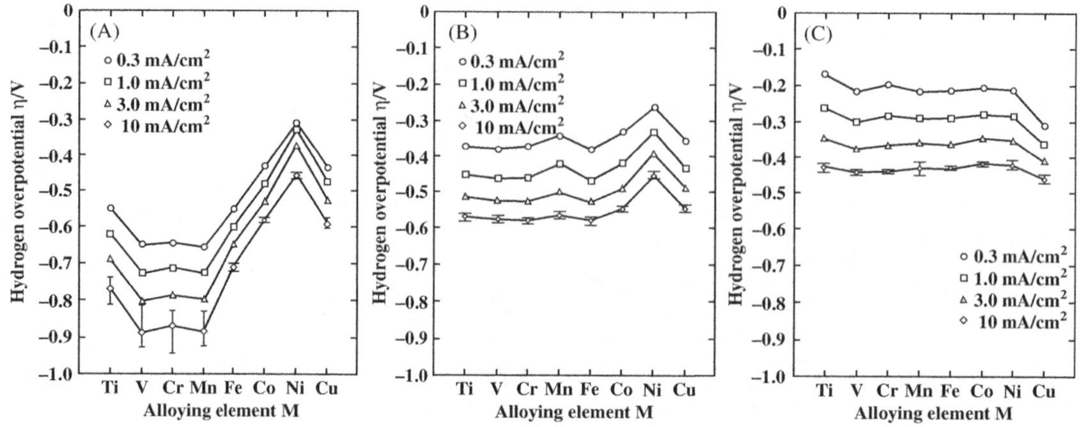

FIGURE 7-3 Hydrogen overpotential for (A) Ti alloys, (B) Fe alloys, and (C) Ni alloys measured in 1 N sulfuric acid at 303K.

to prepare a specimen suitable for the polarization measurement. In Fig. 7-2 the more negative and larger η value means the higher hydrogen overpotential. So, Ti has the highest hydrogen overpotential among these 3*d* elements. As the atomic number increases, the hydrogen overpotential becomes lower and reaches a minimum at Ni, and then again increases at Cu. Such a trend is consistent with the previous results [2,3]. The hydrogen overpotential is high in such metals as the affinity with hydrogen is large. For example, Ti and V of the high hydrogen overpotential metals form hydrides such as TiH_2 and VH_2.

The measured hydrogen overpotential for Ti, Fe, and Ni alloys is shown in Fig. 7-3A–C, respectively. It is apparent from Fig. 7-3A that the hydrogen overpotential for Ti alloys changes considerably depending on the alloying elements, M. Namely, the hydrogen overpotential increases by the addition of V, Cr, and Mn, but it decreases by the addition of Fe, Co, Ni, and Cu. The Ni addition is most effective in lowering the hydrogen overpotential among the 3*d* alloying elements. In case of Fe alloys, as shown in Fig. 7-3B, the measured hydrogen overpotential is nearly similar among the alloys containing Ti, V, Cr, and Mn, except for a small decrease for the Mn containing alloy at the lower current densities. All these elements are located on the left side of Fe in the periodic table. In contrast to this, for the alloys containing the right-side elements, Co, Ni, and Cu, there are large changes in the hydrogen overpotential, and their changes are analogous to the results of Ti alloys containing these elements (compare Fig. 7-3A with Fig. 7-3B). In the case of Ni alloys shown in Fig. 7-3C, there are few changes in the hydrogen overpotential for any alloys that contain the left-side elements of Ni in the periodic table, as is similar to the result of the Fe alloys. However, as compared with pure Ni (i.e., M = Ni), the overpotential decreases slightly in some alloys, in particular in the Ti containing alloy. On the other hand, the addition of Cu, a right-side element of Ni, increases the hydrogen overpotential.

For comparison, the hydrogen overpotential data of every alloy at 10 mA/cm^2 are plotted in Fig. 7-4, together with the data of pure metals taken at the same current density. It is seen

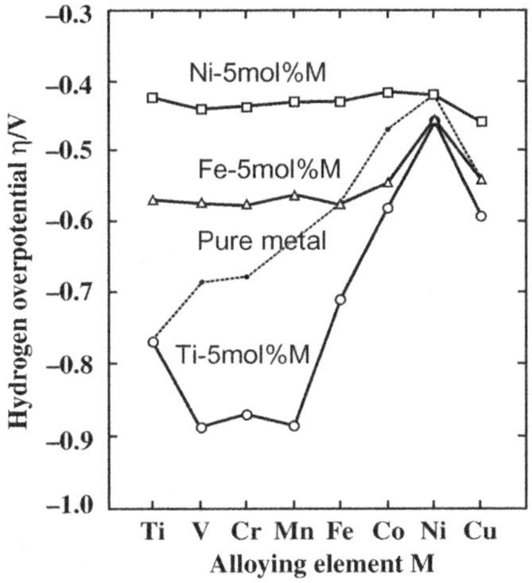

FIGURE 7-4 Comparison of the hydrogen overpotential values at 10 mA/cm^2 for Ti, Fe, and Ni alloys with those for pure metals at the same current density.

from this figure that the hydrogen overpotential change with alloying elements in Ti alloys resembles the change with pure metals, even though there is a small difference in the value of the hydrogen overpotential between them. Similarly, for Fe alloys, the hydrogen overpotential observed in Co, Ni, and Cu reflects the characteristic of each pure metal. Such a feature is also seen in Cu for Ni alloys. The Cu addition leads to the increase in the hydrogen overpotential of the alloy.

Judging from the present trend, it is concluded that the hydrogen overpotential for transition metal–based alloys changes following a simple rule. Only if the right-side element, M, of a base metal in the periodic table is added, does the hydrogen overpotential of the alloy change considerably, reflecting the hydrogen overpotential characteristic of pure M metal. On the other hand, the addition of the left-side element induces little change in the hydrogen overpotential of the alloy, even though a small change still occurs in it.

7.2.4 Compositional Dependence on the Hydrogen Overpotential of Ti–Ni Alloys

For the Ni-5 mol% Ti alloy, it is interesting that the overpotential of the alloy is lower than that of the individual element, Ni or Ti, as shown in Fig. 7-4. In contrast to this, a remarkable change is observed in the Ti-5 mol% Ni alloy as shown in Fig. 7-4. So, the compositional dependence on the hydrogen overpotential of Ti–Ni alloys is further examined, and the results are shown in Fig. 7-5. The hydrogen overpotential decreases rapidly with the 1 or 2 mol% Ni addition into Ti, and its changing rate becomes gradually smaller with further Ni addition. It is interesting to note that the alloy containing 8 mol% Ni shows a lower overpotential than pure Ni at the current densities of 3–10 mA/cm^2. In contrast to such a

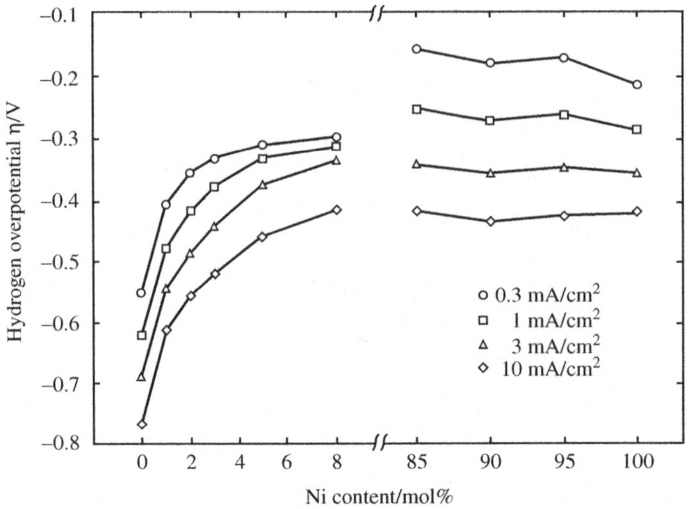

FIGURE 7-5 Effect of Ni content on the hydrogen overpotential for binary Ti–Ni alloys measured in 1 N sulfuric acid at 303K.

remarkable change in the overpotential of Ti-rich alloys, the overpotential of Ni-rich alloys is less changed, but still there is a tendency to decrease with increasing Ti content at low current densities.

7.3 Theoretical Approach to Hydrogen Overpotential of 3*d* Transition Metal Alloys

As pointed out by many investigators, catalytic activity for the hydrogen evolution reaction on the transition metals is related closely to the electronic structures. For alloys, however, there is no well-established concept for understanding hydrogen overpotential.

As explained in Chapter 3, Nickel Alloys,Chapter 4, Iron Alloys, and Chapter 5, Titanium Alloys the electronic structures of 3*d* transition metal–based alloys have been calculated using the DV-Xα cluster method. Here, the calculated results of hcp Ti [19], bcc Fe [20], and Ni_3Al [21] are applied to the present hydrogen overpotential problem. As explained in Chapter 3, Nickel Alloys, the calculated results of Ni_3Al are applicable to the present problem of fcc Ni.

7.3.1 Charge Transfer in Transition Metal Alloys

Following the Mulliken population analysis [22], the ionicity of elements can be calculated. Fig. 7-6 shows the calculated ionicity difference, ΔI, of the 3*d* alloying element in each cluster for (A) hcp Ti, (B) bcc Fe, and (C) Ni_3Al. Here, ΔI is defined as the ionicity difference between the alloying element M and the base metal, both being located at the center of the

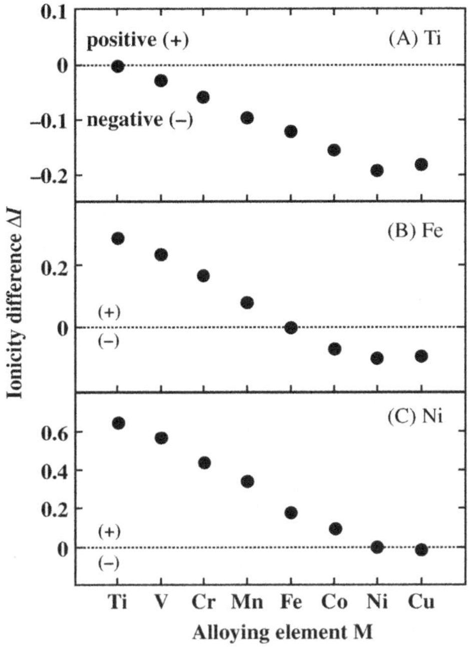

FIGURE 7-6 Ionicity difference between 3*d* alloying element and base-metal element in the alloy cluster of (A) hcp Ti, (B) bcc Fe, and (C) fcc Ni (Ni_3Al).

cluster. For example, in the case of bcc Fe, ΔI is expressed as, $\Delta I = I_M - I_{Fe}$, where I_{Fe} is the ionicity of the central Fe atom in a $FeFe_{14}$ cluster (i.e., pure Fe cluster) and I_M is the ionicity of the central M atom in a MFe_{14} cluster (see Fig. 4-1) [20]. It is found from Fig. 7-6 that the ionicity difference, ΔI, changes systematically with the atomic number of M and also with the electronegativity difference between M and the base metal.

In the case of Ti alloys, the ionicity differences are negative for all of the 3*d* alloying elements as shown in Fig. 7-6A. This result indicates that the electron transfer occurs from Ti to the alloying element, M. The amount of transferred electrons increases as the elements are far away from Ti in the periodic table. For Fe alloys, as shown in Fig. 7-6B, the left-side elements of Fe in the periodic table, Ti, V, Cr, and Mn, have a positive ionicity difference, whereas the right-side elements, Co, Ni, and Cu, have a negative ionicity difference. Therefore, if the former left-side elements are added into Fe, electron transfer occurs in the direction from those alloying elements to the base metal, Fe. The direction of electron transfer is the reverse for the latter right-side elements. For Ni alloys, as shown in Fig. 7-6C, only Cu, which is located just on the right-side of Ni, has a negative ionicity difference, and all other elements have positive ionicity difference, as might be expected from their electronegativity. Thus, these results indicate that the alloying element, M, has a negative charge only if M is located on the right side of the base metal in the periodic table.

7.3.2 Electrocatalytic Activity and Charge Transfer of Alloys

In an acid solution, the overall reaction of hydrogen evolution is expressed as,

$$2H_3O^+ + 2e^- \rightarrow H_2 + 2H_2O \tag{7.1}$$

The initial step of the reaction is probably the discharging process of the hydrated proton existing in the electrolyte. In this process, an electron must transfer to the proton from the metal electrode surface. Therefore, this reaction will be expected to be more active in the excess electron region than the deficient electron region on the electrode surface. In other words, the electronegative sites are more favorable than the electropositive sites for this reaction.

Here, Ti lies at the extreme left position in the periodic table among the 3*d* elements of interest in this study. Therefore, all the alloying elements in Ti accept electrons from the surrounding Ti atoms, as shown in Fig. 7-6A. Thus, there are excess electrons near the alloying element M in Ti. Such electronegative sites near the alloying element on the surface of Ti alloy will act as a source of the electron supply to the proton on the surface of Ti alloy, as illustrated schematically in Fig. 7-7A. Consequently, hydrogen evolution reaction will occur

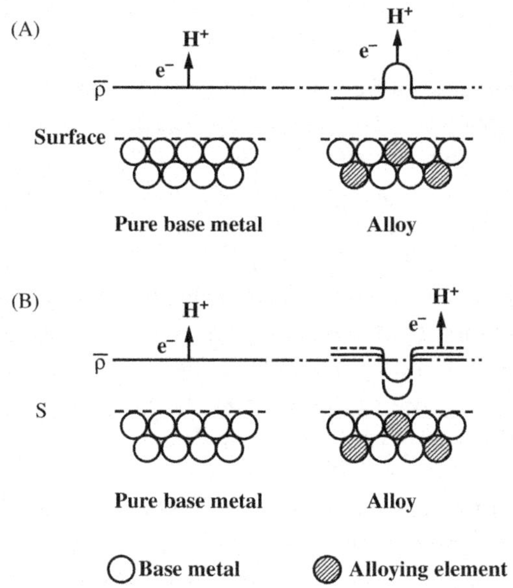

FIGURE 7-7 Schematic illustration of the charge densities ($\overline{\rho}$) on the surface of pure metal and alloy. Charge distributions are modified as a result of charge transfer between atoms. Case (A) and case (B) correspond to the alloy containing alloying elements that are positioned on the right side and the left side of the base metal in the periodic table, respectively. Depending on the type of alloying elements, there are either excess electrons (A) or deficient electrons (B) near the alloying element.

preferentially near the alloying element. As a result, the hydrogen overpotential of each alloying element reflects the measured hydrogen overpotential of the Ti alloy in a straightforward way.

These considerations are applicable to the other alloys containing the right-side elements, Co, Ni, and Cu for Fe alloys and Cu for Ni alloys. Namely, since excess electrons are located preferably near these alloying elements M in the alloys as shown in Fig. 7-6B–C, such elements provide active sites for the hydrogen evolution reaction. As a result, the hydrogen overpotential of the alloy changes following that of the alloying element as shown in Fig. 7-4.

By contrast, the hydrogen overpotential of alloys containing the left-side elements is less sensitive to the alloying elements. Such examples are seen in Fe alloys containing Ti, V, Cr, and Mn and in Ni alloys containing Ti, V, Cr, Mn, Fe, and Co. This result can be also explained by considering the direction of electron transfer between atoms. As already explained in Fig. 7-6B–C, all of these elements have positive charges and tend to transfer electrons to the base metal atoms. Therefore, the hydrogen evolution reaction will occur more easily near the negatively charged base metal atoms than near the positively charged alloying elements. This will account for the scarce change in the hydrogen overpotential for Fe- and Ni-based alloys containing the left-side elements in the periodic table.

However, it is important to note here that there are sometimes more electrons near the base metal atoms in the alloy than in the pure metal because of the electron transfer from the surrounding alloying elements. In such a case, the hydrogen evolution reaction near the base metal atoms will be more active in the alloy than in the pure base metal, as shown schematically in Fig. 7-7B. This effect will be more pronounced by the addition of those elements that can provide more electrons to the base metal. In other words, the addition of the elements with large positive charges (for example, Ti in Ni) is most effective in lowering the hydrogen overpotential of the alloy. This is illustrated in Fig. 7-7B using a dotted line. Needless to say, the alloy composition also affects the total amount of the transferred charges on the base metal, and hence the hydrogen overpotential shows a compositional dependence, as is shown in Fig. 7-5. With increasing Ti concentration, the excess charges increase near the base metal (Ni in the present case), and hence the overpotential decreases for the Ti-rich Ni alloys.

Machida et al. [15–17] investigated the electrocatalytic activity for the hydrogen evolution of several Ni-based alloys, Ni–Ti, Ni–Zr, Ni–V, and Ni–La. From the similarity in the hydrogen evolution kinetics between these alloys and pure Ni, they concluded that Ni is responsible for the electrocatalytic activity of these alloys. Every alloying element treated in their study is positioned on the left side of Ni in the periodic table, and is more electropositive than Ni. Therefore, their results support the present approach to the hydrogen overpotential of alloys. Thus, the direction to which charge transfer takes place is very important in understanding the hydrogen overpotential change with alloying.

Such a charge transfer effect will appear on the hydrogen overpotential for the alloys containing the right-side elements as well as the left-side elements. One example is seen in Fig. 7-5. For Ti–Ni alloys, Ni is the right-side element of Ti in the periodic table.

For this alloy system the hydrogen overpotential changes remarkably with the addition of a small amount of Ni into Ti, which is attributable to the increase in the number of active sites with increasing fraction of Ni atoms on the electrode surface. Interestingly, as explained earlier, a Ti-8 mol% Ni alloy has the lower hydrogen overpotential than pure Ni at the current densities, 3–10 mA/cm^2. This result is interpreted as due to the acceleration effect of the excess electrons on the hydrogen evolution reaction, because relatively large numbers of electrons possibly transfer from the Ti atom to the Ni atom, owing to the large electronegativity difference between them in this alloy. Such a charge transfer will enhance the hydrogen evolution reaction near Ni atom sites in the alloy more than in pure Ni.

7.3.3 Electrocatalytic Synergetic Effect

The lowering of the hydrogen overpotential of alloy below the respective values of the constituent elements has also been observed in various Co [12] and Ni alloys [13–15]. This is known as the electrocatalytic synergetic effect. A qualitative approach was proposed to the appearance of such an effect in transition metal–based alloys [11] using the Engel–Brewer valence bond theory [23]. According to this approach, the synergism in the electrocatalysis for the hydrogen evolution arises in case the metals in the left half of the transition metal series are alloyed with the metals in the right half. The former have empty or half-filled vacant *d*-orbitals, whereas the latter metals have internally paired *d*-electrons, which are not available for bonding in pure metals. In the Engel–Brewer theory, strong bonding is considered to be achieved by the occupancy of vacant *d*-orbitals of the left-half metals with the electrons donating from nonbonding orbitals of the right-half metals, resulting in the formation of stable intermetallic compounds. Thus, attention was directed toward the similarity between the alloy stability and the electrocatalytic activity for hydrogen evolution. Then, it was suggested that mutual pairing and sharing of each *d*-orbital electron provide electronic configurations suitable for proton adherence and transference [11]. However, the theory is too conceptual without any concrete explanation about the hydrogen evolution reaction on metals and alloys. Furthermore, according to the present calculated ionicity differences between atoms in various alloys, the electron transfer from the right-half to the left-half metals as suggested by the Engel–Brewer theory never occurs as far as the crystal structure of base metal is kept unchanged with alloying. The formation of any intermetallic compound with new crystal structure is not related to the synergetic effect of the alloy.

Therefore, it is not reasonable to understand the electrocatalytic activity for the hydrogen evolution of alloys following the Engel–Brewer theory. In contrast to this, the present approach based on the electronic structure calculations provides a more comprehensive interpretation for the hydrogen evolution reaction of transition metal–based alloys. The synergetic effect can be explained qualitatively by considering the charge transfer between constituent atoms in the alloy.

7.3.4 Fermi Energy Level and Hydrogen Overpotential for Ti Alloys

In addition to the ionicity difference, the Fermi energy level, E_f, shows an interesting correlation with the hydrogen overpotential of Ti alloys. The E_f level is calculated with the hcp Ti cluster containing 3*d* elements (see Fig. 5-1B), and the results are shown in Fig. 7-8 together with the hydrogen overpotential for Ti alloys at 1 mA/cm^2. In this case, every alloying element is more electronegative than the base metal Ti, so the active site for the hydrogen evolution reaction is always near the alloying element. In this situation, the hydrogen overpotential correlates with the E_f. Namely, the alloy with the lower E_f has the higher overpotential, whereas the alloy with the higher E_f has the lower overpotential. The Fermi energy (E_f) is a sort of chemical potential for electrons. The work function decreases with increasing E_f. Therefore, as the E_f increases, electrons will transfer from the metal surface to a proton more easily, resulting in the lower overpotential for the higher E_f alloy. This result indicates that the Fermi energy level may also be a dominant factor in controlling the activity for the hydrogen evolution of the alloy.

A similar discussion has already been given as to the corrosion problem of Ti alloys, as shown in Fig. 5-12 in Chapter 5, Titanium Alloys. All the alloying elements M discussed in Chapter 5, Titanium Alloys, are Re, Os, Ir, Pt, Au, Tc, Ru, Rh, Pd, and Ag, and the charge transfer always takes place in the direction from Ti to M. On the other hand, the alloying elements M discussed in this chapter are 3*d* transition metals, and the direction of the charge transfer changes depending on M in the Fe alloys and Ni alloys. But only in the case of the Ti alloys, does the charge transfer always take place from Ti to M, and the hydrogen overpotential correlate with the Fermi energy level, as shown in Fig. 7-8.

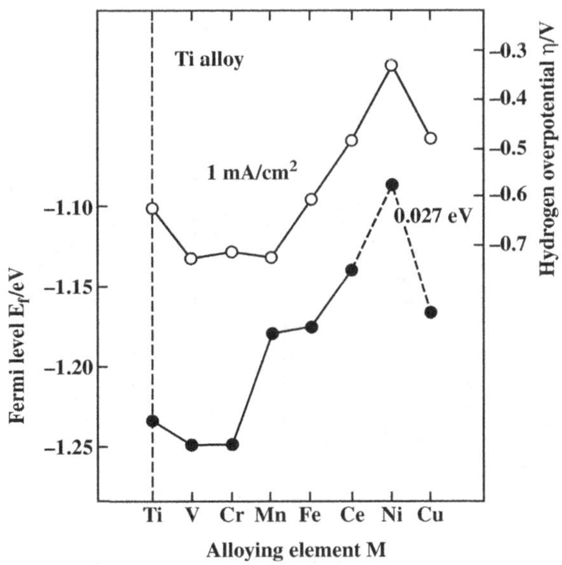

FIGURE 7-8 Comparison of the hydrogen overpotential with the Fermi energy level for Ti alloys.

7.4 Experiments of Hydrogen Overpotential for Intermetallic Compounds, TiAl, FeAl, and NiAl Containing 3*d* Transition Metals

To further confirm the present concept, similar experiments and calculations are carried out with the intermetallic compounds, TiAl, FeAl, and NiAl containing 3*d* transition metals [24].

7.4.1 Specimen Preparation and Hydrogen Overpotential Measurement

Intermetallic compounds of TiAl, FeAl, and NiAl containing 3*d* transition metals are prepared by high purity raw metals, and their compositions are set to be $Ti_{51}Al_{47}M_2$, $Fe_{51}Al_{47}M_2$, and $Ni_{51}Al_{47}M_2$, respectively [24]. Here, M's are 3*d* alloying elements, M = Ti, V, Cr, Mn, Fe, Co, Ni, and Cu. All these compounds have Al-deficient compositions, so that every alloying element, M, is assumed to occupy preferably the Al site in each compound. The hydrogen overpotential is evaluated by measuring cathodic polarization curves in a deaerated 1 N sulfuric acid solution at 303K, which is the same condition as used in the transition metal alloys.

7.4.2 Hydrogen Overpotential for Intermetallic Compounds

The measured hydrogen overpotential for pure compounds, $Ti_{53}Al_{47}$, $Fe_{53}Al_{47}$, and $Ni_{53}Al_{47}$, is shown in Fig. 7-9A. Here, the average data of three measurements are plotted for each

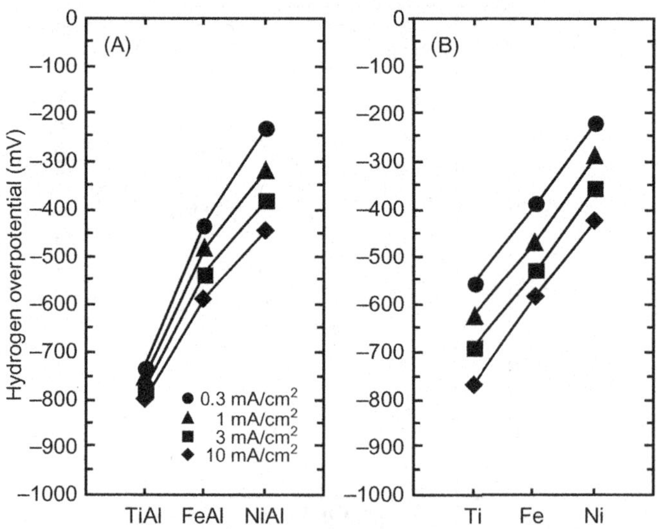

FIGURE 7-9 Comparison of the hydrogen overpotential between (A) TiAl, FeAl, and NiAl and (B) pure Ti, Fe, and Ni.

current density. The overpotential is the largest for TiAl among these compounds, and it changes in the order, TiAl > FeAl > NiAl. For comparison, the results on the hydrogen overpotential obtained for pure Ti, Fe, and Ni [18] are also shown in Fig. 7-9B. It is evident from these figures that the trend of the hydrogen overpotential for compounds is quite similar to that for pure transition metals existing in the base intermetallic compounds.

The measured hydrogen overpotential for alloyed compounds, TiAl-M, FeAl-M, and NiAl-M is shown in Fig. 7-10A–C, respectively. In Fig. 7-10A, the data of the hydrogen overpotential for pure 3*d* metals taken at 1 mA/cm^2, are also plotted using a dotted line (see Fig. 7-2). As shown in Fig. 7-10A, the hydrogen overpotential of alloyed TiAl varies considerably with alloying elements. This variation is very similar to that of pure 3*d* transition metals. For FeAl, as shown in Fig. 7-10B, the addition of Ti, V, Cr, and Mn, all of which are the left-side elements of the base transition metal (in this case Fe) in the periodic table, does not give any remarkable changes in the hydrogen overpotential of FeAl. In contrast to this, significant changes in the hydrogen overpotential are observed in the compounds containing the right-side elements, Co, Ni, and Cu. In the case of NiAl, as shown in Fig. 7-10C, there are scarce changes in the hydrogen overpotential for the compounds containing the left-side elements of the base transition metal (in this case, Ni) in the periodic table. On the other hand, the addition of Cu, which is a right-side element of Ni, increases the hydrogen overpotential of the compound.

The results shown in Fig. 7-10A–C for TiAl, FeAl, and NiAl compounds, respectively, are comparable with the results shown in Fig. 7-3A–C for Ti, Fe, and Ni alloys, respectively. There is a clear resemblance in the hydrogen overpotential change with alloying elements between compounds and alloys, even though there is a slight difference in the value of the hydrogen overpotential between them. Thus, it can be said that the hydrogen overpotential for these intermetallic compounds also changes following a simple rule that is explained in the transition metal alloys.

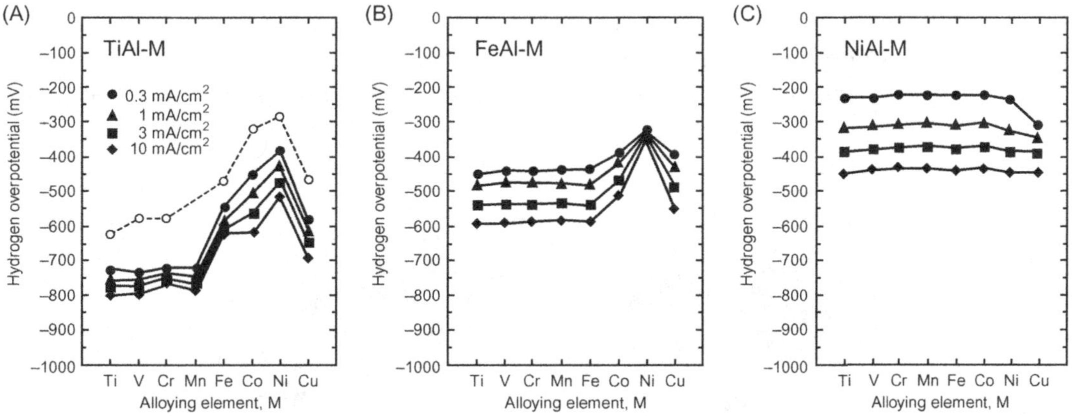

FIGURE 7-10 Alloying effect on the hydrogen overpotential for (A) TiAl, (B) FeAl, and (C) NiAl.

7.5 Theoretical Approach to Hydrogen Overpotential of Intermetallic Compounds

The electronic structures of pure and alloyed compounds are calculated by the DV-Xα cluster method. The cluster models used are shown in Fig. 7-11A–B, where $Ti_8Al_{10}M$, Fe_8Al_6M, and Ni_8Al_6M clusters are made from the crystal structure of each compound. TiAl has the $L1_0$ (AuCu) type structure, and FeAl and NiAl have the B2 (CsCl) type structure. In these clusters, the alloying element is assumed to occupy an Al site by preference because of the deficiency of Al content in the compound.

The ionicity of element M in each cluster is obtained from the Mulliken population analysis [22]. The calculated ionicity differences, ΔI, are shown of the 3*d* alloying elements in Fig. 7-12. Here, ΔI is defined as the ionicity difference between the alloying element M and the base transition metal, as explained earlier. If the value of ΔI is negative, the electron transfers from the base metal atom to the alloying element. If the value of ΔI is positive, the electron transfers in a reverse way. As shown in Fig. 7-12A, all the 3*d* alloying elements have negative ΔI in TiAl. In FeAl, as shown in Fig. 7-12B, the left-side elements of base transition metal Fe, in the periodic table, Ti, V, Cr, and Mn, have a positive ΔI, whereas the right-side elements, Co, Ni, and Cu, have a negative ΔI. In case of NiAl shown in Fig. 7-12C, only Cu, which is the right-side element of the base transition metal Ni, has a negative ΔI. These results indicate that the alloying element, M, has a negative charge only in the case that M is located in the right side of the base transition metal in the periodic table.

Such charge transfer behavior is related consistently to the hydrogen overpotential change with the alloying element M in the compounds. For example, in case of FeAl, Fe is more electronegative than Al. Therefore, for FeAl containing the left-side alloying elements, the most negatively charged Fe atom among the constituent elements is active for the hydrogen evolution reaction, resulting in the scarce change in the hydrogen overpotential, as shown in Fig. 7-10B. By contrast, Co, Ni, and Cu are more electronegative than Fe and Al, judging from the ionicity difference shown in Fig. 7-12B. So, these three elements act as the more preferable site for the hydrogen evolution reaction as compared with the base metals of Fe and Al. Consequently, the hydrogen overpotential for the compounds containing Co, Ni, and Cu exhibits a similar tendency to that for each alloying element, as shown in

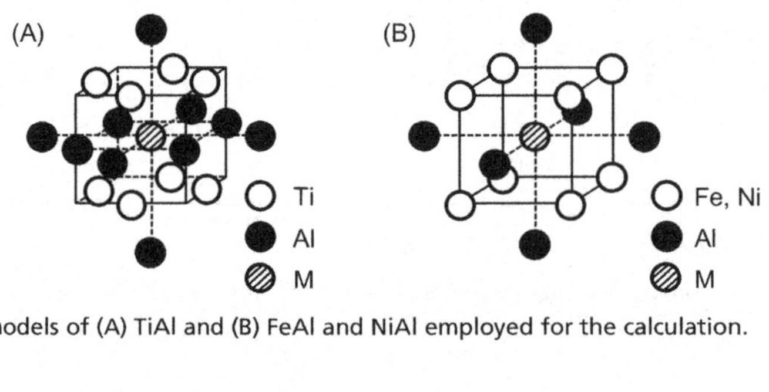

FIGURE 7-11 Cluster models of (A) TiAl and (B) FeAl and NiAl employed for the calculation.

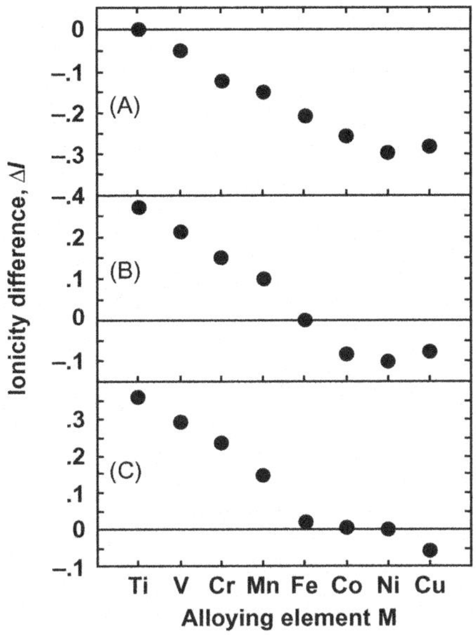

FIGURE 7-12 Ionicity difference between the 3*d* alloying element and the base metal element in (A) TiAl, (B) FeAl, and (C) NiAl.

Fig. 7-10B. For alloyed TiAl and NiAl compounds, the observed hydrogen overpotential can be explained in a similar manner. It is simply noted here that Fig. 7-10A–C for the intermetallic compounds quite resembles Fig. 7-3A–C for the transition metal alloys. Thus, it is confirmed that the concept proposed for understanding the hydrogen overpotential of transition metal alloys is valid for the intermetallic compounds.

7.6 Hydrogen Overpotential of Transition Metal–Based Ternary Alloys

Every compound treated here consists of three metals, namely, a base transition metal, Al, and an alloying element M. This implies that the proposed concept is valid not only for binary alloys, but also for ternary alloys. To confirm this point, Fe-19.1 mol% (18 mass%) Cr-1 mol% M ternary alloys are prepared, and the hydrogen overpotential is measured with them in the same way as explained previously. Here, M's are the 3*d* transition metals, M = Ti, V, Cr, Mn, Fe, Co, Ni, Cu. The base Fe-18 mass% Cr alloy has the composition of a typical ferritic stainless steel. All the constituent elements in the present ternary alloys are 3*d* transition metals. In these alloys, the base metal, Fe, is more electronegative than another base metal, Cr. Therefore, it is presumed that the addition of the more electropositive elements than Fe, in other words, the left-side elements of Fe in the periodic table, will not give any significant change in the hydrogen overpotential of the ternary alloys. Only in case that

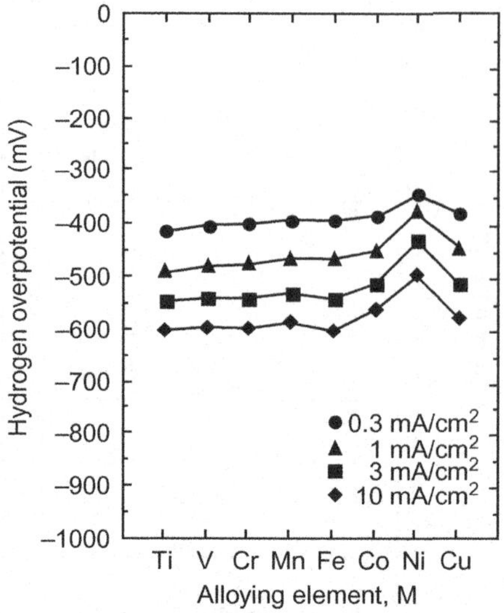

FIGURE 7-13 Alloying effect on the hydrogen overpotential for Fe-19.1 mol% (18 mass%) Cr-based ternary alloys.

the right-side elements, Co, Ni, and Cu, are alloyed, will the characteristic of these alloying elements reflect the hydrogen overpotential of ternary alloys. Fig. 7-13 is the experimental result of the hydrogen overpotential for Fe–Cr–M ternary alloys. A significant change in the hydrogen overpotential is observed only in the Co, Ni, and Cu containing alloys in agreement with the present presumption. Also, this trend is quite similar to the observation of the binary Fe alloys (see Fig. 7-3B) and of the FeAl compounds (see Fig. 7-10B).

7.7 Application of the Present Concept to Nanoelectrocatalysts

Using the present concept, Ezaki et al. [25] developed cathode materials with high catalytic activity for the hydrogen evolution reaction by using metal ultrafine particles (UFPs) of Ni and Mo. These UFPs were prepared by feeding raw metal powders into plasma flames in a reaction chamber. The particle diameter is about 100 nm for Ni UFPs and 30 nm for Mo UFPs. The Ni UFPs sintered at 573K are extremely porous and have an enormously large specific surface area, 5.6 m^2/g. The hydrogen overpotential of the Ni UFPs measured in a 1 N sulfuric acid solution at 303K is as low as that of Pt metal. Further decrease in the hydrogen overpotential is observed when 10 mol% Mo UFPs are mixed with the Ni UFPs, as shown in Fig. 7-14 [25]. As Mo is more electronegative than Ni, the electron transfers from Ni to Mo. As a result, following the mechanism shown in Fig. 7-7A, the electron rich Mo site becomes so active that the hydrogen overpotential further decreases in the Ni-10 mol% Mo UFPs mixture.

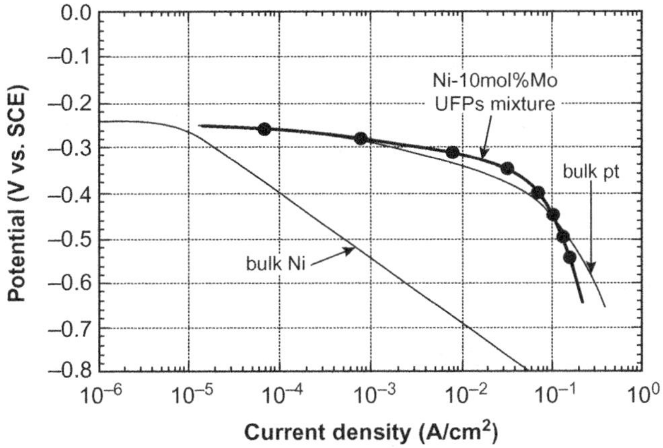

FIGURE 7-14 Cathodic polarization curves for sintered Ni-10 mol% Mo UFPs mixture, bulk Pt, and bulk Ni. *SCE*, saturated calomel electrode; *UFP*, ultrafine particles.

Recently many researchers have been devoted to the development of nanoelectrocatalysts (e.g., molybdenum compounds such as Mo_2C, Mo_2N, and MoS_2 supported on carbon nanotube–graphene hybrid) to produce hydrogen efficiently [26]. A simple concept presented here is applicable to such molybdenum compounds, too. For example, in the case of Mo_2C, C probably operates as an active site for hydrogen evolution reaction since C is more electronegative than Mo. The Mo composition of Mo_2C is so high that a large number of electrons transfer from Mo to C. Such excess electrons existing at the C site in Mo_2C may have a synergetic effect on lowering hydrogen overpotential, following the mechanism shown in Fig. 7-7A. Owing to this synergetic effect, Mo_2C is more active for hydrogen evolution reaction than the carbon nanotube and graphene. Instead the carbon nanotube–graphene hybrid support plays an important role in facilitating the electron transfer to Mo_2C [26].

Another example is a nano-Pt_3Co/C cathode catalyst in a polymer electrolyte fuel cell (PEFC) [27]. In Pt_3Co the electron transfer occurs from Co to Pt owing to the electronegativity difference between them, and the synergetic effect probably emerges following the mechanism indicated by a dotted electron density curve in Fig. 7-7B. Namely, the relatively high electron density around the Pt atom in Pt_3Co activates the cathode reaction, and the rate constants of the surface events on the Pt_3Co/C catalyst are significantly higher than those on the pure Pt/C catalyst [27].

Finally, it is summarized that the hydrogen overpotential for 3*d* transition metal–based alloys varies with alloying elements following a simple rule. In the case of the alloying element positioned on the right side of the base metal in the periodic table, the hydrogen overpotential of the alloy reveals the characteristic of the alloying element rather than the base metal. On the other hand, in the case of the left-side elements to be added into the base metal, the hydrogen overpotential scarcely changes with alloying or decreases slightly as compared with that for the base metal. Such a variation in the hydrogen overpotential can be explained consistently by considering the electronic structures of alloys. In particular, the direction of electron transfer between atoms is primarily important. It is also confirmed that the concept is valid for TiAl, FeAl, and NiAl intermetallic compounds, Fe-18 mass% Cr-based ternary alloys and binary Zr alloys [28] as well. This concept is essentially expressed in terms of the electronegativity difference between constitute elements in alloys and compounds. Therefore, it may not be a great contrast to any previous treatment using the electronegativity [6,7,10], although the present concept is constructed firmly on the basis of the electronic structure calculations. This concept helps us to design new electrocatalytic materials for the hydrogen evolution reaction.

References

[1] J.O.'M. Bockris, Electrolytic polarisation-I. The overpotential of hydrogen on some less common metals at high current densities. Influence of current density and time, Trans. Faraday Soc 43 (1947) 417–429.

[2] H. Kita, Periodic variation of exchange current density of hydrogen electrode reaction with atomic number and reaction mechanism, J. Electrochem. Soc. 113 (1966) 1095–1111.

[3] M.H. Miles, M.A. Thomason, Periodic variations of overvoltages for water electrolysis in acid solutions from cyclic voltammetric studies, J. Eectrochem. Soc. 123 (1976) 1459–1461.

[4] B.E. Conway, J.O.'M. Bockris, Electrolytic hydrogen evolution kinetics and its relation to the electronic and adsorptive properties of the metal, J. Chem. Phys. 26 (1957) 532–541.

[5] G. Belanger, A.K. Vijh, Effect of the state of the surface on the electrocatalysis of the anodic oxidation of hydrazine, Surf. Technol. 5 (1977) 81–84.

[6] S. Trasatti, Work function, electronegativity, and electrochemical behaviour of metals: II. Potentials of zero charge and "electrochemical" work functions, J. Electrochem. Soc. 33 (1971) 351–378.

[7] S. Trasatti, Work function, electronegativity, and electrochemical behaviour of metals: III. Electrolytic hydrogen evolution in acid solutions, J. Electroanal. Chem. 39 (1972) 163–184.

[8] A.T. Kuhn, C.J. Mortimer, G.C. Bond, J. Lindley, A critical analysis of correlations between the rate of the electrochemical hydrogen evolution reaction and physical properties of the elements, J. Electroanal. Chem. 34 (1972) 1–14.

[9] R. Parsons, The rate of electrolytic hydrogen evolution and the heat of adsorption of hydrogen, Trans. Faraday Soc. 54 (1958) 1053–1063.

[10] S. Trasatti, Electronegativity, work function, and heat of adsorption of hydrogen on metals, J. Chem. Soc. Faraday Trans. 1 68 (1972) 229–236.

[11] M.M. Jakšić, Electrocatalysis of hydrogen evolution in the light of the Brewer-Engel Theory for bonding in metals and intermetallic phases, Electrochim. Acta 29 (1984) 1539–1550.

[12] C.M. Lačnjevac, M.M. Jakšić, J. Res. Inst. Catalysis 31 (1983) 1551–1556.

[13] D.E. Brown, M.N. Mahmood, M.C.M. Man, A.K. Turner, Preparation and characterization of low overvoltage transition metal alloy electrocatalysts for hydrogen evolution in alkaline solutions, Electrochim. Acta 29 (1984) 1551–1556.

[14] W.W. Carlin, Cathode electrocatalyst, US Patent 4,010, 085 (1977).

[15] K. Machida, M. Enyo, K. Oguro, M. Nakane, Electrodic properties and surface characteristics of Ni-V alloys for water electrolysis cathode materials, Bull. Chem. Soc. Jpn. 57 (1984) 2809–2815.

[16] K. Machida, M. Enyo, I. Toyoshima, K. Miyahara, K. Kai, K. Suzuki, Amorphous Ni-Ti and Ni-Zr alloys for water electrolysis cathode materials, Bull. Chem. Soc. Jpn. 56 (1983) 3393–3399.

[17] K. Machida, M. Enyo, G. Adachi, J. Shiokawa, The hydrogen electrode reaction characteristics of thin film electrodes of Ni-based hydrogen storage alloys, Electrochim. Acta 29 (1984) 807–815.

[18] H. Ezaki, M. Morinaga, S. Watanabe, Hydrogen overpotential for transition metals and alloys, and its interpretation using an electronic model, Electrochim. Acta 38 (1993) 557–564.

[19] M. Morinaga, N. Yukawa, T. Maya, K. Sone, H. Adachi, Theoretical design of titanium alloys, Proc. Sixth World Conf. on Titanium, Cannes, France, June 6–9, Société Française de Métallurgie, Paris, 1988, pp. 1601–1606.

[20] M. Morinaga, N. Yukawa, H. Adachi, Alloying effect on the electronic structure of BCC Fe, J. Phys. F: Met. Phys. 15 (1985) 1071–1084.

[21] M. Morinaga, N. Yukawa, H. Adachi, Alloying effect on the electronic structure of Ni_3Al (γ'), J. Phys. Soc. Jpn. 53 (1984) 653–663.

[22] R.S. Mulliken, Electronic population analysis on LCAO–MO molecular wave functions. I, II, III, IV, J. Chem. Phys. 23 (1955) 1833–1840, 1841–1846, 2338–2342, 2343–2346.

[23] L. Brewer, Bonding and structures of transition metals, Science 161 (1968) 115–122.

[24] H. Ezaki, M. Morinaga, S. Watanabe, J. Saito, Hydrogen overpotential for intermetallic compounds, TiAl, FeAl and NiAl, containing 3d transition metals, Electrochim. Acta 39 (1994) 1769–1773.

[25] H. Ezaki, T. Nambu, M. Morinaga, M. Udaka, K. Kawasaki, Development of low hydrogen overpotential electrodes utilizing metal ultra-fine particles, Int. J. Hydrogen Energy 21 (1996) 877–881.

[26] D.H. Youn, S. Han, J.Y. Kim, J.Y. Kim, H. Park, S.H. Choi, et al., Highly active and stable hydrogen evolution electrocatalysts based on molybdenum compounds on carbon nanotube-graphene hybrid support, ACS Nano 8 (2014) 5164–5173.

[27] N. Ishiguro, T. Saida, T. Uruga, S. Nagamatsu, O. Sekizawa, K. Nitta, et al., ACS Catal. 2 (2012) 1319–1330.

[28] H. Shibutani, T. Higashijima, H. Ezaki, M. Morinaga, K. Kikuchi, An electronic approach to the hydrogen overpotential for Zr alloys, Electrochim. Acta 43 (1998) 3235–3239.

8 Crystal Structure Maps for Intermetallic Compounds

8.1 Parameters for Structure Maps

There are a variety of crystal structures in intermetallic compounds [1–6]. In 1940 a pioneering work for the structural separation of intermetallic compounds was carried out by Hume-Rothery using the average number of valence electrons per atom (e/a), and these compounds are called electron compounds [7]. Following the Hume-Rothery rule, the average value of e/a is 3/2 (=1.5) for bcc (A2) structure (e.g., AgZn, Cu_3Al), 21/13 (=1.615) for γ-brass ($D8_2$) structure (e.g., Ag_5Zn_8, Cu_5Zn_8), and 7/4 (=1.75) for hcp (A3) structure (e.g., $AgZn_3$) [8]. The Hume-Rothery rule is often discussed on the basis of the tight binding theory of alloys [8]. Since then, great efforts have been made to put in order crystal structures on two- or three-dimensional structure maps using various parameters such as the electronegativity [9,10], the pseudopotential radius [11–14], and the Mendeleev number [15,16]. However, despite such great efforts several difficulties still remain in these approaches as explained later. These days total energy calculations are commonly used for treating the structural stability of the compounds, but the structure map is still very convenient to us in view of the design of intermetallic compounds.

Here, two electronic parameters, *Bo* and *Md*, are used for representing the nature of the chemical bond between atoms in intermetallic compounds [6]. These parameters are calculated by the DV-Xα molecular orbital method [17–19] as explained in Chapter 2, Theory for Alloy Design. Using these electronic parameters, new crystal structure maps are made for aluminide, silicide, and several transition metal compounds [6]. Structural modifications of a few aluminides by alloying will be discussed with the aid of the structure maps. In addition, the ductility problem of TiAl compound will be treated with the aid of electron theory.

8.2 Calculation of Electronic Parameters

8.2.1 Cluster Models

The selection of the cluster model is most essential in the calculation. In general any crystal structure may be divided into a number of small polyhedrons. According to the mathematical polyhedron models [20], there are five simple polyhedrons called the five Platonic solids as shown in Fig. 8-1A–E. Plato is the famous Greek philosopher who left the imprint of his thought deeply fixed in *Euclid's Elements* [20]. They are tetrahedron, octahedron, hexahedron (cube), icosahedron, and dodecahedron. Each of these is a regular polygon of the same kinds

A Quantum Approach to Alloy Design. DOI: https://doi.org/10.1016/B978-0-12-814706-1.00008-X

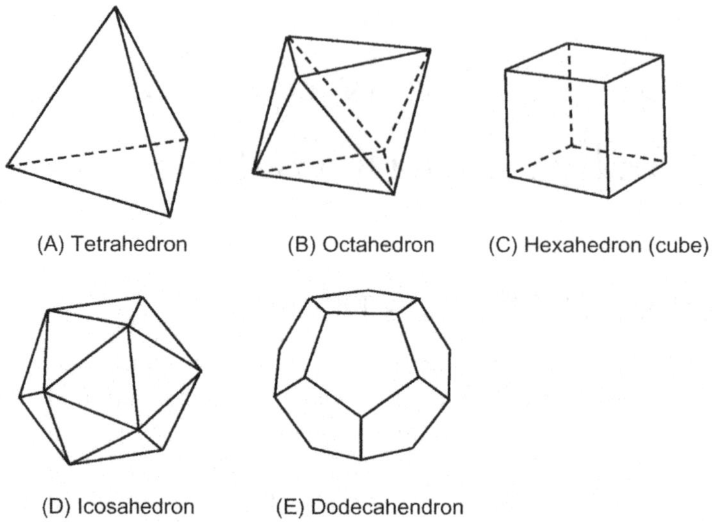

FIGURE 8-1 Five Platonic solids (Plato, "Euclid's Elements").

for all the faces. All the faces are equilateral triangles in the tetrahedron, octahedron, and icosahedron, but squares in the hexahedron (cube) and equilateral pentagons in the dodecahedron. Among them the icosahedron is in simplicity next to the tetrahedron and the octahedron, and this is actually observed in quasi-crystals. It is said that the dodecahedron is in some ways the most attractive of the five Platonic solids [20], but it has never been observed in any material. It is indeed a challenge to find materials made of the dodecahedron. As the dodecahedron has fivefold rotation axes as does the icosahedron, it never appears in crystalline solids. The C_{60} fullerene has the shape of a truncated icosahedron, and all the faces are either hexagons or pentagons [20]. To the author's limited knowledge, there may be a high potential of finding the dodecahedron in the C or C-related materials, if it could exist.

However, in case of crystalline solids the other three polyhedrons are important structural units. Among them an octahedral model is chosen here to express atomic interactions as simply as possible and also to increase the applicability of the results to a variety of crystal structures of the compounds. The actual cluster model used is shown in Fig. 8-2A. Such an octahedron is often seen in most compounds. For example, as shown in Fig. 8-2B–D, there is such an octahedral structural unit, M_2X_4, in B2 (CsCl-type), $L1_0$ (CuAu-type), and $L1_2$ (Cu_3Au-type) structures [6].

In case of aluminides, X is Al and M is the transition metal in the M_2X_4 cluster model shown in Fig. 8-2A, where M's are Sc, Ti, V, Cr, Mn, Fe, Co, Ni, Cu (3*d* transition metals), Y, Zr, Nb, Mo, Tc, Ru, Rh, Pd, Ag (4*d* transition metals), and La, Hf, Ta, W, Re, Os, Ir, Pt, Au (5*d* transition metals). The point group of this cluster is D_{4h}. The *a*-axial length is fixed at 0.2864 nm, twice as large as the atomic radius of Al, and the *c*-axial length varies depending on the atomic radius of M in the way that the M atom sphere contacts with the Al atom sphere as if they are rigid-body spheres. In case of the silicide, a similar cluster model, M_2Si_4, is used

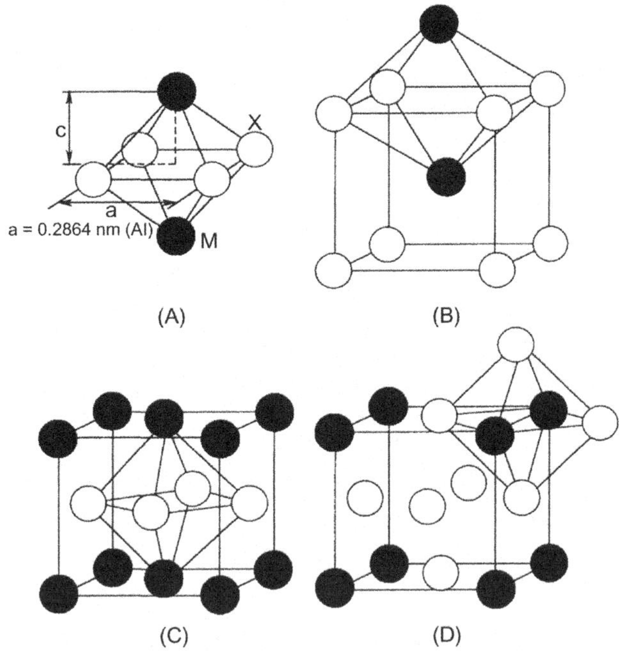

FIGURE 8-2 (A) Octahedral cluster model, M_2X_4, and typical crystal structures containing octahedral clusters; (B) B2-type; (C) $L1_0$-type; and (D) $L1_2$-type structures.

except for the a-axial length of 0.2638 nm, twice as large as the atomic radius of Si. In the case of transition metal compounds, the base metals (X) selected are Ti, Fe, Ni, Nb, and Mo, and similar calculations are carried out with the M_2X_4 cluster for various transition metals, M.

To show the validity of the results of the octahedral model, further calculation is performed with a large $MoSi_2$-based cluster model containing the octahedron, tetrahedron, and hexahedron in it, as is explained later.

8.2.2 *d*-Orbital Energy Level, *Md*

The calculated energy level structures are shown in Fig. 8-3 for the M_2Al_4 cluster where the M's are 3*d* transition metals. The 3*d* component level of M appears in the broad Al *s, p* band, which extends from −8 to 4 eV as shown in Fig. 8-3. The energy of the Fermi level, E_f, is indicated by an arrow (→) in each energy level structure.

A large fraction of the 3*d* component appears, for example, in the b_{2u} level, because of the D_{4h} symmetry of the cluster. In fact, the $2b_{2u}$ level consists of the 3*d* component of more than 90%, and some other minor components. As shown in Fig. 8-3, the height of the $2b_{2u}$ level changes monotonously with the order of elements, M, in the periodic table. Also, the $3b_{2u}$ level exists in the energy range of several electron volts higher than the $2b_{2u}$ level. So, the average *d*-orbital energy level, *Md,* is calculated by taking a weighted average of the

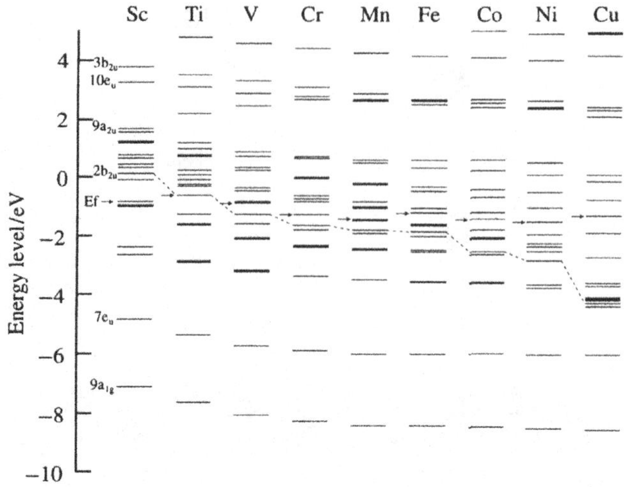

FIGURE 8-3 Energy level structures for the M_2Al_4 cluster containing 3*d* transition metals, M.

energies for the $2b_{2u}$ and $3b_{2u}$ levels using a weighting factor of the 3*d* component in each level. The *Md* level calculated in this way varies with elements, M, as shown in Fig. 8-4A. For comparison, the atomic radius and the electronegativity of elements are shown in Fig. 8-4B.

As explained in earlier chapters, the *Md* level changes periodically with the order of elements in the periodic table. Also, it tends to increase with increasing atomic radius of the element, M [21]. This trend of increasing *Md* level with the atomic radius is due simply to the weakening of the attractive Coulomb interaction between the *d* electrons and a nucleus, since the average *d*-orbital radius increases with the atomic radius of M. As a result, the *Md* level shifts to the high energy side as the atomic radius of M increases. The exception is seen in the elements of Cu, Rh, Pd, Ag, Ir, Pt, Au, all of which have the *d* orbitals nearly or completely filled with the electrons. In particular, in the case of Cu, Ag, and Au, the *d* orbitals are completely filled with the electrons, so that the *d*-orbital levels are further stabilized, resulting in the appearance of their *Md* levels in the very low energy range, as shown in Fig. 8-5A. Thus, the character of the *d* electrons in M reflects well this *Md* parameter, although this parameter is obtained from a very small cluster model shown in Fig. 8-2A. As the *Md* parameter is related to the atomic size, the use of this parameter may guarantee the structural classification even for size-controlling compounds such as the Laves phases.

In addition, as shown in Fig. 8-4B, the *Md* value decreases with increasing electronegativity of the element, M [22]. This correlation is also reasonable, since it is known that the energy level obtained by the Xα method represents the electronegativity value itself [17–19]. The electronegativity is concerned with the charge transfer between atoms, and hence this parameter is a sort of measure to show the ionic interaction between atoms. There is a general trend that a larger atom has a lower electronegativity value. Thus, *Md* is an atomic size parameter as well as a chemical parameter to represent the ionic interaction between M and X atoms through the charge transfer between them.

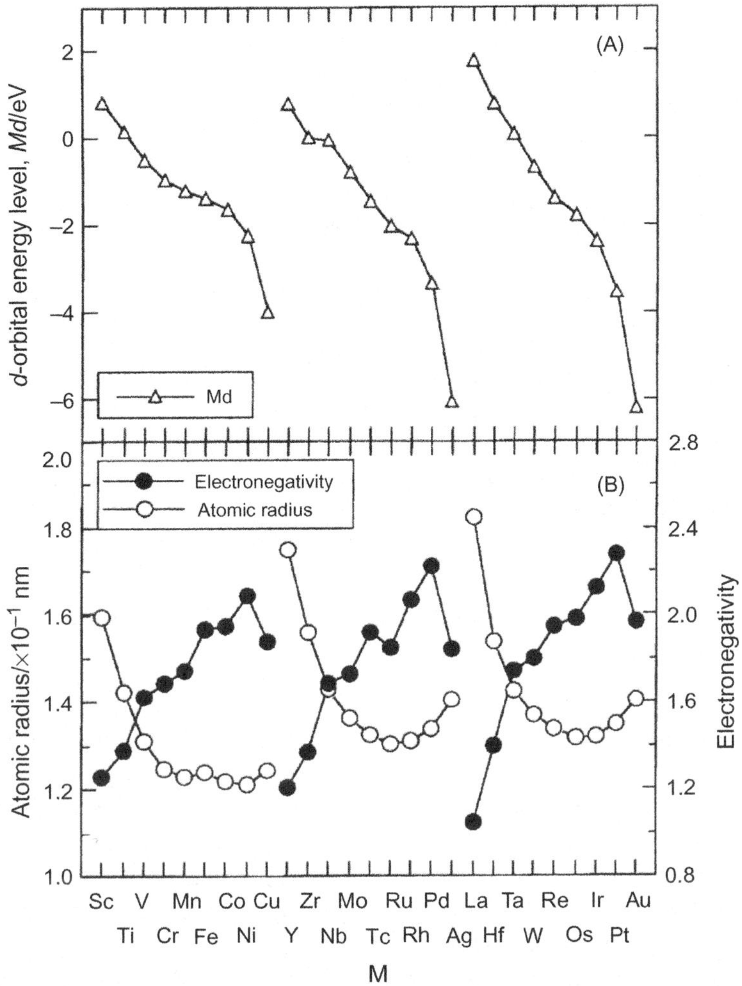

FIGURE 8-4 Changes in (A) the *d*-orbital energy level for the M_2Al_4 cluster, (B) the atomic radius and the electronegativity of elements, M.

The level other than the b_{2u} level may be used, but the other molecular level also changes in a similar way as does the b_{2u} level as shown in Fig. 8-5. This is also the case for the a_{1g}, a_{2u}, e_g, and e_u levels, although they are not shown in the figure.

8.2.3 Bond Order, *Bo*

As explained previously, following the Mulliken population analysis [23–26], the overlap population of electrons between atoms is obtained, and called the bond order. The bond order, *Bo*, is a parameter to show the overlapping of electron clouds between atoms, and

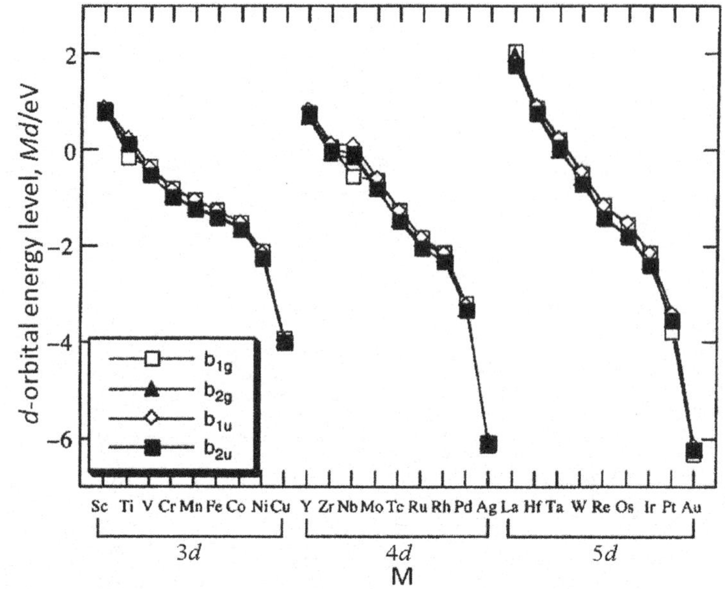

FIGURE 8-5 A comparison of the *d*-orbital energy level among the b_{1g}, b_{2g}, b_{1u}, and b_{2u} levels for the M_2Al_4 cluster.

hence this is a measure of the covalent bond strength between atoms. The bond orders are calculated between M *d* electrons and Al 3*p* electrons in the M_2Al_4 cluster and between M *d* electrons and Si 3*p* electrons in the M_2Si_4 cluster. This is because the *p*–*d* covalent bond is most important to represent the chemical bond between simple metals (Al, Si) and transition metals, M [5]. The calculated results are shown in Fig. 8-6A. Both the bond orders change periodically with the order of elements in the periodic table. There are peaks in the bond order curve near the 6A group elements, Cr, Mo, W in the respective series of 3*d*, 4*d*, 5*d* transition metals. For comparison, the melting temperatures are shown of MAl, M_3Al, and MAl_3 compounds in Fig. 8-6B. The melting temperature changes in a similar way as does the M–Al bond order, except for Cr, Mn, and Fe in the series of the 3*d* transition metals, where there is a steep drop in the melting temperatures of the compounds. Such a drop around Mn is, however, observed even in the melting temperatures of pure 3*d* transition metals.

It is stressed here that the concept of the covalent bonding, ionic bonding, and atomic size, all of which are important factors to determine crystal structures, and it is well condensed in the two calculated parameters, the bond order, *Bo*, and the *d*-orbital energy level, *Md*. Furthermore, both of these two electronic parameters change following the order of elements in the periodic table. This is very important since the crystal structures of compounds also change following the periodic table of elements. In fact, the same or similar crystal structure appears in the same group or its neighborhood in the periodic table. In this sense, these *Bo* and *Md* are indeed suitable parameters for constructing structure maps. For convenience, all the calculated values of the *Bo* and the *Md* parameters are listed in Table 8-1.

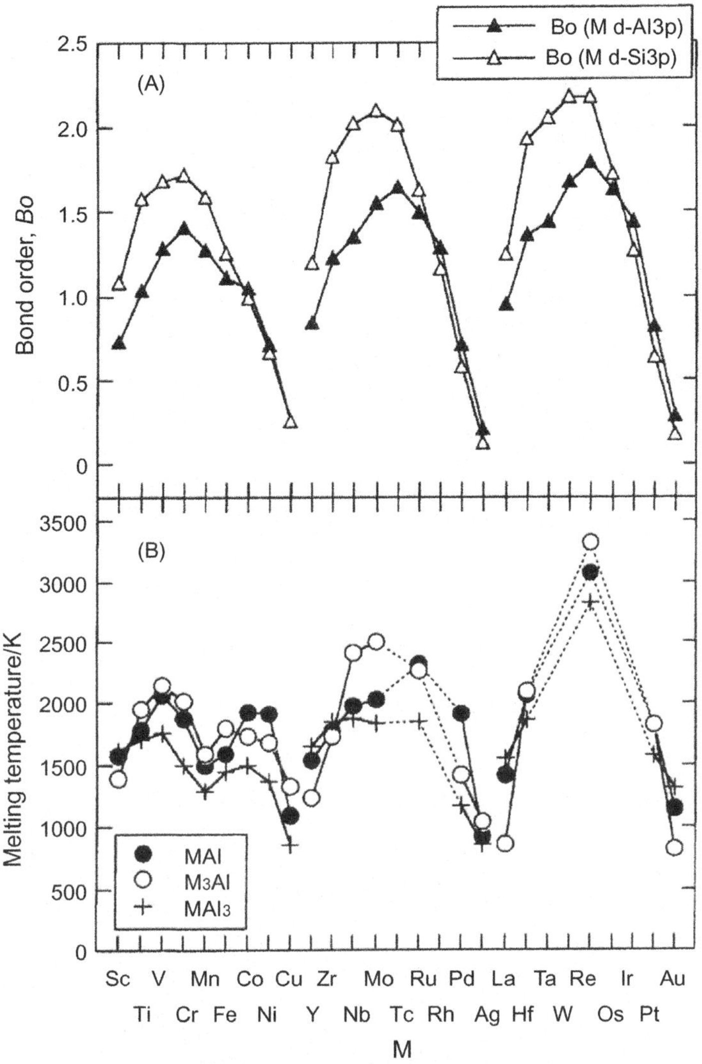

FIGURE 8-6 Changes in (A) the M–Al and the M–Si bond orders and (B) the melting temperatures of aluminides with M.

8.3 Crystal Structure Maps for Aluminide and Silicide

8.3.1 Aluminide

The structure maps are constructed using the bond order (*Bo*) and the *d*-orbital energy level (*Md*). One example is shown in Fig. 8-7 for the aluminide, MAl. This map is drawn using the *Bo* and the *Md* values listed in Table 8-1. The crystal structure data of intermetallic

Table 8-1 *Md* and *Bo* Values for Elements, M, in M_2X_4 and $M_1Mo_{12}Si_{18}$ Clusters

Element	M_2Al_4 Cluster		M_2Si_4 Cluster		M_2Ti_4 Cluster		M_2Fe_4 Cluster		M_2Ni_4 Cluster		M_2Nb_4 Cluster		M_2Mo_4 Cluster		$M_1Mo_{12}Si_{18}$ Cluster	
	Md	*Bo* (*d-3p*)	*Md*	*Bo* (*d-3p*)	*Md*	*Bo* (*d-3d*)	*Md*	*Bo* (*d-3d*)	*Md*	*Bo* (*d-3d*)	*Md*	*Bo* (*d-4d*)	*Md*	*Bo* (*d-4d*)	*Md*	*Bo* (*d-3d*)
Sc	0.802	0.731	−0.184	1.082	2.239	0.591	1.012	0.612	0.333	0.471	2.141	0.767	1.733	0.838		
Ti	0.144	1.034	−0.541	1.573	1.476	0.809	0.164	0.913	−0.825	0.640	1.138	1.101	0.758	1.138	4.524	0.642
V	−0.528	1.282	−0.581	1.684	0.989	1.086	−0.389	0.959	−0.991	0.682	0.776	1.409	0.344	1.392	3.901	0.673
Cr	−0.996	1.405	−1.131	1.719	0.385	1.189	−0.932	0.987	−1.457	0.629	0.080	1.442	−0.225	1.434	3.508	0.652
Mn	−1.225	1.274	−1.433	1.585	0.115	1.113	−1.405	0.825	−1.738	0.482	−0.342	1.340	−0.581	1.218	3.112	0.599
Fe	−1.399	1.115	−1.883	1.250	−0.293	0.969	−1.625	0.528	−2.013	0.278	−0.700	1.064	−1.099	1.020	2.853	0.481
Co	−1.646	1.044	−2.177	0.992	−0.679	0.698	−1.909	0.307	−2.278	0.112	−1.208	0.863	−1.433	0.745	2.669	0.367
Ni	−2.238	0.711	−2.883	0.665	−1.284	0.466	−2.481	0.162	−2.607	−0.041	−1.831	0.589	−1.858	0.414	2.491	0.243
Cu	−4.023	0.253	−4.379	0.256	−2.399	0.138	−3.590	−0.050	−3.820	−0.101	−2.576	0.151	−3.115	0.111		
Y	0.784	0.842	0.042	1.196	2.632	0.660	1.703	0.643	0.337	0.607	2.565	0.879	2.279	0.947		
Zr	0.000	1.222	−0.137	1.826	1.817	0.944	0.849	0.872	−0.457	0.787	1.735	1.256	1.240	1.327	5.654	0.873
Nb	−0.081	1.346	−0.250	2.027	1.325	1.240	0.005	1.100	−0.790	0.848	1.111	1.599	0.685	1.648	5.079	0.939
Mo	−0.819	1.550	−1.020	2.096	0.470	1.351	−0.692	1.067	−1.280	0.711	0.332	1.720	0.065	1.568	4.184	0.928
Tc	−1.474	1.645	−1.519	2.012	−0.062	1.339	−1.463	1.046	−1.759	0.627	−0.463	1.652	−0.713	1.511		
Ru	−2.031	1.494	−1.767	1.625	−0.612	1.079	−1.977	0.706	−2.287	0.384	−1.023	1.353	−1.486	1.304	3.577	0.654
Rh	−2.336	1.280	−2.751	1.161	−1.473	0.834	−2.653	0.455	−2.855	0.135	−1.878	1.020	−2.133	0.859		
Pd	−3.338	0.714	−4.030	0.575	−2.411	0.423	−3.231	0.024	−3.440	−0.082	−2.713	0.484	−2.870	0.358		
Ag	−6.084	0.213	−6.221	0.123	−4.505	0.054	−5.646	−0.107	−5.994	−0.101	−4.670	0.056	−5.078	0.025		
La	1.771	0.952	1.796	1.244	2.756	0.682	1.847	0.828	0.486	0.578	2.769	0.956	2.508	0.999		
Hf	0.762	1.354	−0.056	1.930	2.034	1.051	0.993	1.057	−0.273	0.838	1.953	1.326	1.467	1.408	5.530	0.985
Ta	0.055	1.432	−0.432	2.054	1.589	1.291	−0.027	1.144	−0.714	0.910	1.396	1.677	0.930	1.743	5.078	1.046
W	−0.699	1.672	−0.8009	2.177	0.890	1.412	−0.484	1.135	−0.949	0.755	0.684	1.824	0.219	1.826	4.611	0.995
Re	−1.403	1.784	−1.414	2.170	0.116	1.430	−1.304	1.114	−1.583	0.604	−0.260	1.771	−0.512	1.623	3.931	0.817
Os	−1.783	1.626	−2.198	1.717	−0.563	1.158	−1.839	0.864	−2.090	0.418	−0.885	1.474	−1.350	1.438		
Ir	−2.373	1.432	−2.815	1.262	−1.455	0.919	−2.468	0.407	−2.694	0.148	−1.842	1.143	−2.086	0.971		
Pt	−3.562	0.820	−4.237	0.640	−2.719	0.582	−3.414	0.006	−3.541	−0.103	−2.993	0.706	−2.993	0.407		
Au	−6.210	0.282	−6.316	0.168	−4.557	0.080	−5.733	−0.101	−5.950	−0.138	−4.726	0.083	−5.128	0.041		
Al					1.105	0.358	−0.095	0.360	−0.740	0.291	0.937	0.479	0.566	0.488		
Si					−0.278	0.334	−1.325	0.414	−1.762	0.327	−0.408	0.473	−0.679	0.660		

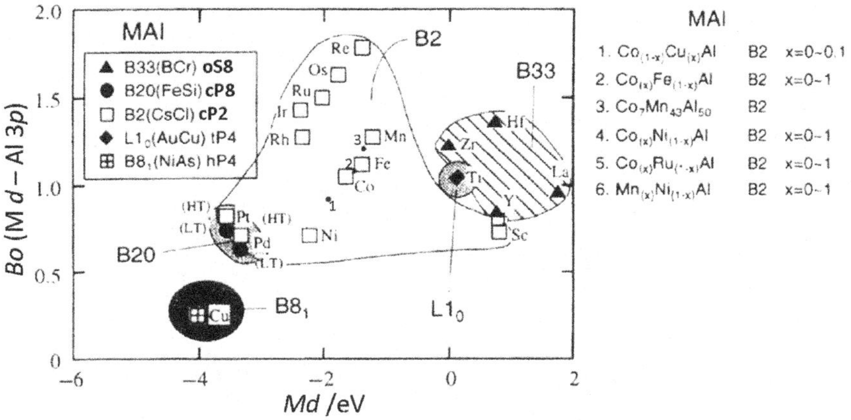

FIGURE 8-7 Structure map for MAl compounds. Crystal structures are denoted using the Structurbericht symbol or the Pearson symbol.

compounds are taken from the Pearson's Handbook [1]. Each crystal structure is distinguished from the others by using different symbols on the map. In addition to binary compounds, ternary compounds, $(M_{1-x}Y_x)$ Al, nos. 1–6, are also located on the map, simply by taking the compositional average of the *Bo* and the *Md* parameters, namely, $Bo = (1-x)(Bo)_M + x\,(Bo)_Y$ and $Md = (1-x)(Md)_M + x\,(Md)_Y$.

The crystal structures appear to be well separated on this map. For example, TiAl with the $L1_0$-type structure is located between the B33-type and the B2-type compounds on the MAl map. PtAl and PdAl take the B20-type structure at low temperatures and the B2-type structure at high temperatures. In response to this, the B20-type region lies at the edge of the B2-type region. All the ternary compounds of nos. 1–6 are located in the B2-type region. Among them the positions of nos. 4–6 are not indicated, because for example, in case of no. 4 ($Co_xNi_{1-x}Al$, $x = 0-1$), both CoAl and NiAl take the B2-type structure as shown in the MAl map. This is also the case in no. 5 ($Co_xRu_{1-x}Al$, $x = 0-1$) and no. 6 ($Mn_xNi_{1-x}Al$, $x = 0-1$).

The structure map for M_3Al is shown in Fig. 8-8. The A15-type region is located in a high-*Bo* region on the map. The $D0_{19}$-type region lies between the A15-type and the $L1_2$-type regions. Both the $L1_2$-type structure (space group, $Pm\bar{3}m$) and the $D0_{19}$-type structure ($P6_3/mmc$) appear in La_3Al, so La_3Al is located in the both regions. Co_3Al has the $L1_2$-type structure. Ternary compounds of nos. 1–19 have the $D0_3$-type structure. The $D0_3$-type region and the $L1_2$-type region partly overlap one another around Co_3Al on the map. The $D0_{19}$-type structure may also appear in alloyed Co_3Al according to a recent publication [27]. The region around Co_3Al is interesting in view of structural modification, since several structures fall on there.

The structure map for MAl_3 is shown in Fig. 8-9. There is the $D0_{22}$-type region near the $D0_{23}$-type region on the map, because of the structural resemblance between them. In fact, $HfAl_3$ takes both the $D0_{22}$-type and the $D0_{23}$-type structures, depending on the temperature. There is a clear separation in the crystal structures on this map.

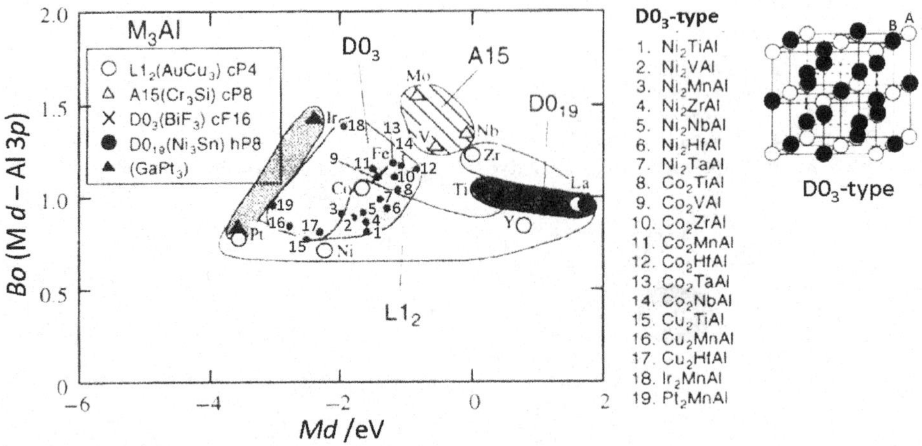

FIGURE 8-8 Structure map for M_3Al compounds.

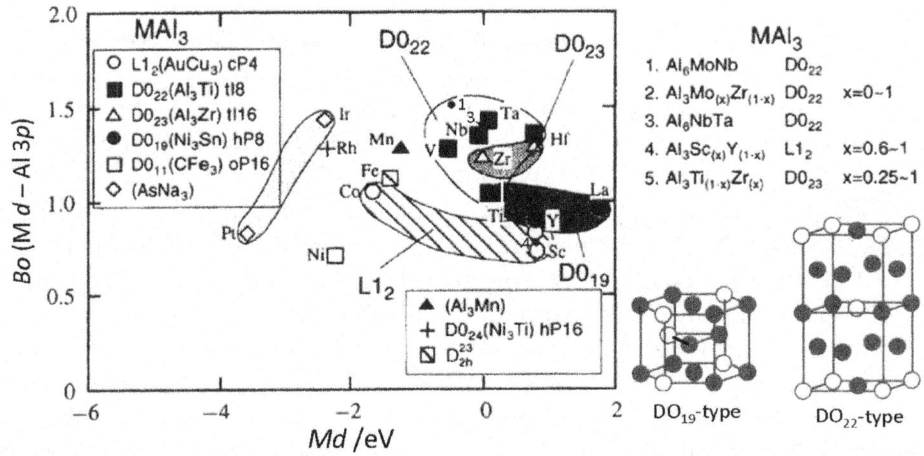

FIGURE 8-9 Structure map for MAl_3 compounds.

8.3.2 Silicide

The structure map is shown in Fig. 8-10 for the silicide, MSi. There is a good separation of the crystal structures on the map. For the ternary compounds, for example, no. 2 ($Cr_{1-x}Ni_xSi$) has the B20-type structure in the compositional range of $x = 0-0.25$. CrSi has the B20-type structure, but NiSi has the B31-type structure, so that the compositional range of $Cr_{1-x}Ni_xSi$ is restricted to $x = 0-0.25$. Also, no. 5 ($Cu_3Hf_7Si_{10}$) has the B33-type structure despite that HfSi has the B27-type structure.

The structure map is shown in Fig. 8-11 for the silicide, M_3Si. The A15-type structure region is adjacent to the PTi_3-type structure region, since both the structures appear, for

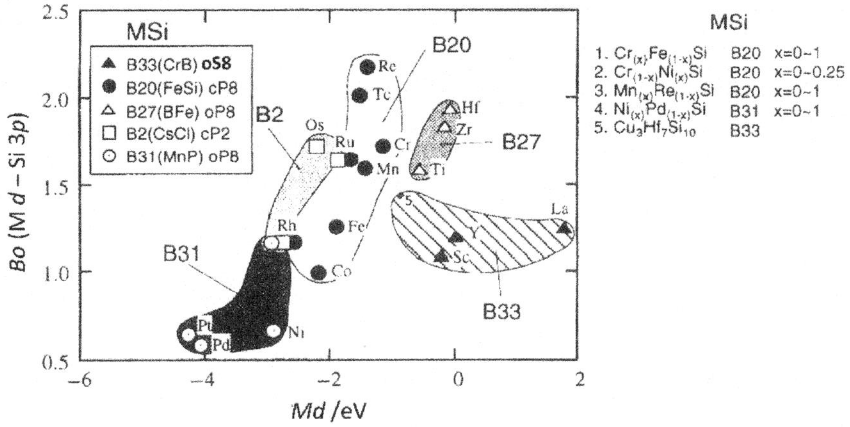

FIGURE 8-10 Structure map for MSi compounds.

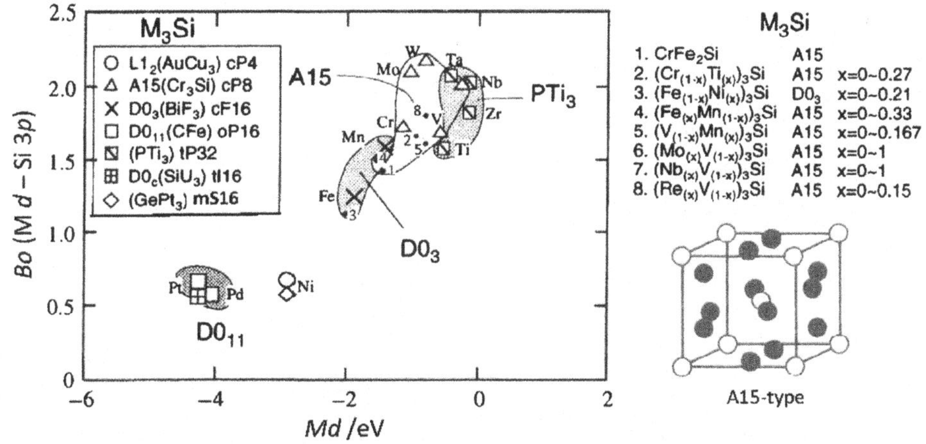

FIGURE 8-11 Structure map for M_3Si compounds.

example, in Nb_3Si. The ternary compounds of nos. 1–8 have the A15-type structure except for no. 3 $(Fe_{1-x}Ni_x)_3Si$ ($x = 0$–0.21) with the $D0_3$-type structure. Here, Ni_3Si with the $GePt_3$-type or the $L1_2$-type structure is located in the lower *Bo* region than Fe_3Si with the $D0_3$-type structure. The Fe-rich no. 3 compound is located near Fe_3Si, while keeping the $D0_3$-type structure.

The structure map is shown in Fig. 8-12 for the silicide, MSi_2. For example, the $C11_b$-type structure (e.g., $MoSi_2$ [28]) is located in a high-*Bo* region, and adjacent to the C40-type structure region on the map. This is reasonable since even for pure $MoSi_2$ the C40-type structure is found when it is rapidly quenched from the melt, even though this may be a metastable phase [29]. It is also known that in case a (1 1 0) stacking fault is introduced into the $C11_b$-type structure, the crystal structure becomes identical with the C40-type structure.

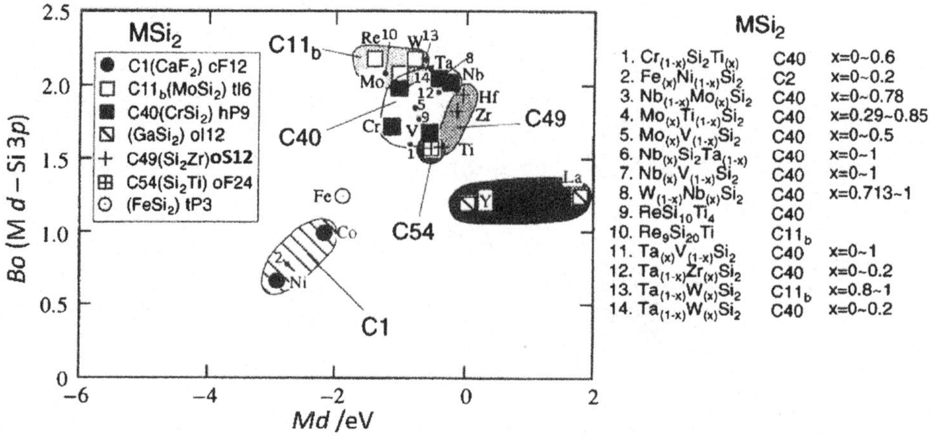

FIGURE 8-12 Structure map for MSi_2 compounds.

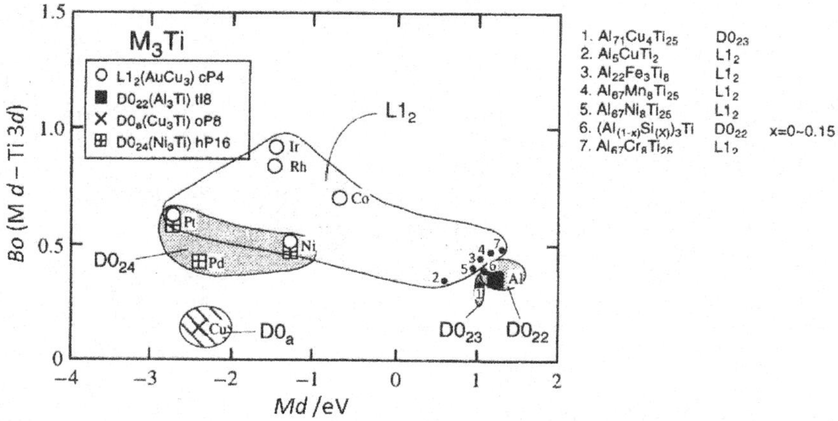

FIGURE 8-13 Structure map for M_3Ti compounds.

8.4 Crystal Structure Maps for Transition Metal Compounds

A series of structure maps is shown in Fig. 8-13 for M_3Ti, Fig. 8-14 for MNi, Fig. 8-15 for MNi_3, Fig. 8-16 for MFe, Fig. 8-17 for MNb_3, and Fig. 8-18 in MMo. It is apparent that any crystal structures are separated clearly on these structure maps. A brief explanation will be given below focusing on the ternary compounds to get an idea for structural modification by alloying.

For M_3Ti shown in Fig. 8-13, ternary compounds of nos. 1–7 are composed of Al, Ti, and M, and all of them have Al-rich compositions. Al_3Ti has the $D0_{22}$-type structure, whereas the

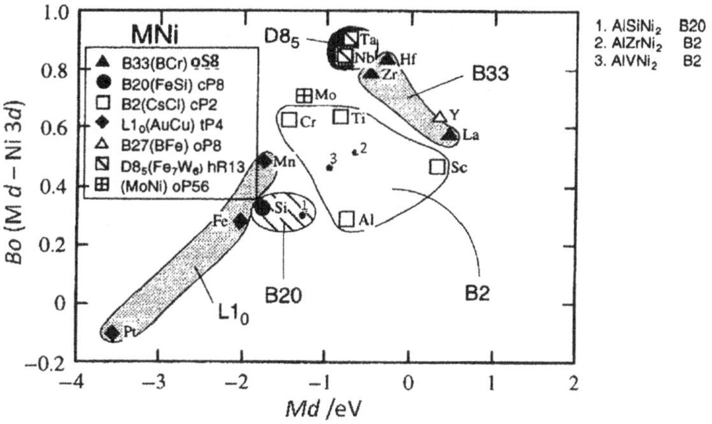

FIGURE 8-14 Structure map for MNi compounds.

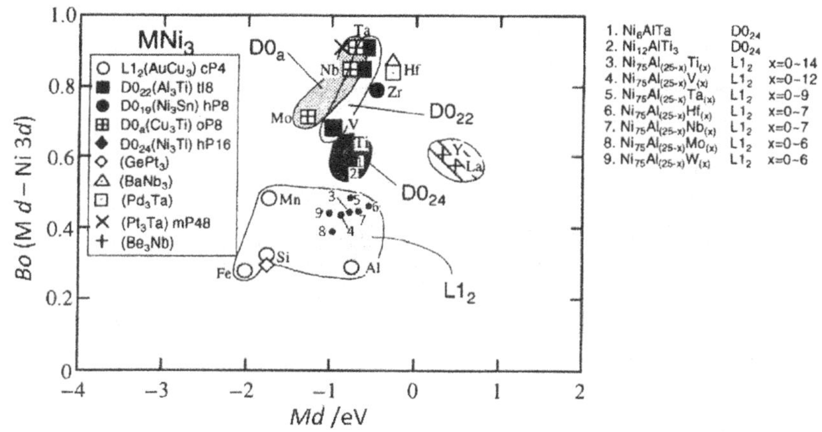

FIGURE 8-15 Structure map for MNi_3 compounds.

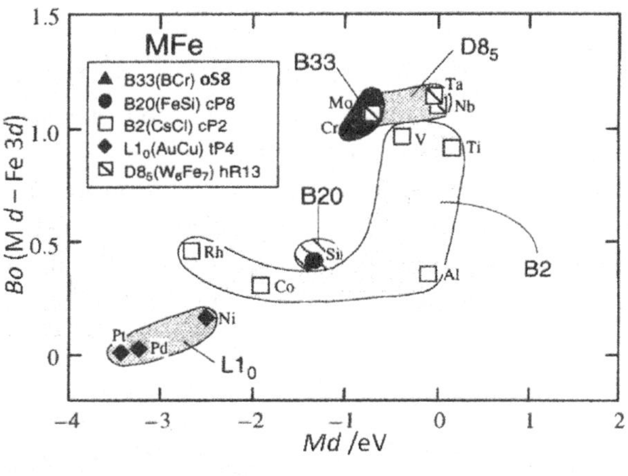

FIGURE 8-16 Structure map for MFe compounds.

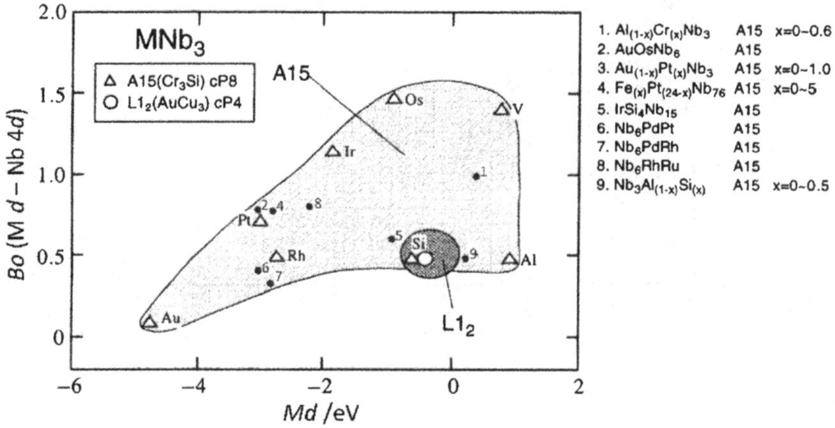

FIGURE 8-17 Structure map for MNb_3 compounds.

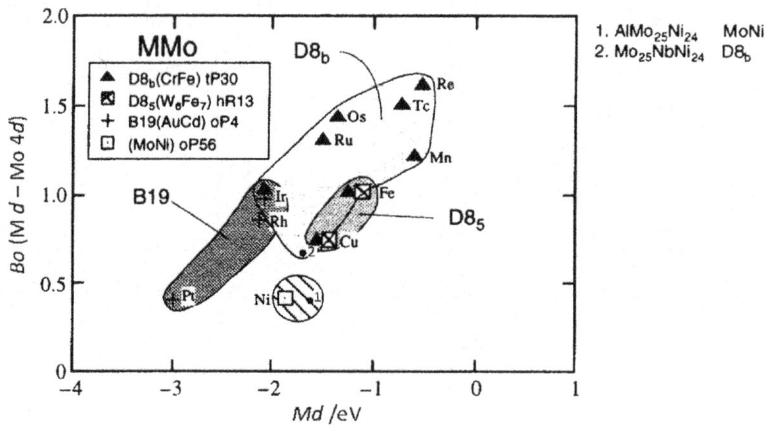

FIGURE 8-18 Structure map for MMo compounds.

compounds of nos. 1–7 have either the $D0_{22}$-type (no. 6) or $D0_{23}$-type (no. 1) or $L1_2$-type (the others). So, these structure regions are adjacent to each other on the map. This information is useful for the structural modification of Al_3Ti.

Also, as shown in Fig. 8-14 for MNi, ternary compounds of nos. 1–3 are composed of Al, Ni, and M. The crystal structure of AlNi is the B2-type, and the crystal structures of nos. 1–3 are either the B2-type (no. 2 and no. 3) or B20-type (no. 1). So, these structure regions are adjacent to each other on the map. In addition, as shown in Fig. 8-15 for MNi_3, every ternary compound contains Al and has either the $L1_2$-type structure or the $D0_{24}$-type structure. So, the $L1_2$-type region exists near the $D0_{24}$-type region.

Similarly as shown in Fig. 8-17 for MNb_3, an A15-type region is extended over the map. All the ternary compounds of nos. 1–9 have the A15-type structure. Finally, as shown in

Fig. 8-18 for MMo, the ternary compound, no. 1 has the Ni-rich composition, so it has the same crystal structure as NiMo. But no. 2 has the $D8_b$-type structure. It is apparent that the NiMo-type region is close to the $D8_b$-type region on the map. Thus, the crystal structures for ternary transition metal compounds are traceable using the present structure maps.

8.5 Cluster Model Dependence on Alloying Parameters

Two electronic parameters, *Md* and *Bo*, are determined using an octahedral cluster model shown in Fig. 8-2A. However, the results may be dependent on the cluster model. As explained before, for crystalline solids, tetrahedron, octahedron, and hexahedron (cube) are important structural units. As shown in Fig. 8-19A, there are all three polyhedrons in the C11b-type structure of $MoSi_2$ [1], even though each polyhedron is slightly distorted from the regular one because of the tetragonal symmetry of the unit cell. Therefore, to show the validity of the structure maps derived from the octahedral cluster model, further calculation is carried out with a $MoSi_2$ cluster model containing three types of polyhedrons in it.

8.5.1 *Bo* and *Md* Parameters for $MoSi_2$ Cluster

The cluster model used for the calculation is shown in Fig. 8-19B [30]. In this $M_1Mo_{12}Si_{18}$ cluster, four Mo atoms and eight Si atoms are added besides the atoms existing in the unit cell shown in Fig. 8-19A. This is because the second nearest neighbor interactions are probably important in this structure. The transition metal, M, is substituted for the Mo atom at a center of the cluster, where M's are Ti, V, Cr, Mn, Fe, Co, and Ni (3*d* transition metals),

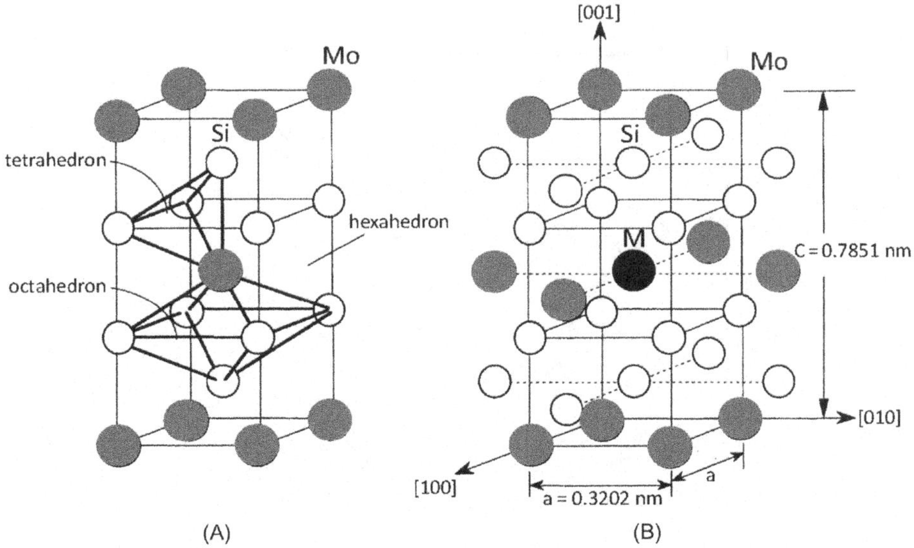

FIGURE 8-19 (A) The crystal structure of $MoSi_2$ and (B) the cluster model, $M_1Mo_{12}Si_{18}$, used in the calculation.

Zr, Nb, Mo, and Ru (4*d* transition metals), and Hf, Ta, W, and Re (5*d* transition metals). The lattice parameters used are $a = 0.3202$ nm and $c = 0.7851$ nm, the same values as in the bulk [1].

The values of *Bo* and *Md* are calculated and the detailed derivation of these parameters are given elsewhere [30]. The calculated results are shown in Fig. 8-20A–B, using the symbols of solid triangles for *Md* in (A) and of solid circles for *Bo*. For comparison, the results of the *Md* parameter obtained from the calculation of the M_2Si_4 cluster are shown by using the open triangles in (A). *Md* parameter changes monotonously with the order of elements in

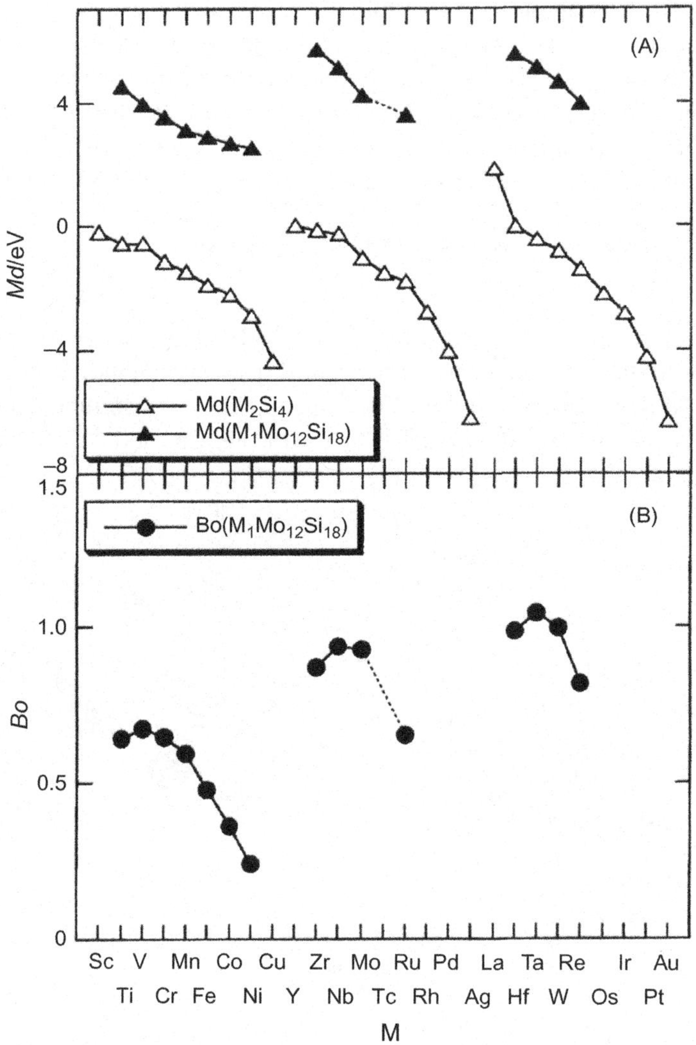

FIGURE 8-20 Changes in (A) the *d*-orbital energy level and (B) the bond order with M for $M_1Mo_{12}Si_{18}$ cluster.

the periodic table. Also, as shown in Fig. 8-20B, the calculated bond orders between Si 3*p* and M *d* electrons change periodically with the order of elements in the periodic table. This trend resembles the results of the octahedral cluster shown in Fig. 8-6A.

8.5.2 Structure Map of Silicide

A structure map is constructed using the *Bo* and the *Md* parameters for a large cluster model of $M_1Mo_{12}Si_{18}$. The results for di-silicide, MSi_2, are shown in Fig. 8-21, and compared with the results shown in Fig. 8-12 for the octahedral cluster model.

As shown in Fig. 8-21, $MoSi_2$ compound with the C11b-type structure is located in a high-*Bo* region on the map. Therefore, this compound has a stronger chemical bond than other compounds, which yields the higher melting temperature, 2303K [21]. Also, there is the C40-type region near the C11b-type region. Therefore, the structure can be modified to the C40-type by adding those elements (i.e., Cr, V, Nb, Ta) into $MoSi_2$, which forms the C40-type compounds. This is confirmed experimentally [3,31]. The C49-type region seems to extend to the C54-type region of $TiSi_2$, since the C49-type structure is observed in some cases in $TiSi_2$ [32]. Also, both the C1-type region and the C49-type region are located far away from the $C11_b$-type region. Therefore, continuous structural modification from the $C11_b$-type to the C49 type or the C1 type may be difficult by alloying [3]. Not only the structural separation, but also solid solubility of elements in $MoSi_2$ can be understood with the aid of this *Bo*–*Md* map [28]. Therefore, the structure map shown in Fig. 8-21 is probably highly reliable.

The structure map shown in Fig. 8-12 resembles the map shown in Fig. 8-21 as a whole. Both Y and La are not calculated using a large cluster model of $M_1Mo_{12}Si_{18}$. So, the $GaSi_2$-type region relevant to YSi_2 and $LaSi_2$ is not included in Fig. 8-21. Thus, the general trend of the chemical bond between M and Si atoms is reproduced well in Fig. 8-12, despite the use

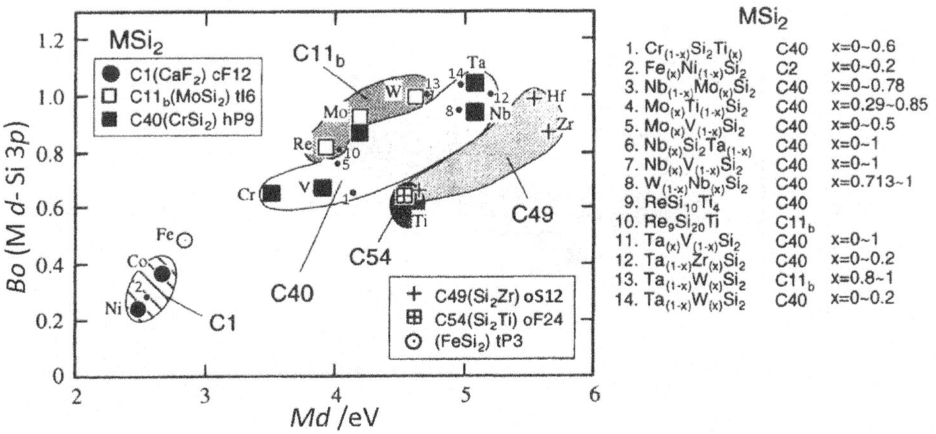

FIGURE 8-21 Structure map for MSi_2 compounds obtained from the calculation of $M_1Mo_{12}Si_{18}$ cluster.

of the small octahedral cluster model in the calculation. In other words, the *Bo* and *Md* parameters obtained from the small octahedral cluster are usable for constructing structure maps of the compounds.

8.6 Comparison With Other Structure Maps

8.6.1 Miedema's Electronegativity and Electron Density at the Wigner–Seitz Atomic Cell Boundary

Miedema proposed two parameters to estimate the heat of formation of intermetallic compounds [10]. One is the electronegativity, ϕ^*, and the other is the electron density at the boundary of the Wigner–Seitz atomic cell, n_{ws}. The structure map for M_3Al is made using these parameters and the result is shown in Fig. 8-22. The regions of the $D0_{19}$-type, the $D0_3$-type, and the $L1_2$-type structures are mixed together on the map. The separation of crystal structures is poorer on this map, as compared with the map shown in Fig. 8-8.

8.6.2 Zunger's Pseudopotential Radius

Zunger calculated the pseudopotential radii of *s* electrons, r_s, and also of *p* electrons, r_p [11]. Using these r_s and r_p, dual parameters are obtained for the AB compound as,

$$R_\sigma^{AB} = \left|\left(r_s^A + r_p^A\right) - \left(r_s^B + r_p^B\right)\right| \tag{8.1}$$

$$R_\pi^{AB} = \left|\left(r_s^A - r_p^A\right) + \left(r_s^B - r_p^B\right)\right| \tag{8.2}$$

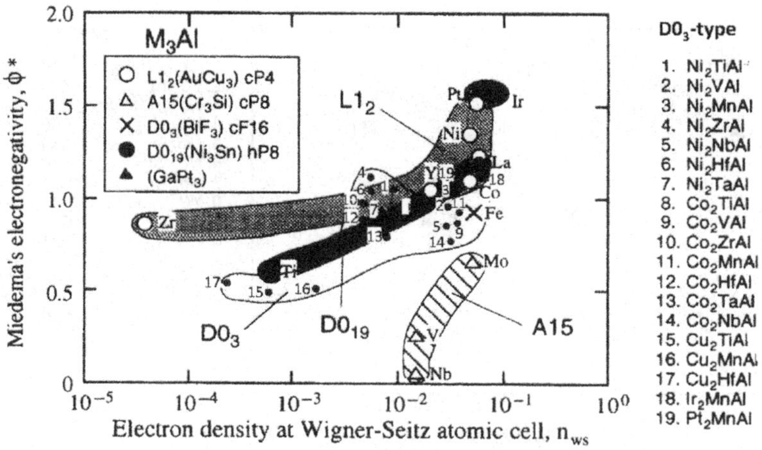

FIGURE 8-22 Structure map for M_3Al compounds made of the Miedema parameters.

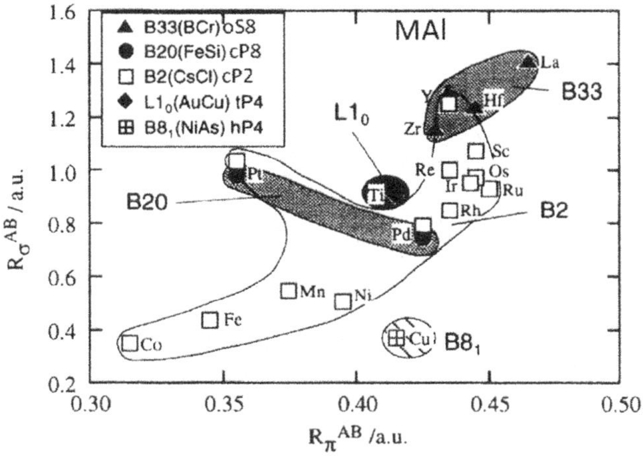

FIGURE 8-23 Structure maps for MAl compounds made of the Zunger parameters (1 a.u. = 0.0529 nm).

Here, R_{σ}^{AB} is a measure to show the difference between the total effective core radii of A and B atoms (i.e., size mismatch). R_{π}^{AB} is a measure to show the sum of the orbital nonlocality of the *s* and *p* electrons on each site [11]. However, any covalent bonding character of the *d* electrons in transition metal compounds is not taken into account for these parameters. By using these parameters the structure map for MAl is made as shown in Fig. 8-23. The B2-type region is very broad and overlapped with the B20-type and the B33-type regions. Thus, the separation of the crystal structure is poorer on this map than the *Bo–Md* structure map shown in Fig. 8-7. As explained by Zunger [11], these parameters are useful for structural separation in some compounds (e.g., binary octet compounds, $A^{N}B^{(8-N)}$), but not in various compounds listed in Table III in his paper [11]. It may be a cause of the trouble that the values of the pseudopotential radii are assigned to pure elements and hence any alloying effect is not involved in them.

8.6.3 The Mendeleev Number

Pettifor proposed the Mendeleev number for the structural classification of intermetallic compounds [15,16]. The magnitude of the Mendeleev number for each element changes as if it forms a one-dimensional chain lying on the periodic table. For binary A*m*B*n* compounds, a series of structure maps is constructed using two Mendeleev numbers (M_A, M_B) [15,16]. For example, the AB_3 structure map is made as shown in Fig. 8-24. Following the previously used notations, AB_3 compounds are expressed as M_3Al for aluminide (i.e., A = Al) and M_3Si for silicide (i.e., A = Si), where M's (i.e., B = M) are the transition metals.

However, several ill predictions are seen on the map shown in Fig. 8-24. For example, either the $L1_2$-type or the $D0_{19}$-type region is separated into two different regions on the map. In addition, the compositional average of the Mendeleev number is taken for the

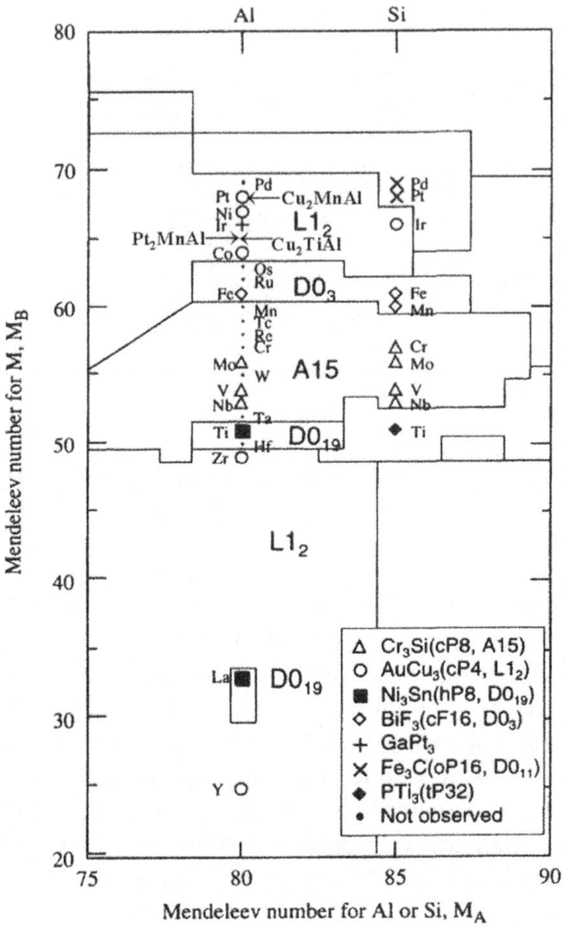

FIGURE 8-24 AB_3 structure map made of the Mendeleev number for aluminides and silicides containing transition metals.

structural classification of pseudobinary $(A_xX_{1-x})_m(B_yY_{1-y})_n$ systems, namely, $M_A = x(M_A) + (1-x)(M_X)$ and $M_B = y(M_B) + (1-y)(M_Y)$. However, there is an inconsistency in some ternary systems. For example, ternary compounds, Cu_2TiAl (no. 15), Cu_2MnAl (no. 16), and Pt_2MnAl (no. 19) are presented on the M_3Al map shown in Fig. 8-8. All these compounds have the $D0_3$-type structure, but they are located in the $L1_2$-type region on the map shown in Fig. 8-24.

The main reason for these discrepancies is probably the lack of two-dimensional periodicity in the single Mendeleev parameter, despite the fact that any compounds having the same crystal structure tend to appear in the same group or near the group in the two-dimensional periodic table. The use of two parameters instead of the single parameter is required to represent the characteristics of the two-dimensional periodic table of elements.

On the other hand, there are no such discrepancies on the *Bo*–*Md* structure map shown in Fig. 8-8. For example, Cu_2TiAl (no. 15), Cu_2MnAl (no. 16), and Pt_2MnAl (no. 19) are positioned in the $D0_3$-type region, being consistent with the experiment. Also, neither the $L1_2$-type region nor the $D0_{19}$-type region ever splits into different regions on this map, as La_3Al has the $L1_2$-type structure, too. Any compounds with the same crystal structure gather together on the map. This is because, as presented in Figs.8-4 and 8-6, two electronic parameters of *Md* and *Bo* reflect well the nature of the chemical bond between atoms in the compounds, while showing a periodical change in the magnitude, following the order of elements in the two-dimensional periodic table.

8.7 Structural Modification

Here, a possibility of structural modification by alloying will be discussed with the aid of structure maps. Presented are two examples, Nb_3Al and Al_3Ti. Any intermetallic compounds have more than two sublattices in the crystal lattice. In case a third element is added into the binary compound, it is first necessary to take into account the substitutional site of the element. For this reason, special attention is directed toward the substitutional problem of the third element in the following discussion.

8.7.1 Nb_3Al Compound

The Nb_3Al compound with the A15-type crystal structure is one of the promising materials for high-temperature applications, because of its high melting temperature, 2333K (2060°C). However, the low ductility at room temperature is a large barrier to the practical use. A main problem is to find out how to change the A15-type structure to the more simple $L1_2$-type structure by alloying.

In the M_3Al structure map shown in Fig. 8-8, the A15-type region is located in a high-*Bo* region. This indicates that the A15-type compound has stronger chemical bonding than the $L1_2$-, $D0_{19}$-, and $D0_3$-type compounds. Therefore, to modify the structure to the $L1_2$-type one, it may be preferable to substitute the lower *Bo* element for the Nb atom. In fact, Zr_3Al with the $L1_2$-type structure is located just below the A15-type region. Also, the Nb–Zr binary phase diagram [33] shows an all-proportional solid solution above 1250K without forming any intermetallic compounds between them. This implies that Zr atoms are substitutable for Nb atoms. The Zr substitution for Nb may lead to the formation of the $L1_2$-type phase, although it is not certain whether the microstructure consists of the $L1_2$-type phase alone or of the $L1_2$-type and the A15-type phases coexisting in the compound.

On the other hand, in the MNb_3 structure map shown in Fig. 8-17, Nb_3Al is located in the highest *Md* region. Including Nb_3Al, there are many A15-type compounds, but the $L1_2$-type compound is limited only to Nb_3Si. However, a ternary compound, $Nb_3Al_{1-x}Si_x$ ($x = 0$–0.5, no. 9), keeps the A15-type structure. Therefore, it is probably difficult to modify the structure of Nb_3Al by substituting third elements for Al atoms.

Thus, we need to first modify the A15-type structure of Nb_3Al to the $L1_2$-type structure by substituting Zr atoms for Nb atoms, and then to substitute some transition metals (e.g., Ti) for Al atoms. This is because the Al *p*–Nb *d* interaction probably causes poor ductility of the compound as explained later. The substitution of the transition metal atoms for Al atoms in the compound is effective in lowering the *p*–*d* interaction and in increasing the *d*–*d* interaction.

8.7.2 Al_3Ti Compound

The Al_3Ti compound is characteristic of the low density and excellent oxidation resistance. However, this compound is very brittle at room temperature, partially due to the tetragonal $D0_{22}$-type structure as shown in Fig. 8-9. To solve this problem, a trial is made to modify its crystal structure to the cubic $L1_2$-type as illustrated in Fig. 8-2D [2].

For the structure map shown in Fig. 8-9, the Al_3Ti compound is positioned in a lower *Bo* part of the $D0_{22}$-type region. Both the $D0_{19}$-type and the $L1_2$-type regions extend just below the position of Al_3Ti. Therefore, the substitution of the lower *Bo* elements for Ti atoms may be able to modify the crystal structure, although there has been no previous report that the $D0_{22}$-type structure is transformed into either the $L1_2$-type or the $D0_{19}$-type structure by alloying.

On the other hand, the M_3Ti structure map is shown in Fig. 8-13. The Al_3Ti compound is located in the highest *Md* region, but the $L1_2$-type region extends widely over a lower *Md* region than the $D0_{22}$-type region. Therefore, it is considered that the substitution of the lower *Md* elements for Al atoms is beneficial to the structural modification. In fact, it is known that the $D0_{22}$-type structure is modified to the $L1_2$-type one by the addition of Cu, Fe, Mn, Ni, Cr into Al_3Ti, as shown in the ternary compounds of nos. 2, 3, 4, 5, 7. It is important to note that all these ternary compounds are located near the mother compound, Al_3Ti. In other words, the crystal structure may be modified only in case the locations of the mother and alloyed compounds are close to each other on the structure map. This is because constituent elements in the compounds neighboring on the structure map have similar *Bo* and *Md* values. Therefore, the heat of formation of the compounds is probably not so different between the mother and alloyed compounds. The crystal structure can be modified by alloying as shown here. However, it is a different matter altogether whether the structurally modified compound is ductile or brittle.

8.8 Ductility of Intermetallic Compound, TiAl

There has been renewed interest in the titanium aluminides, in particular in a lightweight TiAl, stemming from increasing demands for high temperature structural materials [34,35]. However, it still remains a question why TiAl is brittle at room temperature. Despite the complex and manifold origins of mechanical properties, certain mechanical behavior may be related with some specific features of the electronic structures [36,37]. Presented here is a

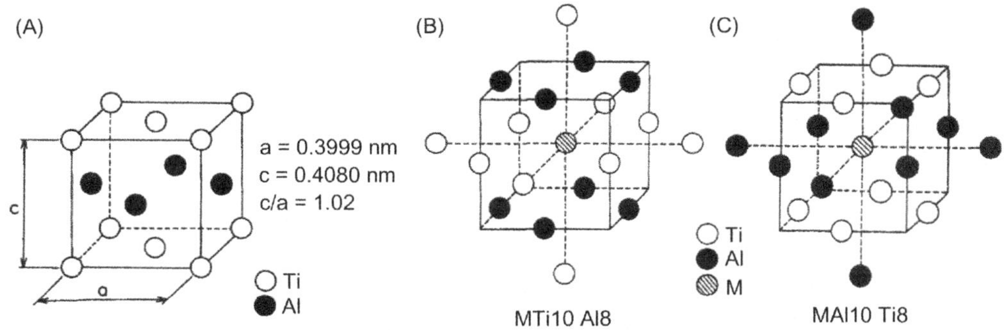

FIGURE 8-25 Crystal structure of (A) TiAl, and cluster models of (B) $MTi_{10}Al_8$, and (C) $MAl_{10}Ti_8$, where M is an alloying element.

piece of study concerning the alloying effect on the ductility of TiAl. This study is based on the electronic structure calculations by the DV-Xα method [5].

8.8.1 Cluster Models

TiAl has the $L1_0$-type structure as shown in Fig. 8-25A. Following this crystal structure, two types of cluster models are used for the calculation. One is the $MTi_{10}Al_8$ cluster model, where an alloying element, M, is substituted for Ti atom, and the other is the $MAl_{10}Ti_8$ cluster model, where an alloying element, M, is substituted for Al atom. The alloying elements are 3*d* elements (Ti, V, Cr, Mn, Fe, Co, Ni, Cu), 4*d* elements (Zr, Nb, Mo), and nontransition elements (Mg, Al, Si).

8.8.2 Bond Order

The bond order is calculated to estimate the change of the chemical bond strength with alloying or substitution of M for Al or Ti [5]. As far as the alloying element M is a transition metal, the chemical bonds between M and Al and between M and Ti are attributable mainly to the M *d*–Al *p* and the M *d*–Ti *d* covalent bonding, respectively. Therefore, the corresponding bond orders are calculated, and the results are summarized in Fig. 8-26A for the $MTi_{10}Al_8$ cluster, and (B) for the $MAl_{10}Ti_8$ cluster. For comparison, the bond orders of several nontransition metals are calculated in the same way as those of transition metals. In the figure, Total means the sum of the M *d*–Al *p* and the M *d*–Ti *d* bond orders. Also, Ti in $MTi_{10}Al_8$ (i.e., M = Ti), shown in Fig. 8-26A, and Al in $MAl_{10}Ti_8$ (i.e., M = Al), shown in Fig. 8-26B, refer to the results of pure TiAl without any alloying elements.

As shown in Fig. 8-26A, in case M is substituted for Ti, the M *d*–Ti *d* bond order is smaller than the M *d*-Al *p* bond order, since an M atom is surrounded by four Ti atoms and eight Al atoms in the first nearest neighbors. For 3*d* transition elements, the M *d*–Ti *d* bond order curve shows a small peak at V and then decreases with increasing atomic number.

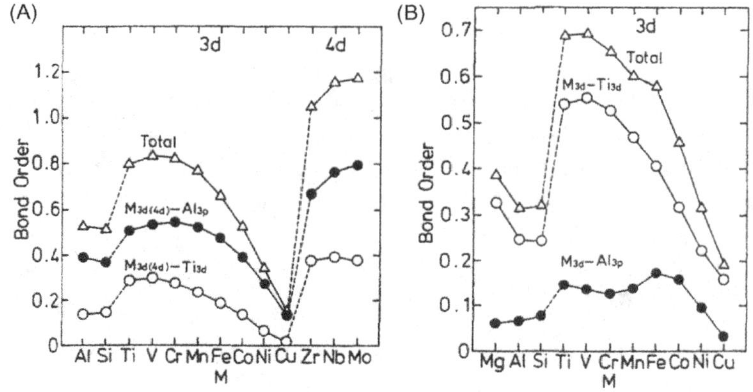

FIGURE 8-26 Bond order changes with alloying element, M. (A) $MTi_{10}Al_8$ cluster with the substitution of M for Ti, and (B) $MAl_{10}Ti_8$ cluster with the substitution of M for Al.

The M *d*–Al *p* bond order curve shows a broad peak at Cr. For the 4*d* elements, both bond orders are higher than those for the 3*d* elements. On the other hand, the Al substitution for Ti lowers both the bond orders.

On the other hand, as shown in Fig. 8-26B, in case M is substituted for Al, the M *d*–Ti *d* bond order is much larger than the M *d*–Al *p* bond order, since an M atom is surrounded by eight Ti atoms and four Al atoms at the first nearest neighbors. For 3*d* transition elements, the M *d*–Ti *d* bond order change with M is similar to the result shown in Fig. 8-26A. The Ti substitution for Al yields a large increase in the M *d*–Ti *d* bond order. On the other hand, the M *d*–Al *p* bond order change with M is a little different from that of Fig. 8-26A. Its magnitude for Ti, V, Cr, and Mn is slightly larger than that for Al.

8.8.3 Alloying Effect on the Ductility Improvement of TiAl

TiAl appears to be an intrinsically brittle compound and undergoes a cleavage-type fracture. The Al *p*–Ti *d* covalent bonds are very strong in TiAl and this probably contributes to a large portion of the cohesive energy of TiAl on the analogy of NiAl [38].

On the other hand, there are many ductile compounds with the $L1_2$-type structure [39]. For example, a single crystal of Ni_3Al is ductile and its electronic structure is very similar to that of pure Ni [40], where *d*–*d* covalent interactions are most dominant as explained in Chapter 3, Nickel Alloys. Also, other $L1_2$-type compounds, Co_3Ti, Cu_3Pd, Ni_3Mn, and Ni_3Fe, are more ductile [41]. Their constituent elements are only transition metals and hence *d*–*d* covalent interaction is the most dominant part of the cohesive energy.

Thus, it is likely that the ductility is related partially to the nature of chemical bonds between atoms. The *p*–*d* interactions are enhanced with increasing composition of nontransition metals (Al in the present case). The *p–d* covalent bonds with the lower coordination number are very directional. High directionality of the bonds increases the shear strength and gives a large barrier to the shear deformation [42].

According to the study by Rice and Thomson [43], brittle and ductile behavior may be governed by the stress distribution near the tip of an atomically sharp crack. A material will, in principle, fail in a brittle manner if the local tensile stress operating at the crack tip reaches the cohesive strength before the local shear stress reaches the shear strength. Such a situation may be expected to occur when the shear strength is high owing to the directional *p–d* interaction. Therefore, the enhancement of the *d–d* interaction probably leads to the improvement of the ductility of TiAl.

There are two ways to do this. One is the substitution of transition elements for Al atoms. For example, as shown in Fig. 8-26B, Ti, V, Cr, and Mn substitutions strongly enhance the M *d*–Ti *d* bonding, while keeping a relatively small increase in the M *d*–Al *p* interaction. In fact, the improvement of ductility due to the addition of V and Mn into TiAl has been confirmed experimentally [44]. There are many deformation twins in the TiAl alloyed with Mn [44]. Also, the Ti-rich TiAl is more ductile than the Al-rich TiAl [44]. The Cr addition to TiAl will also improve the ductility.

In contrast, the substitution of 3*d* transition metals for titanium atoms is not so effective in improving ductility. This is because, as is evident from Fig. 8-26A, such substitution does not result in the increase of the *d–d* interaction. However, for the 4*d* transition metals the *d–d* interaction is strongly enhanced, but at the same time the *p–d* interaction is also enhanced. This balance may cause TiAl to be more ductile in some way, even though this improvement is supposed to be restricted.

The other way to improve the ductility is to substitute nontransition atoms for Al atoms. As shown in Fig. 8-26B, for example, Mg has the smaller M *d*–Al *p* and the larger M *d*–Ti *d* interaction than Al. This meets the conditions for improving ductility. On the contrary, Si does not follow the conditions and hence its addition will be harmful to the ductility.

Thus, the existence of directional Al *p*–Ti *d* covalent bonds causes TiAl to be brittle. The addition of those elements into TiAl that weaken *p–d* interactions but enhance *d–d* interactions is most effective in improving the ductility.

Needless to say, a variety of factors affect mechanical properties [45]. For instance, the axial ratio change and the atomic disordering with alloying may be important [44]. The stacking fault energies and shear constants may also affect mechanical properties of the compounds [46]. However, the electronic factor appears to be one of the crucial factors to control the ductility of TiAl.

Finally, it is summarized that new crystal structure maps are constructed by using two electronic parameters, *Md* and *Bo*. There is a better separation of the crystal structures for various compounds on the *Bo–Md* maps, as compared with the previously proposed maps. The *Bo–Md* maps are useful for the prediction of crystal structures not only for binary compounds but also for ternary compounds. Also, the ductility problem is treated in TiAl in view of the electronic bonding between atoms.

References

[1] P. Villars, L.D. Calvert, Pearson's Handbook of Crystallographic Data for Intermetallic Phases, vols. 1–3, ASM, Metal Park, OH, 1985.

[2] H. Mabuchi, K. Hirukawa, Y. Nakayama, Formation of structural $L1_2$ compounds in $TiAl_3$-base alloys containing Mn, Scr. Metal. 23 (1989) 1761–1765.

[3] Y. Harada, Y. Funato, M. Morinaga, A. Ito, Y. Sugita, Solid solubility of ternary elements and their effects on microstructure of $MoSi_2$, J. Jpn. Inst. Met 58 (1994) 1239–1247.

[4] K. Aoki, O. Izumi, Ductilization of $L1_2$ intermetallic compound Ni_3Al by microalloying with boron, Mater. Trans., JIM 31 (1990) 443–448.

[5] M. Morinaga, J. Saito, N. Yukawa, H. Adachi, Electronic effect on the ductility of alloyed TiAl compound, Acta Metall. Mater. 38 (1) (1990) 25–29.

[6] Y. Harada, M. Morinaga, J. Saito, Y. Takagi, New crystal structure maps for intermetallic compounds, J. Phys.:Condens. Matter 9 (1997) 8011–8030.

[7] W. Hume-Rothery, P.W. Reynolds, G.V. Raynor, Factors affecting the formation of 3/2 electron compounds in alloys of copper, silver, and gold, J. Inst. Metals 66 (1940) 191–207.

[8] Z.M. Ziman, Principles of the Theory of Solids, second ed., Cambridge University Press, Cambridge, 1972.

[9] T.B. Massalski, Phase diagrams in materials science, Metall. Trans. A 20 (1985) 1295–1323.

[10] A.R. Miedema, F.R. Deboer, P.F. Dechatel, Empirical description of the role of electronegativity in alloy formation, J. Phys. F: Metal Phys. 3 (1973) 1558–1576.

[11] A. Zunger, Systematization of the stable crystal structure of all AB-type binary compounds: a pseudopotential orbital-radii approach, Phys. Rev. B 22 (1980) 5839–5872.

[12] P. Villars, A. Three-Dimensional, Structural stability diagram for 998 binary AB intermetallic compounds, J. Less Common Met 92 (1983) 215–238.

[13] P. Villars, A. Three-Dimensional, Structural stability diagram for 1011 binary AB_2 intermetallic compounds: II, J. Less Common Met 99 (1984) 33–43.

[14] P. Villars, Three-dimensional structural stability diagrams for 648 binary AB_3 and 389 binary A_3B_5 intermetallic compounds: III, J. Less Common Met 102 (1984) 199–211.

[15] D.G. Pettifor, The structures of binary compounds. I. Phenomenological structure maps, J. Phys. C: Solid State Phys 19 (1986) 285–313.

[16] D.G. Pettifor, Structure maps for pseudobinary and ternary phases, Mater. Sci. Technol 4 (1988) 675–691.

[17] J.C. Slater, Quantum Theory of Molecules and Solids, vol. 4, McGraw-Hill, New York, 1974.

[18] F.W. Averill, D.E. Ellis, An efficient numerical multicenter basis set for molecular orbital calculations: Application to $FeCl_4$, J. Chem. Phys. 59 (1973) 6412–6418.

[19] H. Adachi, M. Tsukada, C. Satoko, Discrete variational X-alpha cluster calculations, 1. Application to metal clusters, J. Phys. Soc. Jpn. 45 (1978) 875–883.

[20] M.J. Wenninger, Polyhedron Models, Cambridge University Press, Cambridge, 1971.

[21] E.T. Teatum, K.A. Gshneidner Jr., J.T. Waber, Department of Commerce LA-2345 (1968).

[22] R.E. Watson, L.H. Bennet, Transition metals: d-band hybridization, electronegativities and structural stability of intermetallic compounds, Phys. Rev. B 18 (1978) 6439–6449.

[23] R.S. Mulliken, Electronic population analysis on LCAO-MO molecular wave functions. I, J. Chem. Phys 23 (1955) 1833–1840.

[24] R.S. Mulliken, Electronic population analysis on LCAO-MO molecular wave functions. II. Overlap populations, bond orders, and covalent bond energies, J. Chem. Phys 23 (1955) 1841–1846.

[25] R.S. Mulliken, Electronic population analysis on LCAO-MO molecular wave functions. III. Effects of hybridization on overlap and gross AO populations, J. Chem. Phys 23 (1955) 2338–2342.

[26] R.S. Mulliken, Electronic population analysis on LCAO-MO molecular wave functions. IV. Bonding and antibonding in LCAO and valence-bond theories, J. Chem. Phys 23 (1955) 2343–2346.

[27] S.K. Makineni, A. Samanta, T. Rojhirunsakool, T. Alam, B. Nithin, A.K. Singh, et al., A new class of high strength high temperature cobalt based $\gamma - \gamma'$ Co–Mo–Al alloys stabilized with Ta addition, Acta Mater 97 (2015) 29–40.

[28] D.A. Berztiss, R.R. Cerchiara, E.A. Gulbransen, F.S. Pettit, G.H. Meier, Oxidation of $MoSi_2$ and comparison with other silicide materials, Mater. Sci. Eng. A 155 (1992) 165–181.

[29] V.N. Svechnikov, Y.A. Kocherzhinskii, L.M. Yupko, Phase diagram of titanium–silicon system, Dokl. Akad. Nauk SSSR 193 (1970) 393.

[30] Y. Harada, M. Morinaga, A. Ito, Y. Sugita, Electronic states of alloying elements in $MoSi_2$ compound, J. Alloys Compd. 236 (1996) 92–101.

[31] W.J. Boettinger, J.H. Perepezko, P.S. Frankwicz, Application of ternary phase diagrams to the development of $MoSi_2$-based materials, Mater. Sci. Eng. A 155 (1992) 33–44.

[32] R. Beyers, D. Coulman, P. Merchant, Titanium disilicide formation on heavily doped silicon substrates, J. Appl. Phys 61 (1987) 5110–5117.

[33] H. Okamoto, Phase Diagrams for Binary Alloys, second ed., ASM International, Materials Park, OH, 2010.

[34] R.L. Fleischer, High-strength, high-temperature intermetallic compounds, J. Mater. Sci 22 (1987) 2281–2288.

[35] A. Khataee, H.M. Flower, D.R.F. West, New titanium–aluminum–X alloys for aerospace applications, J. Mater. Eng 10 (1988) 37–44.

[36] M.E. Eberhart, D.D. Vvedensky, Localized grain-boundary electronic states and intergranular fracture, Phys. Rev. Lett 58 (1987) 61–64.

[37] D.D. Vvedensky, M.E. Eberhart, Toward a microscopic basis for mechanical behaviour, Phil. Mag. Lett 55 (1987) 157–161.

[38] D. Hackenbracht, J. Kübler, Electronic magnetic and cohesive properties of some nickel–aluminium compounds, J. Phys. F. Metal Phys 10 (1980) 427–440.

[39] T. Takasugi, O. Izumi, Electronic and structural studies of grain boundary strength and fracture in Ll_2 ordered alloys II. On the effect of third elements in Ni_3Al alloy, Acta Metal 33 (1985) 1259–1269.

[40] M. Morinaga, N. Yukawa, H. Adachi, Alloying effect on the electronic structure of Ni_3Al (γ'), J. Phys. Soc. Jpn 53 (1984) 653–663.

[41] T. Takasugi, O. Izumi, N. Masahashi, Electronic and structural studies of grain boundary strength and fracture in $L1_2$ ordered alloys I. On binary A_3B alloys, Acta Metal 33 (1985) 1247–1258.

[42] M.E. Eberhart, R.M. Latanision, K.H. Johnson, Overview no. 44: the chemistry of fracture: A basis for analysis, Acta Metal 33 (1985) 1769–1783.

[43] J.R. Rice, R. Thomson, Ductile versus brittle behaviour of crystals, Philos. Mag 29 (1974) 73–97.

[44] K. Hashimoto, H. Doi, K. Kasahara, T. Tsujimoto, T. Suzuki, Effects of third elements on the structures of TiAl-based alloys, J. Jpn Inst. Metals 52 (1988) 816–825.

[45] G. Hug, A. Loiseau, A. Lasalmonie, Nature and dissociation of the dislocations in TiAl deformed at room temperature, Phil. Mag. A, Phys. Condens. Matter Def. Mech. Prop. 54 (1986) 47–65.

[46] D.G. Pettifor, Theoretical predictions of structure and related properties of intermetallics, Mater. Sci. Technol 8 (1992) 345–349.

9

A Universal Relation between Electron Density Minima and Atomic or Ionic Radii

9.1 Calculation of Electronic Structures

A series of electronic structure calculations is performed using the DV-$X\alpha$ molecular orbital method [1–5]. For the calculation of solids, a relatively large cluster model is made by using experimental data on the crystal structure. For example, it is a Fe_{59} cluster model for pure iron with the bcc structure and a $Mg_{32}O_{32}$ cluster model for MgO with the rock salt structure.

9.2 Spatial Electron Density Distribution

One example of the calculated results is illustrated in Fig. 9-1A, C for MgO. The rock salt structure is shown in (A). Also, as shown in Fig. 9-1B, any clear change is not observable in the normal plot of the electron density, $\rho(r)$, along the line linking between O and Mg nuclei. So, the natural logarithm of the electron density, log $\rho(r)$, is taken to detect a very small change in the $\rho(r)$ curve. As is evident from Fig. 9-1C, log $\rho(r)$ decreases monotonously with the distance, r, from the oxygen nucleus, but the curve is separated into two regions in the oxygen side, and each region has its own slope. The slope varies with the principal quantum number, n, which is a quantum number to control the orbital energy and the electron distribution in the molecule. In the case of the oxygen shown in Fig. 9-1, $n = 1$ for the inner region and $n = 2$ for the outer region.

The slopes are calculated using the fitting functions given in Fig. 9-2. These functions for $n = 1$ and $n = 2$ are set following the radial distribution functions, Rnl(r), of hydrogen-like atom, since $\rho(r)$ is proportional to the square of Rnl(r). In Fig. 9-2, R10(r), R20(r), and R21(r) are the radial distribution functions for 1s electrons, 2s electrons, and 2p electrons, respectively. The slope expressed as ξ is about 15.75 for the inner region composed mainly of O 1s electrons, and about 6.61 for the outer region composed mainly of O 2s, 2p electrons. It is known that η $(=2Z/n)$ is the slope expected from the radial distribution function of a hydrogen-like atom consisting of only one electron, where Z is the atomic number. As $Z = 8$ for oxygen, the η values are 16 for $n = 1$ and 8 for $n = 2$, each value being close to the respective value, ξ, of oxygen described previously.

Such slopes are obtained for other elements, B, C, N, and F, and the results are summarized in Fig. 9-3. Every value is close to the η $(=2Z/n)$ value. These elements, however,

A Quantum Approach to Alloy Design. DOI: https://doi.org/10.1016/B978-0-12-814706-1.00009-1

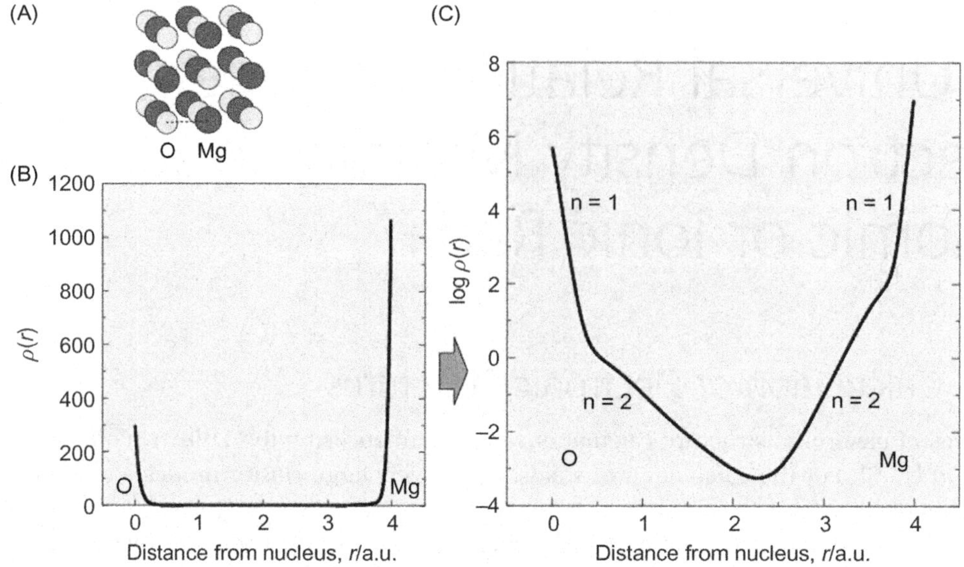

FIGURE 9-1 Representation of electron density distribution in MgO. (A) Crystal structure of MgO, (B) electron density, $\rho(r)$, along the line linking between O and Mg nuclei, and (C) natural logarithm of the electron density, log $\rho(r)$. Here, the unit of electron density is electrons / $(\text{a.u.})^3$, (1 a.u. = 0.0529 nm).

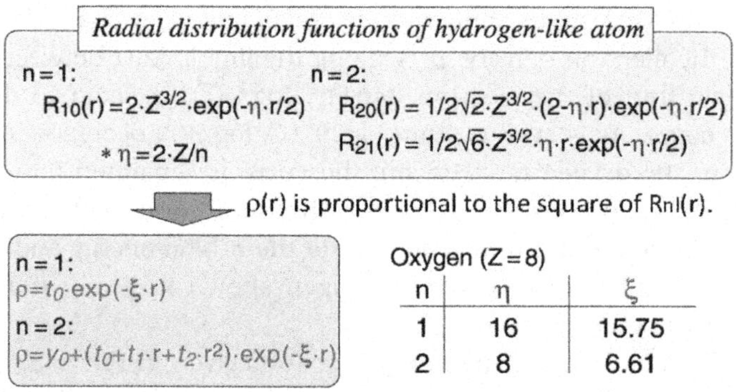

FIGURE 9-2 Radial distribution functions of hydrogen-like atom, where R10(r), R20(r), and R21(r) are the functions for 1*s*, 2*s*, and 2*p* electrons, respectively. Two fitting functions for $n = 1$ and $n = 2$ are used for determining the ξ parameter, and the ξ values for oxygen are compared with the η values for the hydrogen-like atom.

contain many electrons, and hence the screening of nuclear charges by the outer electrons will occur inevitably, resulting in the deviation of the slopes from the η $(=2Z/n)$ values of the hydrogen-like atom. Despite this screening effect, the calculated values, ξ, are similar to those of the hydrogen-like atom containing only one electron. The similarity is particularly great in the case of $n = 1$ for 1*s* electrons. To confirm this further, the preexponential term, t_0, is calculated using an equation, $\rho(r) = t_0\exp(-\xi r)$ for 1*s* electrons as shown in Fig. 9-2.

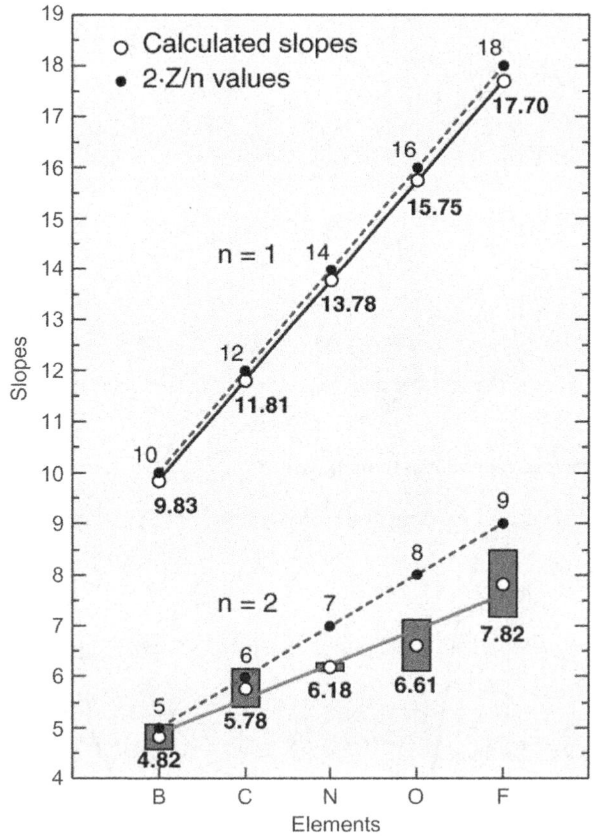

FIGURE 9-3 Calculated slopes of elements, B, C, N, O, and F, and comparison with the $\eta = 2Z/n$ values for the radial distribution functions of the hydrogen-like atom.

According to the radial distribution function, R10(r), for 1*s* electrons in the hydrogen-like atom, t_0 should be proportional to Z^3. This is indeed satisfied as shown in Fig. 9-4.

It is also found that the extent of the region ($n = 2$) for O-2*s*, 2*p* electrons is dependent largely on the neighboring metal atom in various oxides. For example, the result of log $\rho(r)$ in Al_2O_3 is indicated by a dotted curve in Fig. 9-5 together with the result of MgO. The region of $n = 2$ is rather limited compared with that in MgO. However, it is stressed here that the slope still remains unchanged. Even the valence electrons such as O-2*s*, 2*p* keep their own slope, instead of spreading over the space in a random manner. In addition, the extent of $n = 2$ region means the oxygen's sphere of influence. So, its extension probably increases as the effective number of electrons increases at the oxygen site. Therefore, the extension becomes larger when the oxygen has a larger and more negative ionicity. In this sense, the extent of $n = 2$ region is a measure of the ionicity as well as the ionic size. This is also the case of other elements shown in Fig. 9-3.

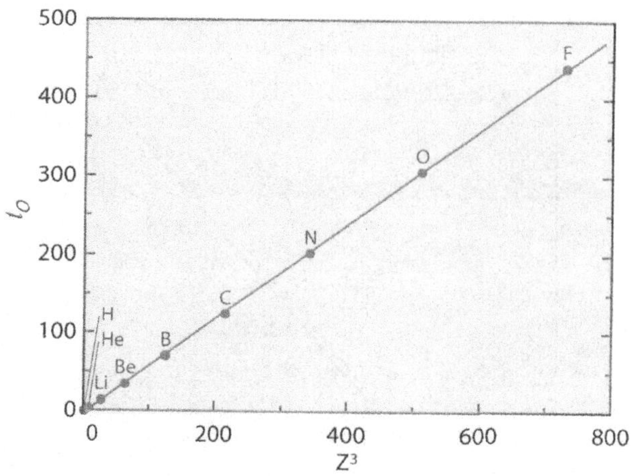

FIGURE 9-4 Correlation of the pre-exponential term, t_0, with Z^3.

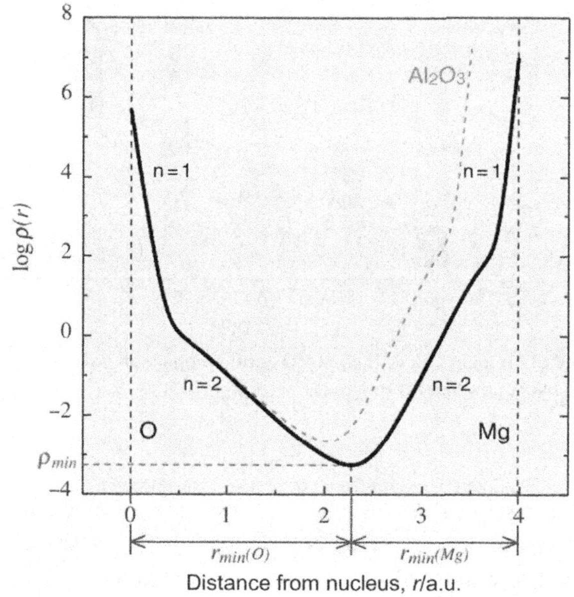

FIGURE 9-5 Representation of natural logarithm of the electron density, log $\rho(r)$, along the line linking between O and Mg nuclei. Minimum electron density, ρ_{min}, and ionic radii, $r_{min}(O)$ for O and $r_{min}(Mg)$ for Mg, are illustrated in the figure. The dotted curve is the result of Al_2O_3.

9.3 A Universal Relation between Electron Density Minima and Atomic or Ionic Radii

Next, as illustrated in Fig. 9-5, we define the minimum electron density, ρ_{min}, and the atomic or ionic radii, r_{min}, at the position of showing ρ_{min}. The values of ρ_{min} and r_{min} are calculated for gases, water and solids of over 130 species, and the results are summarized in Fig. 9-6. In this figure the vertical axis is log (ρ_{min}/Z^3) and the horizontal axis is $2(Z/n)r_{min}$. This

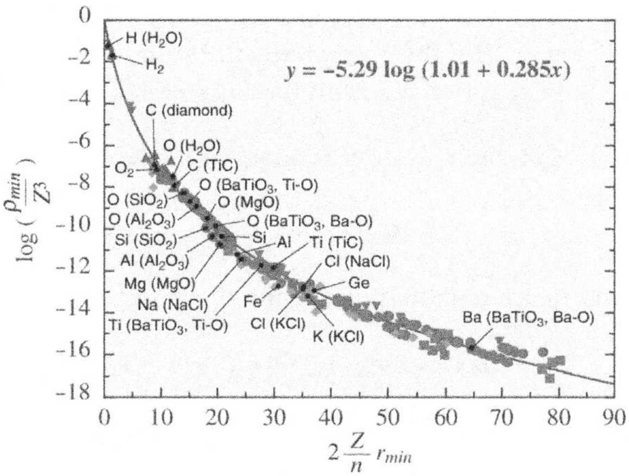

FIGURE 9-6 A universal relation between the minimum electron density, ρ_{min}, and the atomic or ionic radii, r_{min}. Notations used in the figure: for example, H_2 denotes H in H_2 gas. Ti (TiC) and C (TiC) denote Ti and C in the first nearest neighbor Ti-C pair in TiC, respectively. Ti (or O) ($BaTiO_3$, Ti-O) denotes Ti (or O) in the first nearest neighbor Ti-O pair in $BaTiO_3$. The others are denoted in a similar way.

coordinate is chosen because in the case of the radial distribution functions of hydrogen-like atoms, $\rho(r)/Z^3$ is a function of only one variable, $2(Z/n)$r.

In these plots, the value of n is set with the help of the calculated electron density around ρ_{min}. For example, the value of n is set to be 2 for both O and Mg, because the region of $n=3$ for Mg-*3s, 3p* electrons are missing in the electron density shown in Fig. 8-1. This is probably due to the occurrence of charge transfer from Mg to O in MgO. For most pure solids the value of n is equal to the principal quantum number of the outermost electrons of the neutral atom in the ground state. For example, n is 4 for bcc iron and 2 for diamond.

It is surprising that every type of matter falls on one curve, irrespective of gases, water, and solids, despite the great difference in the nature of the chemical bond among them. For example, some of the matter named in Fig. 9-6 are gases (H_2, O_2), water (H_2O), diamond (C), metals (Fe, Al), semiconductors (Si, Ge), oxides (MgO, Al_2O_3, SiO_2, $BaTiO_3$), alkali halide crystals (NaCl, KCl), and metal compounds (TiC, ZrN). Thus, this curve is indeed a universal curve that is common to all matter. By the least-squared analysis, it is expressed in a simple formula as,

$$\log\left[\rho_{\min}/Z^3\right] = \mathrm{a} \bullet \log\left[\mathrm{b} + \mathrm{c} \bullet 2(Z/n)\mathrm{r}_{\min}\right] \tag{9.1}$$

Here, the fitting parameters are $\mathrm{a} = -5.29 \pm 0.07$, $\mathrm{b} = 1.01 \pm 0.04$, and $\mathrm{c} = 0.285 \pm 0.010$.

9.4 Determination of Ionic Radius in Compounds

Eq. (9.1) is useful, for example, for determining ionic radii. Let's explain this using an AB binary compound with the first nearest neighbor A-B distance, r_0. If $2(Z/n)\ \mathrm{r}_{\min} = 2(Z_A/n_A)\ \mathrm{r}_A$ for A and $2(Z/n)\ \mathrm{r}_{\min} = 2(Z_B/n_B)\ \mathrm{r}_B$ for B, then $\mathrm{r}_A + \mathrm{r}_B = \mathrm{r}_0$. Also, $\rho_{\min}$ is the minimum electron density at the r_A position from A nucleus and at the r_B position from B nucleus in the A-B connecting line (see Fig. 9-1 for the case of MgO). Under this condition, the following linear equation between r_A and r_B is derived from the Eq. (9.1),

$$2(Z_A/n_A)r_A - \alpha \bullet 2(Z_B/n_B)r_B = (b/c) \bullet (\alpha\text{-}1), \tag{9.2}$$

$$\alpha = \exp\left[(1/a)\log\left(Z_B^3/Z_A^3\right)\right]$$

As $r_A + r_B = r_0$, r_A and r_B are obtained by solving these simultaneous equations.

$$r_A = \left[r_0 \bullet \alpha \bullet 2(Z_B/n_B) + (b/c) \bullet (\alpha - 1)\right]/\beta, \tag{9.3a}$$

$$r_B = \left[r_0 \bullet 2(Z_A/n_A) - (b/c) \bullet (\alpha - 1)\right]/\beta, \tag{9.3b}$$

$$\beta = 2(Z_A/n_A) + \alpha \bullet 2(Z_B/n_B).$$

Thus, the ionic radii, r_A and r_B, are calculated from a given r_0, using Eqs. (9.3a) and (9.3b). Then, ρ_{min} is calculated from Eq. (9.1). The ionic radii have been proposed by many investigators [6], but strictly speaking they are not able to be defined correctly, since the ionic radius, r_A, is dependent on the neighboring B ion, as shown in Eq. (9.3a).

Thus, there is a universal relation that is common to every material. However, the physical meaning of the coefficients, a, b, and c, in Eq. (9.1) still remains unknown.

References

[1] J.C. Slater, The Calculation of Molecular Orbitals, Wiley, New York, 1979.

[2] F.W. Averill, D.E. Ellis, Efficient numerical multicenter basis set for molecular-orbital calculations-application to $FeCl_4$, J. Chem. Phys. 59 (1973) 6412–6418.

[3] H. Adachi, M. Tsukada, C. Satoko, Discrete variational Xα calculations. 1. Application to metal clusters, J. Phys. Soc. Jpn. 45 (1978) 875–883.

[4] H. Adachi, DV-Xα method and molecular structure, in: H. Adachi, T. Mukoyama, J. Kawai (Eds.), Hartree-Fock-Slater Method for Materials Science, The DV-Xα Method for Design and Characterization of Materials, Springer, Berlin Heidelberg, 2006, pp. 1–20.

[5] M. Morinaga, N. Yukawa, H. Adachi, Alloying effect on the electronic structure of $Ni_3Al(\gamma')$, J. Phys. Soc. Jpn. 53 (1984) 653–663.

[6] R.D. Shannon, Revised effective ionic-radii and systematic studies of interatomic distances in halides and chalcogenides, Acta Crystallogr. A. 32 (1976) 751–767.

Further Reading

R.F.W. Bader, H. Essen, The characterization of atomic interactions, J. Chem. Phys. 80 (1984) 1943–1960.

10

Atomization Energy Approach to Alloys and Metal Compounds

10.1 Current Molecular Orbital Approach to Hydrogen Storage Alloys

Hydrogen related materials have been investigated experimentally and theoretically [1–8]. Among them hydrogen storage alloy is one of the key materials for the development of future hydrogen energy systems. Hydrogen is stored in the alloy by forming metal hydrides. For example, $LaNi_5$ forms $LaNi_5H_6$ and Mg_2Ni forms Mg_2NiH_4 by hydrogenation [9]. As shown in Table 10-1, various hydrogen storage alloys have been developed, but the total amount of hydrogen stored in any alloy is still low. The target properties for the onboard application of automobiles are expressed as, (the amount of hydrogen stored in alloy) $>$5.5 mass%, and (hydrogen desorption temperature) $<$423K (150°C).

As shown in Table 10-1, most hydrogen storage alloys such as $LaNi_5$ (AB_5-type), $ZrMn_2$ (AB_2-type), TiFe (AB-type), and Mg_2Ni (A_2B-type) consist of hydride forming elements, A, and nonforming elements, B. For example, A's are La, Ca, Zr, Ti, and Mg, and LaH_2, CaH_2, ZrH_2, TiH_2, and MgH_2 are formed by hydrogenation. On the other hand, B's are Ni, Mn, Cr, and Fe. In the case of Ni, NiH is formed in the condition of high pressure and high temperature (e.g., 1 GPa, 435K), but never formed in the ordinary condition [10].

10.1.1 Metal–Hydrogen Interaction

A series of molecular orbital calculations is performed to get a clue to alloy design. For example, cluster models used in the calculation for $LaNi_5H_6$ and Mg_2NiH_4 are shown in Fig. 10-1A, and their contour maps of the electron densities are shown in (B). As is evident from this figure, a high electron-density region around hydrogen extends to the Ni atom, but not to the La atom in $LaNi_5H_6$. This result clearly indicates that hydrogen interacts more strongly with Ni atoms than La atoms in $LaNi_5H_6$ [11], despite a larger affinity of the La atom for hydrogen than the Ni atom in the binary metal–hydrogen system. This is also the case in Mg_2NiH_4 [12], since a relatively high electron-density region around hydrogen extends to the Ni atom, but not to the Mg atom. It is apparent that hydrogen interacts more strongly with Ni atoms than Mg atoms in Mg_2NiH_4 [13]. Similar results are also obtained in a $ZrMn_2$ system [14], and a TiFe system [15]. In any case hydrogen interacts more strongly with the B atom (Ni, Mn, or Fe) than the A atom (La, Mg, Zr, or Ti) without exception.

A Quantum Approach to Alloy Design. DOI: https://doi.org/10.1016/B978-0-12-814706-1.00010-8

Table 10-1 Hydrogen Storage Alloys

Type	Alloy	Amount of Hydrogen (Mass%)
AB_5	$LaNi_5$	1.4
	$CaNi_5$	1.2
AB_2	$ZrMn_2$	1.7
	$TiCr_{1.8}$	2.4
AB	TiFe	1.8
A_2B	Mg_2Ni	3.6
Solid solution	b.c.c V	<3.0

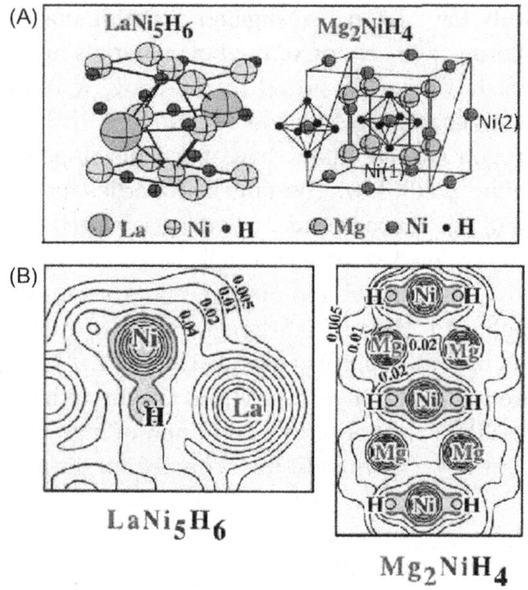

FIGURE 10-1 Cluster models used for molecular orbital calculations of $LaNi_5H_6$ and Mg_2NiH_4, and contour maps of electron densities. The H atom is located near the Ni atom rather than the La or Mg atom in these hydrides.

10.1.2 Role of Hydride Forming and Nonforming Elements

To understand the role of hydride forming and nonforming elements, further calculation is continued with a small cluster model shown in the inset of Fig. 10-2 [16,17]. This is a Ni-based octahedral cluster model, M_2Ni_4H, and a hydrogen atom is placed at the center of the cluster. M's are various elements denoted on the horizontal axis. The Ni–M interatomic distance is varied, depending on the atomic size of M. The Ni–H bond order and M–H bond order are calculated using the Mulliken population analysis [18]. For comparison, either the heat of hydride formation of M metal or the heat of hydrogen dissolution in M metal, ΔH, are drawn in the figure.

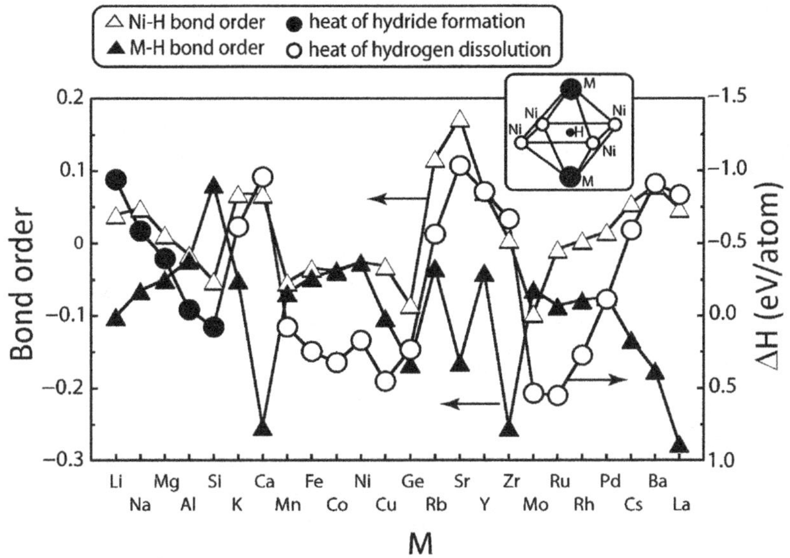

FIGURE 10-2 Bond orders between Ni and H atoms and between M and H atoms in a M_2Ni_4H cluster, and comparison with the heat of hydride formation of M metal or the heat of hydrogen dissolution in M metal, ΔH.

It is apparent from this figure that the Ni–H bond order is large when the M–H bond order is small. In particular, the Ni–H bond order is enhanced when M's are the hydride forming elements such as Na, Mg, Ca, Sr, and La. This means that hydrogen interacts strongly with Ni atoms if the hydride forming elements exist in the neighborhood. On the other hand, when M's are the hydride nonforming elements such as Fe, Co, and Ni, the Ni–H bond order is very small, so that hydrogen will not be absorbed in such an octahedral cluster. Also, it is evident from Fig. 10-2 that the magnitude of the Ni–H bond order correlates well with the heat of hydride formation of M metal or the heat of hydrogen dissolution in M metal, ΔH. Similar results are also obtained for the Ti-based and the Mg-based octahedral clusters [13,16,17]. Thus, there is a general trend that the B–H bond becomes strong only when the hydride forming element, A, exists adjacent to B and H atoms. Otherwise, the B–H bond remains weak or even antibonding, so that it is difficult for hydrogen to be absorbed in the alloy. Thus, both A and B are indeed essential elements in hydrogen storage alloys.

However, it is noticed that the B–H (e.g., Ni–H) interaction in the alloy is still weaker than the A–H (e.g., La–H) interaction in pure metal hydride, AH_2, (e.g., LaH_2), judging from the values of the heat of formation of the hydride, ΔH. As shown in Fig. 10-3C, ΔH is about -209 kJ/mol H_2 for LaH_2 and -30 kJ/mol H_2 for $LaNi_5H_6$. The hydrogen desorption process can be activated readily because of such a weak B–H (i.e., Ni–H) interaction operating in every hydrogen storage alloy. On the other hand, LaH_2 is too stable to activate the hydrogen desorption reaction at moderate temperatures. In this meaning, we may say that the existence of weak hydrogen–metal interactions is characteristic of hydrogen storage alloys [13,16,17].

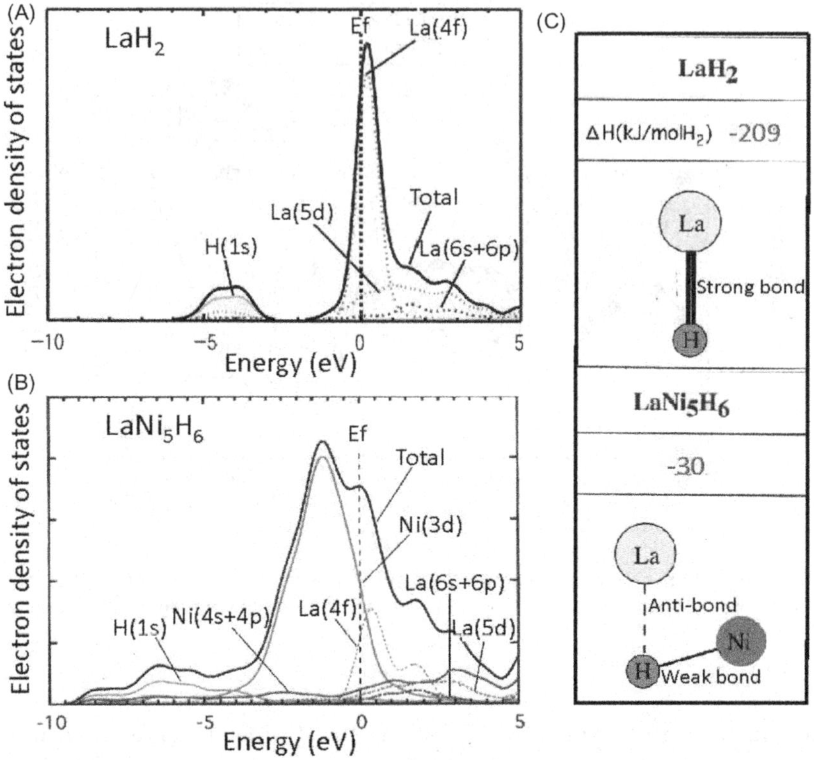

FIGURE 10-3 Electron density of states for (A) LaH_2 and (B) $LaNi_5H_6$, and (C) the heat of hydride formation ΔH for them.

The trend mentioned previously is probably concerned with the fact that the B-H interatomic distance is shorter than the A–H interatomic distance, because of the smaller atomic size of the B element than that of the A element in most cases. Since hydrogen occupies an interstitial site at the center of distorted octahedron or tetrahedron with unequal edge lengths, the B–H interaction with the shorter B–H interatomic distance is inevitably strengthened, whereas the A–H interaction with the longer A–H interatomic distance is weakened. Thus, a polyhedron made of A and B elements provides a unique interstitial spacing for hydrogen in the A–B alloy, being very different from that in pure A metal.

In addition to the geometrical reason, there is another reason, which is the difference in the nature of the chemical bond between LaH_2 and $LaNi_5H_6$. The calculated electron density of states is shown in Fig. 10-3A for LaH_2 and (B) for $LaNi_5H_6$. In the case of (A) LaH_2, the La-4f band is located well above the H-1*s* band, so that the charge transfers from La to H, resulting in the strong ionic interaction operating between La and H atoms in the LaH_2 hydride. On the other hand, in the case of (B) $LaNi_5H_6$, the H-1*s* energy band appears in the energy range much closer to the Ni-3*d* band than the La-4*f* band. In other words, the Ni-3*d* band appears in between the H-1*s* band and the La-4*f* band. As a result, the Ni–H covalent interaction becomes stronger than the La–H interaction. The La–H interaction appears to be antibonding following the Mulliken population analysis [18]. Also, the effective charge of hydrogen in metals is estimated to be about −0.2.

10.1.3 Alloying Effect on the Hydride Stability

The hydride stability changes with the addition of alloying elements into the base alloy. This is explained using an example of Mg_2NiH_4 hydride formed by the hydrogenation of Mg_2Ni. The measured heat of formation of Mg_2NiH_4 is about −64 kJ/mol H_2 [9]. Therefore, this hydride is so stable that the hydrogen desorption temperature is as high as about 573K (300°C). To lower this temperature down to room temperature, the heat of hydride formation, ΔH, should be as low as about −30 kJ/mol H_2. Various third elements are often added into Mg_2Ni to lower the hydride stability. For example, the changes in ΔH with alloying element, M, are measured for the $Mg_2Ni_{0.75}M_{0.25}$ alloy, where M's are alloying elements to be substituted for Ni in Mg_2Ni. The experimental results are shown in Fig. 10-4B. In the figure, Ni for M (i.e., M = Ni) corresponds to pure Mg_2NiH_4. It is apparent from this figure that the Cu addition is most effective in lowering the hydride stability among various M, although the ΔH is still lower than the target value of −30 kJ/mol H_2.

To understand this experimental result, molecular orbital calculations are performed using a cluster model shown in Fig. 10-1A (the right figure) [12]. Alloying element M is substituted for Ni(1) at a center of the cluster. The Ni atom located at the edge of the cluster is denoted as Ni(2). Following the Mulliken population analysis [18], the bond order per bond is calculated for various elements, M. The results are shown in Fig. 10-4A. Among the bond orders shown in the figure, the most remarkable change with M is seen in the M–H

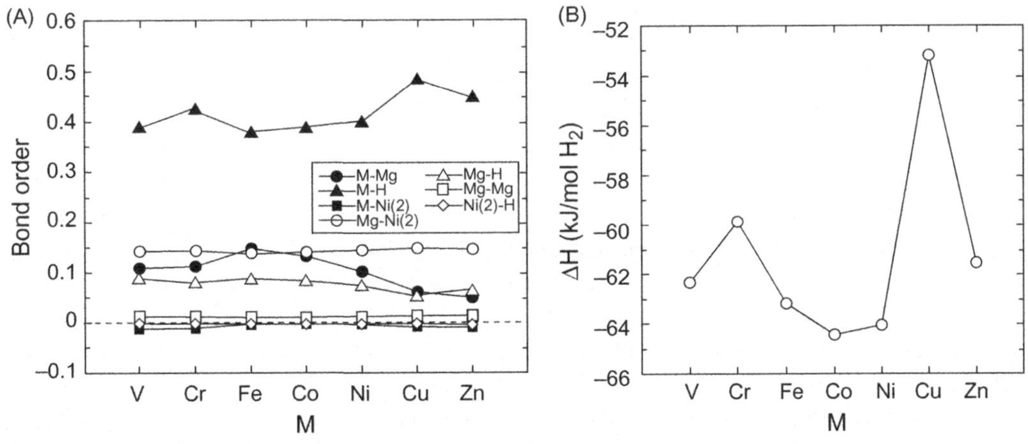

FIGURE 10-4 (A) Bond order change between atoms with alloying element M, which is substituted for Ni atom, and (B) the heat of hydride formation of Mg_2Ni alloyed with M.

bond order indicated by solid triangles and in the M–Mg bond order indicated by solid circles in the figure. These two bond orders vary with M in an approximately reverse way, and their variation quite resembles the ΔH change with M shown in Fig. 10-4B.

For example, in the case of M = Cu, the M–H bond order is higher than the Ni–H bond order in the pure hydride, indicating that the hydride stability increases partially with the M–H bond order. However, there is an attendant decrease in the M–Mg bond order. In pure Mg_2NiH_4 both the Ni–Ni bond and the Mg–Mg bond are rather weak because of the long interatomic distances between them. But instead the Ni–Mg bond is strong enough to stabilize the hydride. Therefore, the significant decrease in the M–Mg bond order with the M (= Cu) addition makes the hydride unstable, resulting in vanishing the stability effect due to the increasing M–H bond order. As a result, the ΔH value changes from -64 kJ/mol H_2 to -53 kJ/mol H_2, as shown in Fig. 10-4B. However, this value is still lower than the target value of $\Delta H = -30$ kJ/mol H_2.

Yukawa et al. have simulated the alloying effect on the hydride stability in various hydrogen storage alloys, for example, $LaNi_5$ (AB_5-type) [11], $ZrMn_2$ (AB_2-type) [14], TiFe (AB-type) [15], and bcc V [19]. Alloying effects on the hydride stability are understood in a reasonable manner [13].

10.1.4 Criterion for Alloy Design

We have been searching for the criterion for the design of new alloys for hydrogen storage. As previously explained in Fig. 10-2, the B-H bond is formed in case the A element exists in the neighborhood of the B element. Therefore, the number of the B–H bonds will increase as the concentration of A element increases in the alloy. In other words, the total amount of hydrogen absorbed in the alloy will increase with increasing A/B compositional ratio. However, when the concentration of A element is too high in the alloy, a disproportionation reaction tends to take place in the alloy.

Here, let's explain this in more detail, assuming that there are three types of tetragonal clusters in the A–B binary alloy as shown in Fig. 10-5A–C. If the A/B compositional ratio is high, or the A–A bond is stronger than the A–B or the B–B bond, the clusters shown in Fig. 10-5A are dominant in the alloy. When hydrogen is introduced into a center of the tetrahedron in the alloy, hydrogen will interact mainly with A elements, resulting in the formation of strong A–H bonds. Also, the disproportionation reaction, AB + H→AH + B, will take place easily in it. Here, AH is so stable that the onset of this reaction is undesirable in view of the activation of the hydrogen desorption process. By contrast, when the A/B compositional ratio is low, or the B–B bond is stronger than the A–A or the A–B bond, the clusters shown in Fig. 10-5C are dominant in the alloy. In this case, it is very difficult for hydrogen to be absorbed in such a B abundant cluster.

Therefore, to absorb and desorb hydrogen smoothly without any disproportionation reaction, the A/B compositional ratio should be controlled within a proper range, depending on the chemical bond strength between A and B elements. In other words, optimum clusters shown in Fig. 10-5B may be formed in the alloy only in case the bond order ratio, 2Bo(A–B)/[Bo(A–A) + Bo(B–B)], takes a suitable value for a given A/B compositional ratio.

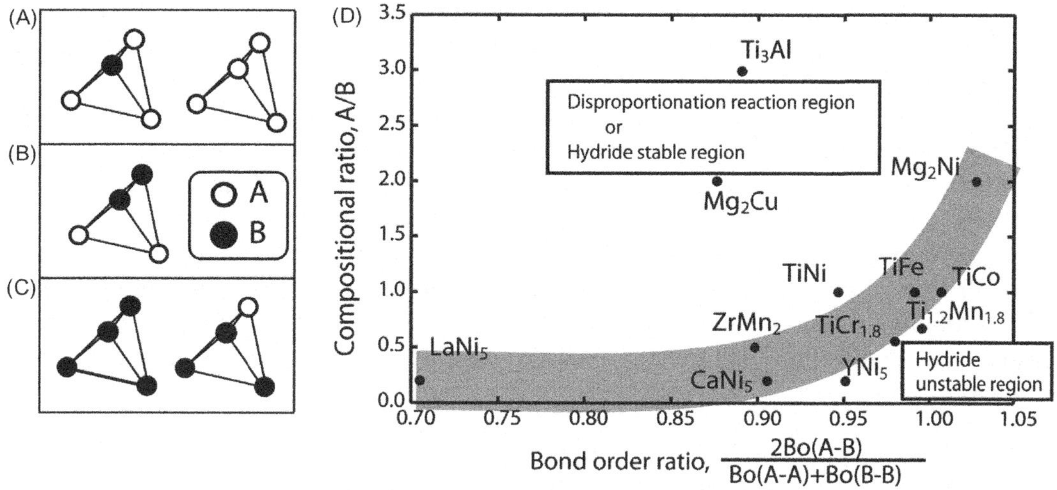

FIGURE 10-5 (A)–(C) Three types of tetrahedral clusters and (D) correlation between the A/B compositional ratio and the bond order ratio.

10.1.5 Search for Hydrogen Storage Alloys

To realize this criterion, a series of calculation is performed to obtain the bond order ratio, 2Bo(A – B)/[Bo(A – A) + Bo(B – B)], for various alloy systems using the tetrahedral cluster, and the results are shown in Fig. 10-5D [20]. Here, this bond order ratio has a physical meaning similar to the ordering energy parameter, Ω_{AB} $(=V_{AB}-1/2(V_{AA}+V_{BB}))$, where V_{AB}, V_{AA}, V_{BB} are the respective bond strengths between atoms given in the subscript. Needless to say, the atomic arrangement shown in Fig. 10-5A or C is present in the clustering-type alloy system, but the one shown in Fig. 10-5B is present in the ordering-type alloy system. The hydrogen storage alloys except for bcc V alloys have the ordered atomic arrangement, so that their characteristics will be represented by using an ordering energy parameter, Ω_{AB}. But instead of Ω_{AB}, the bond order ratio is used here, because it is a parameter to be estimated easily from the calculations, while still holding an approximately similar physical meaning as Ω_{AB}.

As shown in figure Fig. 10-5D, there is a strong correlation between the A/B compositional ratio and the bond order ratio. For example, all the typical hydrogen storage alloys, $LaNi_5$, $ZrMn_2$, TiFe, and Mg_2Ni, are located on a narrow band illustrated in the figure. Here, for Mg_2Ni, the bond order ratio is high, because the Mg–Ni bond is much stronger than the Mg–Mg bond. Also, for $LaNi_5$, the bond order ratio is low, because the La–Ni bond is weaker than the La–La bond. All the alloys located in the upper region above this narrow band tend to decompose during hydrogenation and/or tend to form a stable hydride. One example is the Mg_2Cu system, in which a disproportionation reaction, $Mg_2Cu + 2H_2 \rightarrow 2MgH_2 + Cu$, takes place and a stable hydride, MgH_2, is formed during hydrogenation [9]. Therefore, the region lying above the narrow band is called the disproportionation reaction region or hydride stable region. On the other hand, the alloys located in the region below this narrow band tend to form unstable hydrides. YNi_5 is an example of this. Therefore, this region is called the hydride unstable region.

Similar results are also obtained for the octahedral clusters [17]. Thus, the present result is independent of the clusters used for the calculation. Once A and B elements are chosen, the most suitable A/B compositional ratio is probably determinable from the bond order ratio shown in Fig. 10-5D.

However, despite enormous efforts we are still in a thick mist in view of the search for new hydrogen storage alloys. We may need further information to complement to the present information so far obtained from the current molecular orbital calculations. For this reason, the atomization energy approach is developed as explained in the following sections.

10.2 Atomization Energy Approach to Materials Design

As shown in Fig. 10-6, there are three kinds of materials for hydrogen storage. They are metallic, inorganic, and organic materials. However, several problems still remain in each material as shown in Fig. 10-6.

For example, the total amount of hydrogen stored in metallic materials is usually less than 3 mass%, being much lower than target value of 5.5 mass%. On the other hand, a large amount of hydrogen is stored in both inorganic and organic materials, but the hydrogen desorption temperature is about 573K (300°C), which is much higher than the target

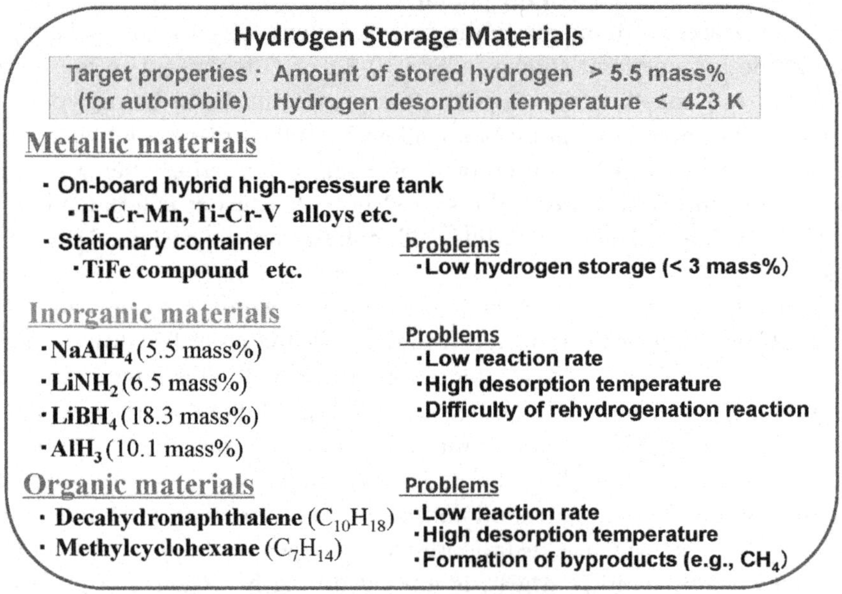

FIGURE 10-6 Three types of hydrogen storage materials, metallic, inorganic, and organic materials.

temperature of 423K (150°C). To get a clue to solving the problems, the nature of the chemical bond between atoms in three materials should be analyzed in an energy scale. A keen insight into the character of these materials may be obtained from such an analysis.

Nakai [3,4] has proposed a new analyzing technique called energy density analysis (EDA). In this method, the total energy of a system calculated by the Kohn–Sham type density functional theory (DFT) [21] is partitioned into atomic energy densities of constituent elements in the material, and the nature of the chemical bond between atoms is expressed in terms of each atomic energy density instead of the total energy. As the feature of each constituent element in the material reflects the atomic energy density, this EDA is very useful for materials design. This is because we can see, so to speak, a face of the individual element in the material through the atomization energy.

10.3 Energy Density Analysis and Atomization Energy

10.3.1 Energy Density Analysis

As explained above, the energy density analysis (EDA) is a method to partition the total energy, *E*, into the respective atomic energy densities of A, E^{A} [3,4]. According to the EDA, the atomic energy density of X atom is evaluated by,

$$E = \sum_{\mathrm{A}}^{\mathrm{atom}} E^{\mathrm{X}} = \sum_{\mathrm{A}}^{\mathrm{atom}} E_{\mathrm{NN}}^{\mathrm{X}} + E_{\mathrm{KIN}}^{\mathrm{X}} + E_{\mathrm{Ne}}^{\mathrm{X}} + E_{\mathrm{CLB}}^{\mathrm{X}} + E_{\mathrm{XC}}^{\mathrm{X}}, \tag{10.1}$$

where $E_{\mathrm{NN}}^{\mathrm{X}}$ is the nuclear–nuclear repulsion energy density, $E_{\mathrm{KIN}}^{\mathrm{X}}$ is the noninteracting kinetic energy density, $E_{\mathrm{Ne}}^{\mathrm{X}}$ is the nuclear–electron attraction energy density, $E_{\mathrm{CLB}}^{\mathrm{X}}$ is the classical Coulomb energy density, $E_{\mathrm{XC}}^{\mathrm{X}}$ is the exchange-correlation energy density.

In Eq. (10.1), for example, $E_{\mathrm{XC}}^{\mathrm{X}}$ is evaluated by the partial sum for the numerical quadrature technique,

$$E_{\mathrm{XC}}^{\mathrm{X}} = \sum_{g}^{\mathrm{grid}} \omega_g(r_g) p_{\mathrm{A}}(r_g) F_{\mathrm{XC}}(r_g). \tag{10.2}$$

Here, $\omega_g(r_g)$ is the weighting factor, $p_{\mathrm{A}}(r_g)$ is the partition function, and $F_{\mathrm{XC}}(r_g)$ is the exchange-correlation functional. The other terms in Eq. (10.1), which are evaluated by the analytical integration with the Kohn–Sham orbitals [21], are partitioned into their energy densities on the analogy of the Mulliken population analysis [18]. A further explanation is given elsewhere [3,4].

It is noted here that the energy density analysis [3,4] deals with the energy density, whereas the Mullkien population analysis [18] deals with the electron density. However, among the five terms shown in Eq. (10.1), an $E_{\mathrm{Ne}}^{\mathrm{X}}$ term that represents the nucleus–electron attraction, is significantly larger as compared with the other terms, and hence the electron density around a nucleus reflects the energy density. In this meaning, we may get information on the electron density from the energy density analysis. In general, the chemical bond between X and Y atoms in a XY molecule is formed as a result of the attractive Coulomb interaction between the X (or Y) nucleus and the electrons in the molecule. The electron density distribution is optimized so as to maximize the attractive Coulomb interactions in it. The interactions between electrons are also taken into account in a proper manner in this optimization process. Thus, the energy density analysis provides us important information on the nature of the chemical bond between atoms in the material in an energy scale.

10.3.2 Atomization Energy

Here, let's define the atomization energy. For example, in case of binary hydride, MH, the energy of the isolated neutral atom, E_M^{atom} for metal M and E_H^{atom} for hydrogen H, are used as a reference. Then, the atomization energies, ΔE_M and ΔE_H, are defined as [1],

$$\Delta E_M = E_M^{atom} - E_M^{hydride}, \tag{10.3}$$

$$\Delta E_H = E_H^{atom} - E_H^{hydride}, \tag{10.4}$$

where, $E_M^{hydride}$ and $E_H^{hydride}$ are the atomic energy densities for M and H in MH, respectively. First, the total energy of a system, E, is calculated by the Gaussian03 [22] under the periodic boundary condition. Then, E is partitioned into $E_M^{hydride}$ and $E_H^{hydride}$, using the energy density analysis [3,4]. Finally ΔE_M and ΔE_H are obtained from Eqs. (10.3) and (10.4). In Fig. 10-7 the calculation steps are illustrated using an example of KH hydride (i.e., M = K). The atomization energy is the energy gained or lost by arranging each atom in a crystalline solid, as compared with the gas state. If ΔE_H is positive, H in KH gains the energy as compared with H in the gas state, whereas if ΔE_K is negative, K in KH loses the energy as compared with K in the gas state. Such an energy transfer takes place between K and H to form the K–H chemical bond in KH.

In case of ternary hydrides, $(M1M2)H_n$, ΔE_M is defined as $(\Delta E_{M1} + \Delta E_{M2})/n$, that is the average atomization energy of M1 and M2 to be counted per hydrogen atom. Even in the other type of ternary hydrides, ΔE_M is defined in a similar way.

Then, the cohesive energy, E_{coh}, of the hydride per hydrogen atom is defined as,

$$\Delta E_M + \Delta E_H = E_{coh}, \tag{10.5}$$

Thus, ΔE_M and ΔE_H are the components of E_{coh}. The advantage of using the atomization energy is to know the contribution of each constituent element to the hydride formation. Such a quantity is never able to be obtained from the total energy calculation alone.

By setting that $y = \Delta E_H$ and $x = \Delta E_M$, we obtain a relation, $y = -x + E_{coh}$. So, E_{coh} is expressed as a point of intersection of this line and y-axis at $x = 0$.

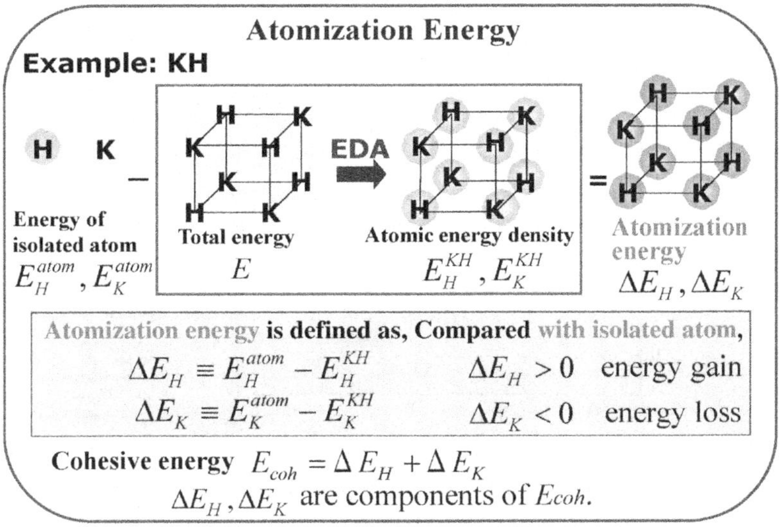

FIGURE 10-7 Definition of atomization energy in case of KH.

For hydrocarbons expressed as a chemical formula, C_mH_n, ΔE_H, and ΔE_C are defined as,

$$\Delta E_C = (E_C^{atom} - E_C^{hydrocarbon}) \times \left(\frac{m}{n}\right), \tag{10.6}$$

$$\Delta E_H = E_H^{atom} - E_H^{hydrocarbon}. \tag{10.7}$$

Here, ΔE_C is the average atomization energy of carbon per hydrogen atom in C_mH_n. For simplicity, the sum of ΔE_H and ΔE_C is hereafter called cohesive energy, E_{coh}, of the hydrocarbon per hydrogen atom, although it exists in either solid, liquid, or gas.

10.4 Atomization Energy Diagram

This energy approach is first applied to a variety of hydrides such as metal hydrides and complex hydrides and hydrocarbons as well, all of which are candidate materials for hydrogen storage [1,5]. As shown in Fig. 10-8, all these hydrides and hydrocarbons are located on an atomization energy diagram, in which ΔE_H is shown in the vertical axis, and ΔE_M is shown in the horizontal axis. In the case of hydrocarbons, the horizontal axis is regarded as ΔE_C instead of ΔE_M. To show the reliability of the calculation, the result of the cohesive energy, E_{coh}, is shown in Table 10-2. The difference between the calculated and experimental values for E_{coh} lies within 0.4 eV for binary metal hydrides and 0.2 eV for hydrocarbons [1]. Thus, the calculation is performed in a reasonable manner.

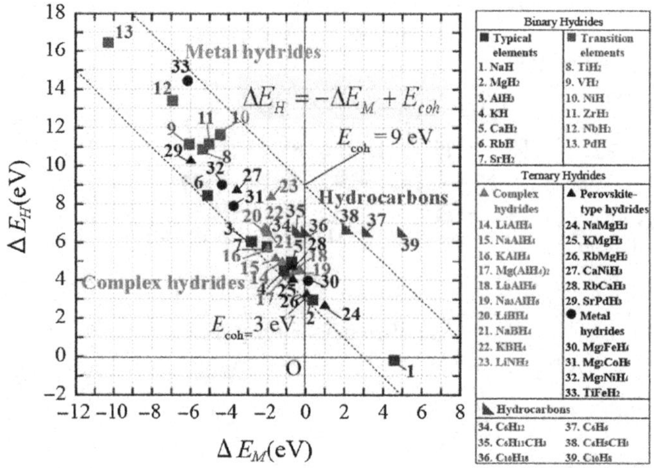

FIGURE 10-8 Atomization energy diagram for hydrides and hydrocarbons.

Table 10-2 Comparison of Cohesive Energy Between Experiment and Calculation for (A) Binary Metal Hydrides and (B) Hydrocarbons

(A) Binary Metal Hydrides (eV per Hydrogen Atom)

	E_{coh} (Cal.)	E_{coh} (Exp.)		E_{coh} (Cal.)	E_{coh} (Exp.)
NaH	4.39	3.96	TiH_2	5.52	5.38
MgH_2	3.34	3.41	VH_2	5.08	5.13
AlH_3	3.22	3.56	NiH	7.20	7.32
KH	3.38	3.78	ZrH_2	6.13	6.29
CaH_2	4.27	4.12	NbH_2	6.48	6.20
RbH	3.34	3.64	PdH	6.17	6.58
SrH_2	3.76	4.04			

(B) Hydrocarbons (eV per Hydrogen Atom)

	E_{coh} (Cal.)	E_{coh} (Exp.)		E_{coh} (Cal.)	E_{coh} (Exp.)
Ethane, C_2H_6	4.95	4.86	Benzene, C_6H_6	9.68	9.49
Ethane, C_2H_4	5.93	5.81	Toluene, C_7H_8	8.80	8.64
Acetylene, C_2H_2	8.45	8.54	p-Xylene, C_8H_{10}	8.26	8.14

All the hydrides and hydrocarbons are compared with each other on this diagram despite the significant difference in the nature of the chemical bond between atoms among them. For example, metal hydrides are located in the upper region of this diagram where ΔE_H is large, whereas complex hydrides are located in the lower region where ΔE_H is small. The hydrocarbons are located near the complex hydrides. The characteristics of hydrocarbons, complex hydrides, and metal hydrides will be explained in this order individually in the following sections.

10.5 Hydrocarbons (C_mH_n)

10.5.1 Atomization Energies for Ethane, Ethylene, Acetylene, and Arene

First of all a simple analysis is given for ethane (C_2H_6), ethylene (C_2H_4), and acetylene (C_2H_2). As already shown in Table 10-2B, there is good agreement in the cohesive energy between the calculation and the experiment.

The chemical bonds between C atoms in these hydrocarbons have been expressed using the concept of a single bond for C_2H_6, a double bond for C_2H_4 and a triple bond for C_2H_2. The atomization energies for carbon, ΔE_C, and for hydrogen, ΔE_H, are shown in Fig. 10-9A. Here, ΔE_C is the atomization energy weighted by a factor, m/n, following Eq. (10.6). As is evident from Fig. 10-9A, ΔE_C increases linearly with the ratio of carbon number (m) to hydrogen number (n), m/n, whereas ΔE_H is nearly constant. For comparison, the bond energies between C atoms, E_{bond} [23], and its normalized values per hydrogen atom, E'_{bond} ($=E_{bond}/n$), are plotted in Fig. 10-9A. E_{bond} values change in the order, $C_2H_6 < C_2H_4 < C_2H_2$, in agreement with the order of ΔE_C. Also, the E'_{bond} line is parallel to the ΔE_C line. Thus, the atomization energy for carbon, ΔE_C, reflects well the bond strength between C atoms in these hydrocarbons.

Next explained here are the arenes, such as benzene (C_6H_6), toluene ($C_6H_5CH_3$), naphthalene ($C_{10}H_8$), *p*-xylene (C_8H_{10}), fluorine ($C_{13}H_{10}$), and coronene ($C_{24}H_{12}$). Arene has a larger m/n ratio as compared with alkane, alkene, and alkine. As shown in Fig. 10-9B, ΔE_H values stay nearly constant and only the ΔE_C values vary linearly with m/n. For example, in the case of benzene, it is known that the C–C bond order is about 1.67, which is larger than the simply averaged value, 1.5 of a single and a double bond owing to the resonance effect [23]. The C–C bond order varies depending on the type of the arene. But even in such an arene, ΔE_C changes linearly with m/n.

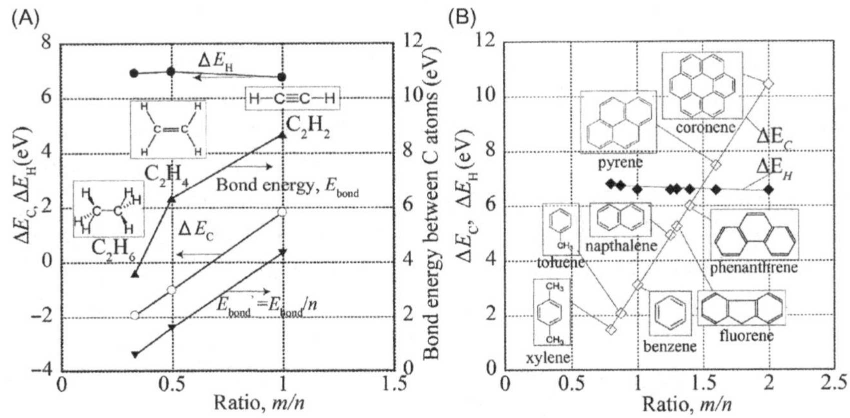

FIGURE 10-9 Correlation of ΔE_C and ΔE_H with the ratio of carbon number to hydrogen number, m/n, in (A) C_2H_6, C_2H_4, C_2H_2 and (B) arene.

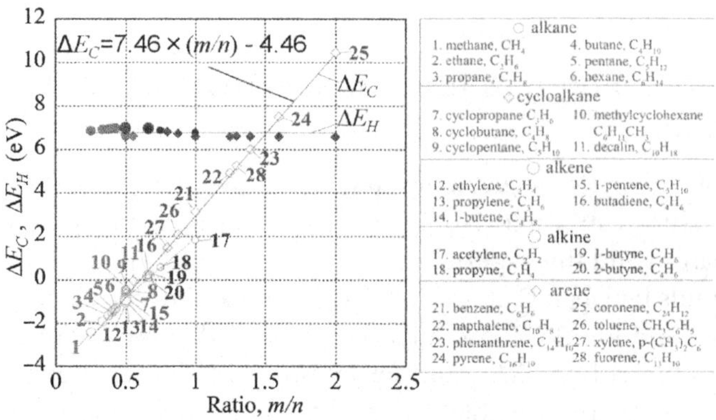

FIGURE 10-10 Atomization energy of carbon and hydrogen in various hydrocarbons.

10.5.2 Correlation of ΔE_C and ΔE_H With *m*/*n* Ratio for Various Hydrocarbons, C_mH_n

In Fig. 10-10 the calculated results are summarized for various hydrocarbons [1]. The atomization energy of C, ΔE_C, increases linearly with m/n, and ΔE_C is expressed as,

$$\Delta E_C\ (\text{eV}) = 7.46 \times \left(\frac{m}{n}\right) - 4.46. \tag{10.8}$$

According to Eq. (10.6), the slope is ($E_C^{\text{atom}} - E_C^{\text{hydrocarbon}}$), and it is nearly constant, 7.46 eV, regardless of the type of hydrocarbons. On the other hand, the atomization energy for hydrogen, ΔE_H is a nearly constant, 6.76 eV, irrespective of the types of hydrocarbons. So, the cohesive energy per one hydrogen atom in hydrocarbons, E_{coh}, is expressed as,

$$E_{coh}\ (\text{eV}) = 7.46 \times \left(\frac{m}{n}\right) + 2.30. \tag{10.9}$$

This beautiful relation is satisfied in all the hydrocarbons. If the cohesive energy is normalized by one molecular unit, it is expressed as $E'_{coh} = 7.46\ m + 2.30\ n$. In either case, the chemical stability of C_mH_n is dependent only on m and n. It is noted here that m is not equal to the number of C—C bonds in every hydrocarbon. For example, in the case of fluorine ($C_{13}H_{10}$), m is 13 and the number of C—C bonds is 15, as shown in Fig. 10-9B. Thus, using the atomization energies, ΔE_C and ΔE_H, we may understand the nature of chemical bonds in the hydrocarbons without using the concept of single or multiple bonds.

10.6 Complex Hydrides

For example, the crystal structure of a complex hydride, $NaAlH_4$, is shown in Fig. 10-11A. It is known that a complex ion, $(AlH_4)^-$ is formed owing to the strong covalent interaction between

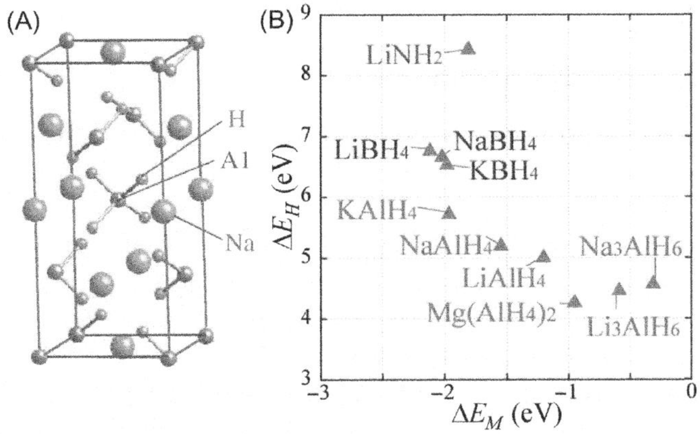

FIGURE 10-11 (A) Crystal structure of $NaAlH_4$ and (B) atomization energy diagram for complex hydrides.

H and Al, and also the ionic interaction is operating between $(AlH_4)^-$ and Na^+ in $NaAlH_4$ [24]. As shown in Fig. 10-11B, complex hydrides are located in a narrow and small ΔE_M range (-2.2 eV $< \Delta E_M < -0.3$ eV), but in a relatively wide ΔE_H range (4.2 eV $< \Delta E_H <$ 8.5 eV). The magnitude of ΔE_H changes in the order, $Mg(AlH_4)_2 < M1AlH_4$ (M1 = Li, Na, K) $< M1BH_4$ (M1 = Li, Na, K) $< LiNH_2$. The cohesive energy, E_{coh}, also changes in the same order. Thus, the H state is well stabilized in compensation for the destabilized M state.

In the case of alanate (Al)-type complex hydrides, $Mg(AlH_4)_2$ and $M1AlH_4$ (M1 = Li, Na, K), the magnitude of ΔE_H changes in the order, $Mg(AlH_4)_2 < LiAlH_4 < NaAlH_4 < KAlH_4$, as shown in Fig. 10-11B. Thus, the H state in $Mg(AlH_4)_2$ is the least stable among these hydrides, in agreement with the experiments [25,26].

In addition, the coordination number effect is observed on the atomization energy diagram. In Na_3AlH_6 (or Li_3AlH_6) an Al atom is surrounded by six hydrogen atoms, whereas in $NaAlH_4$ (or $LiAlH_4$) an Al atom is surrounded by four hydrogen atoms (see Fig. 10-11A). As shown in Fig. 10-11B, ΔE_H is lower in Na_3AlH_6 (or Li_3AlH_6) than in $NaAlH_4$ (or $LiAlH_4$). Thus, the H state is less stable in Na_3AlH_6 (or Li_3AlH_6) than in $NaAlH_4$ (or $LiAlH_4$), as might be expected from the coordination number.

Also, there is a trend that ΔE_H increases with decreasing Al—H distance. For example, it is 0.161 nm for $NaAlH_4$ and 0.175 nm for Na_3AlH_6. The coordination number and the interatomic distance are parameters to show local atomic arrangements. Besides, the material density, which is a parameter to show the atomic arrangements (i.e., crystal structure), also affects the atomization energy [2], as explained later.

In contrast to this approach, the nature of the chemical bond between atoms in complex hydrides has been investigated using the current molecular orbital calculations. For example, Yoshino et al. [24] proposed a new phase stability diagram of complex hydrides. In addition, Yukawa et al. [27] performed a Raman scattering experiment to examine the phase stability of $NaAlH_4$ and Na_3AlH_6. Orimo et al. [28,29] also investigated various complex hydrides for hydrogen storage.

10.7 Metal Hydrides

10.7.1 Atomization Energy Diagram

The ΔE_H vs ΔE_M diagram for metal hydrides is shown in Fig. 10-12C [1], together with the crystal structures of (A) $TiFeH_2$ and (B) Mg_2NiH_4. Mg_2NiH_4 is sometimes grouped into complex hydrides, but here it is grouped into metal hydrides for convenience. The cohesive energy per hydrogen lies in the range, $3\ eV < E_{coh} < 9\ eV$ for metal hydrides.

In case the hydrides have a resemblance in the chemical bonding state, their locations are close to each other on the diagram. For example, binary hydrides of transition elements (e.g., PdH and NbH_2, both shown by solid square (■) in Fig. 10-12C) appear in the higher ΔE_H region than those of typical elements (e.g., NaH and MgH_2, both shown by open square (□) in Fig. 10-12C). This indicates that transition elements could stabilize the hydrogen state remarkably in the binary hydrides. In other words, the hydrogen atom interacts more strongly with transition elements than typical elements.

First, let's consider ternary metal hydrides, $M1M2H_n$, where M2 is a transition metal (e.g., $TiFeH_2$ shown by open circle (○) in Fig. 10-12C). In these hydrides the M2−H interaction is important, since H is located in the neighborhood of M2 (e.g., M2 = Fe) in the crystal lattice as illustrated in Fig. 10-12A. In the case of Mg_2NiH_4 shown by open circle (○) in Fig. 10-12C, the hydrogen atom is located in the neighborhood of Ni as illustrated in Fig. 10-12B, and the interaction is operating between Ni and H in it. In agreement with this, Mg_2NiH_4 is located near NiH rather than MgH_2 on the atomization energy diagram shown in Fig. 10-12C. However, the ΔE_H value is smaller in Mg_2NiH_4 than in NiH, indicating that the H state is destabilized by the presence of Mg in Mg_2NiH_4. (As previously explained, NiH is not formed in the ordinary conditions, but formed in the high pressure and high temperature conditions [10].)

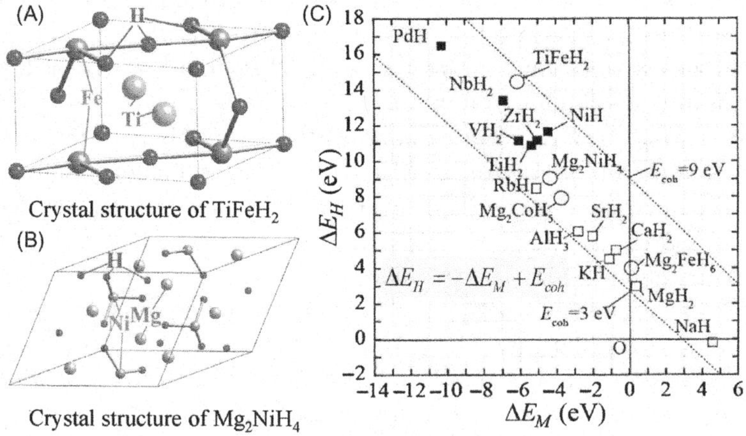

FIGURE 10-12 Crystal structure of (A) $TiFeH_2$ and (B) Mg_2NiH_4, and (C) atomization energy diagram for metal hydrides.

On the other hand, $TiFeH_2$ is located well above TiH_2, so that the ΔE_H value is larger in $TiFeH_2$ than in TiH_2, indicating that the H state is further stabilized by the presence of Fe in the neighborhood instead of neighboring Ti. It is interesting that $TiFeH_2$, which desorbs hydrogen at room temperature, has the largest cohesive energy, 8.3 eV, among the metal hydrides. $TiFeH_2$ are formed from intermetallic compound, TiFe, following the reaction, $TiFe + H_2 \rightarrow TiFeH_2$, and the formation energy is as low as about -30 kJ/mol H_2 [9].

10.7.2 Energy Transfer from Metal to Hydrogen in the Course of Hydrogenation

Mg_2NiH_4 is formed by the hydrogenation of intermetallic compound, Mg_2Ni, as is similar to $TiFeH_2$, which is formed by the hydrogenation of TiFe as described previously. Therefore, further analyses are carried out with these compounds, TiFe and Mg_2Ni. To understand the role of metal atoms in the bond formation, the values of ΔE_{M1} (M1 = Ti or Mg) and ΔE_{M2} (M2 = Fe or Ni) are plotted in Fig. 10-13 for TiFe and Mg_2Ni, together with the values for $TiFeH_2$ and Mg_2NiH_4. As shown in the figure, ΔE_{Ti} for TiFe and ΔE_{Mg} for Mg_2Ni are negative, whereas ΔE_{Fe} for TiFe and ΔE_{Ni} for Mg_2Ni are positive. By hydrogenation, the atomization energy decreases in both M1 and M2 atoms. However, the decrement is more remarkable in ΔE_{M2} than in ΔE_{M1}. This is because hydrogen exists in the neighborhood of M2 and interacts with M2 (M2 = Fe or Ni) in the hydrides. As a result, most of the energy of M2 transfers to the hydrogen and the H state becomes stable, for example, $\Delta E_H = 14.4$ eV for $TiFeH_2$ and $\Delta E_H = 9.0$ eV for Mg_2NiH_4. The energy transfer means the chemical bond formation between H and M2 (M2 = Fe or Ni).

It is interesting to note that there is a large difference between ΔE_{M1} and ΔE_{M2} in both TiFe and Mg_2Ni. This means that when these compounds are formed, a significantly large

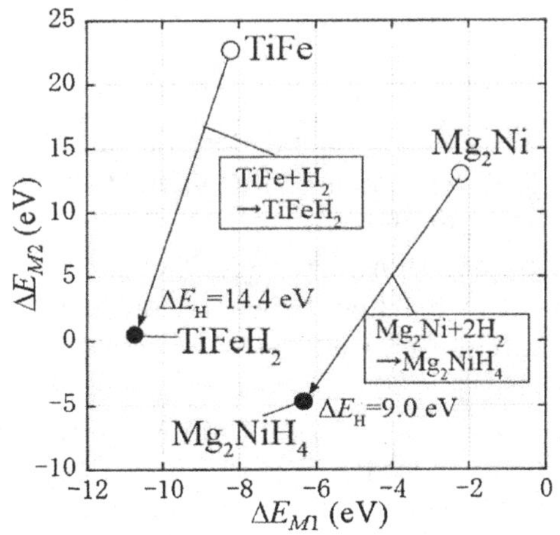

FIGURE 10-13 Change in the atomization energy for M1 and M2 atoms after the hydrogenation reaction of intermetallic compounds.

modification occurs in the electron density distribution between M1 and M2 atoms, which leads to the strong chemical bond formation between atoms in both TiFe and Mg_2Ni intermetallic compounds. This is a reason why the hydrogenation and dehydrogenation reactions take place smoothly without changing alloy compositions in them. Thus, this energy approach can clarify the role of constituent elements in the formation of metal hydrides [1].

10.8 Oxide Catalysts for Dehydrogenation Reaction of MgH_2

The present energy approach is applied to the analysis of the oxide catalysts to promote the dehydrogenation reaction of magnesium hydride (MgH_2) [7]. MgH_2 has been considered as one of the promising candidates for hydrogen storage, because it possesses a high hydrogen capacity of 7.6 mass%. However, the enthalpy of formation of MgH_2, ΔH, is -74 kJ/molH_2, indicating that MgH_2 is a stable hydride. As a result, the temperature of the desorption reaction, $MgH_2 \rightarrow Mg + H_2$, is as high as about 573K. In addition, this desorption reaction proceeds very slowly in the moderate conditions. However, it is found experimentally that some transition metal oxides (e.g., Nb_2O_5) have a large catalytic effect on the hydrogen desorption reaction of MgH_2 [30,31]. However, no quantitative method has been proposed for understanding this catalytic problem yet. This problem is treated using the atomization energy concept [7].

10.8.1 Atomization Energy Diagram for Binary Metal Oxides

The atomization energy diagram is first calculated for binary metal oxides for the use of catalysts. For metal oxides expressed as M_xO_y, the energy of the isolated neutral atom, E_M^{atom} (or E_O^{atom}), is taken as a reference, and the atomization energy, ΔE_M (or ΔE_O), is defined as,

$$\Delta E_M = \left(E_M^{atom} - E_M^{oxide}\right) \times \left(\frac{x}{y}\right), \tag{10.10}$$

$$\Delta E_O = E_O^{atom} - E_O^{oxide}. \tag{10.11}$$

Here, E_M^{oxide} and E_O^{oxide} are the atomic energy densities for M and O in M_xO_y, respectively. Thus, ΔE_M is the energy of M to be counted per O atom. The composition difference among M_xO_y is normalized in this way using Eq. (10.10).

The plots of ΔE_M vs ΔE_O are shown in Fig. 10-14 for binary metal oxides, M_xO_y [2]. Those oxides that are located in the upper right region in Fig. 10-14 have large cohesive energies. Also, the contribution of each element to the cohesive energy of the oxide is understood from this figure. For example, the cohesive energy is nearly the same between NbO_2 and Al_2O_3 (NbO_2: $E_{coh} = 10.1$ eV, Al_2O_3: $E_{coh} = 10.7$ eV). However, they are located in very different positions in Fig. 10-14. NbO_2 has a large ΔE_O value, but a very small ΔE_M value, indicating that the O atoms in NbO_2 make a significant contribution to the cohesive energy. On the other hand, in the case of Al_2O_3, the value of ΔE_O is almost zero. Instead, its ΔE_M value is very large. Thus, the metal atoms in Al_2O_3 contribute mainly to the cohesive energy.

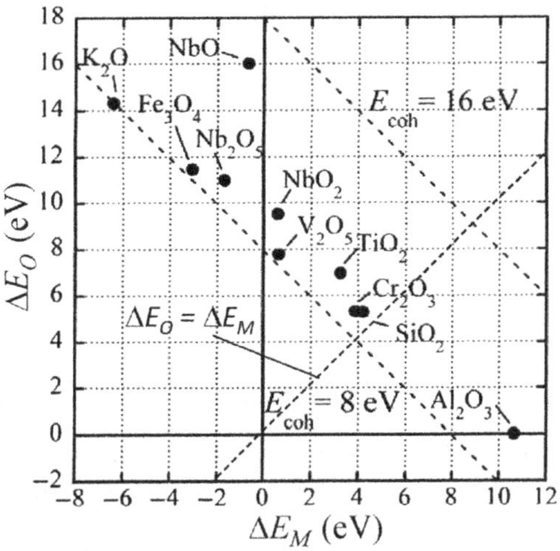

FIGURE 10-14 Atomization energy diagram for binary metal oxides.

Thus, an inequality relation, $\Delta E_O > \Delta E_M$, is satisfied for transition metal oxides like NbO_2, but $\Delta E_O < \Delta E_M$ is satisfied for Al_2O_3. This difference affects the catalytic activity of oxides, as explained later.

10.8.2 Quantitative Analysis of Catalytic Activity of Metal Oxide

Barkhordarian et al. [30] reported the change in the hydrogen desorption rate with catalytic metal oxides as shown in Fig. 10-15A. The hydrogen desorption rate is measured at 573K by mixing 1 mol% catalytic metal oxides into MgH_2. This data is adopted here as a measure of the catalytic activity of each metal oxide. Nb_2O_5 is most active among various oxides as shown in Fig. 10-15A.

To evaluate the catalytic activity quantitatively, the atomization energies are used of constituent elements in metal oxides. As shown in Eqs. (10.10) and (10.11), atomization energies are defined as the values per O atom. However, metal oxides, M_xO_y, are mixed into the MgH_2 system by a mole unit in the experiment. So, the values of the atomization energies are needed to be converted into the values per mole unit. Therefore, each of ΔE_O and ΔE_M is multiplied by the number of O atoms, y, in M_xO_y, which is expressed as $y \times \Delta E_O (= (E_O^{atom} - E_O^{oxide}) \times y)$, and $y \times \Delta E_M (= (E_M^{atom} - E_M^{oxide}) \times x)$.

In Fig. 10-15B, the measured desorption rate is plotted against the atomization energy for O atoms, $y \times \Delta E_O$ [7]. It is apparent that metal oxides that have large $y \times \Delta E_O$ values tend to accelerate the hydrogen desorption rate of MgH_2. But, such a clear correlation is not found between $y \times \Delta E_M$ values and the hydrogen desorption rate [7].

As previously explained about the hydrogenation process of TiFe or Mg_2Ni, the element (e.g., Fe or Ni) with the higher atomization energy interacts more strongly with the hydrogen atom than the other element (e.g., Ti or Mg) with the lower atomization energy [1]. As a

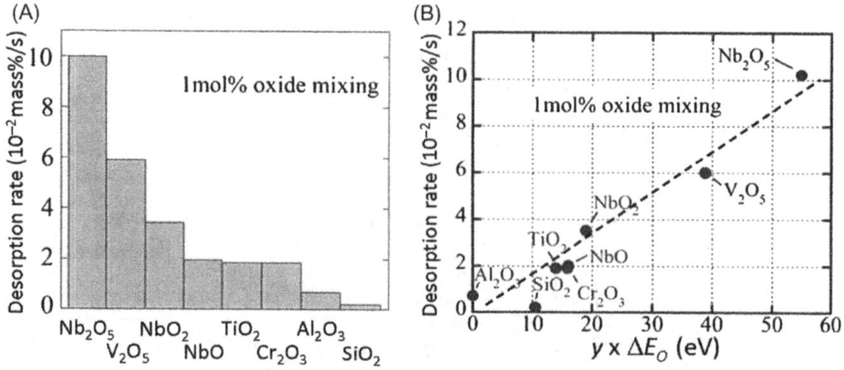

FIGURE 10-15 (A) Catalytic effect of metal oxides on hydrogen desorption rate of MgH_2 at 573K, and (B) evaluation of the desorption rate of MgH_2 using the energy, $y \times \Delta E_O$, of oxide ion per chemical formula, M_xO_y.

result, hydrogen atoms are located near Fe in $TiFeH_2$, and near Ni in Mg_2NiH_4 as shown in Fig. 10-12A and B. This is also the case in metal oxides. The element with the higher atomization energy in the metal oxide must interact more strongly with the hydrogen atom in MgH_2. Even in MgH_2, the hydrogen atom has higher energy than Mg according to the atomization diagram shown in Fig. 10-12C.

As described earlier, in case of transition metal oxides, the atomization energy is higher in the O atom than in the M atom. Therefore, the O atoms in the metal oxides with larger $y \times \Delta E_O$ values are supposed to interact more strongly with hydrogen atoms in MgH_2, as is shown in Fig. 10-15B. The stronger O–H interaction probably leads to the higher catalytic activities of oxides. On the other hand, in the case of Al_2O_3 where the atomization energy of the O atom is low, the catalytic activity is very low as shown in Fig. 10-15A and B. Thus, the strong O–H interaction is operating in the metal oxides with large $y \times \Delta E_O$ values, which works to destabilize MgH_2 and to enhance the catalytic activities of metal oxides [7]. It is stressed here that the catalytically active element in the hydrogen desorption reaction is the O atom, but not the M atom in metal oxides, M_xO_y. The ΔE_O value changes with M in M_xO_y, so that the catalytic activity changes with M_xO_y. Thus, M has nothing to do with the catalytic reaction.

10.8.3 FT-IR Experiment to Prove the Presence of O-H Interaction

As shown in Fig. 10-16A and B, the presence of the O-H bonding is confirmed experimentally in the course of the hydrogen desorption of MgH_2 containing 1 mol% Nb_2O_5 catalyst [8]. A hydrogen desorption curve measured at 573K in vacuum is shown in Fig. 10-16A. On the basis of this result, four specimens (A–D) are prepared for the FT-IR experiment, where (A) is a hydrogen nonreleased specimen, (B) is a 2.3 mass% hydrogen released specimen, (C) is a 5.6 mass% hydrogen released specimen, and (D) is a hydrogen all-released specimen. The theoretical amount of hydrogen released from MgH_2 is 7.6 mass%, but the measured value is

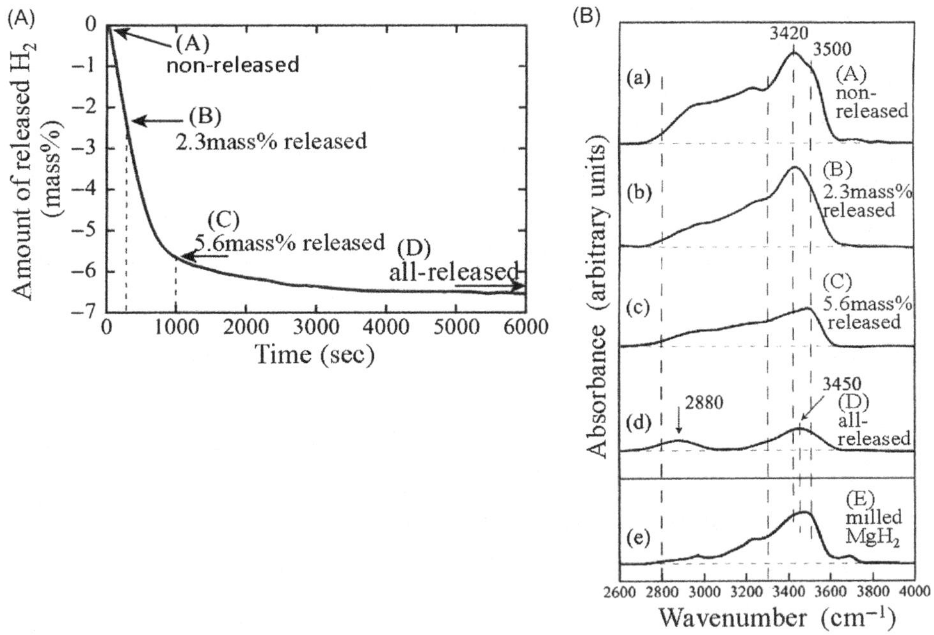

FIGURE 10-16 (A) Hydrogen desorption curve at 573K for MgH_2 with 1 mol% Nb_2O_5, and (B) FT-IR spectra for (A–D) specimens and (E) milled MgH_2 specimen prepared for comparison.

about 6.9 mass%. For comparison, a milled MgH_2 specimen without any oxide catalysts is also prepared and it is denoted as (E).

The FT-IR experiment is carried out with these five specimens of (A)–(E), and the results are shown in Fig. 10-16B [8]. According to previous experiments, the IR absorption frequency of the O–H stretching vibration mode appears in the range of 2800–3800 cm^{-1}, so FT-IR spectra are measured in the frequency range of 2600–4000 cm^{-1}. As shown in Fig. 10-16B, the absorption intensities decrease gradually with decreasing amount of hydrogen retained in the specimen, in agreement with our presumption that the O–H interaction is operating in the hydrogen desorption process. Also, absorption intensities are much larger in the non-released specimen (A) containing 1 mol% Nb_2O_5 catalyst, as compared with the catalyst-free milled MgH_2 specimen (E). This implies that the addition of 1 mol% Nb_2O_5 to MgH_2 actually enhances the O–H interaction. The signals of the O–H vibration mode observed in both the milled MgH_2 specimen (E) and the hydrogen all-released specimen (D) are probably due to the contamination of oxygen introduced during the specimen preparation, since the presence of MgO in the specimen is confirmed from the X-ray diffraction experiment. The further explanation is given in the original paper [8].

Barkhordarian et al. [31] also reported that the mixing of 0.2 mol% Fe_3O_4 strongly enhances the hydrogen desorption reaction of MgH_2. The catalytic effect of Fe_3O_4 is high next to Nb_2O_5, according to their experimental results. By using the ΔE_O value of 11.4 (eV) for Fe_3O_4 shown in Fig. 10-14, the $y \times \Delta E_O$ value is calculated to be 45.6 (eV) because of $y = 4$ in this case. This value is high next to Nb_2O_5 in Fig. 10-15B, in agreement with the present atomization energy analysis. Thus, it is apparent that the element that plays an important role in the catalytic reaction is not the Fe atom, but the O atom in Fe_3O_4. The ΔE_O value changes depending on Fe in Fe_3O_4, but Fe is not an active element in the catalytic reaction. Thus, the present atomization energy analysis is usable for the examination of which element is more catalytically active in the catalyst.

Fe_3O_4 is a main catalyst for the ammonia production in the Harber–Bosch process, $N_2(g) + 3H_2(g) \rightarrow 2NH_3(g)$. The Fe atom that is reduced from Fe_3O_4 by hydrogen has been considered to be a catalytically active element in this reaction. However, to the author's limited knowledge, it may be needed to reexamine the catalytic reaction in the Harber–Bosch process.

10.8.4 A Perspective on Catalytic Problems

It is important to note that the catalytic activities of oxides are evaluated in a quantitative way by using the atomization energy concept, as shown in Fig. 10-15B. In Chapter 7, Hydrogen Overpotential of Alloys and Intermetallic Compounds, a concept was proposed to treat the electrocatalytic problem on the hydrogen overpotential. It is shown that the catalytically active element, X, has excess electrons transferred from other constituent element, Y, owing to the electronegativity difference between X and Y. This concept is presented in a qualitative way, since our attention is mainly directed toward the method of how to find out the catalytically active element. However, this concept may be further extended in a quantitative way if the atomization energy concept is applied to this electrocatalytic problem.

Even in the present case, the catalytic active oxygen is more electronegative than any metal in metal oxides. Therefore, the high atomization energy of oxygen, ΔE_O, is probably interpreted as due to the electron transfer from metal to oxygen in the metal oxide. This is because such excess electrons at the oxygen site lead to the increase in the attractive Coulomb energy between the O nucleus and the surrounding electrons, resulting in the higher atomization energy of oxygen, ΔE_O, as compared with the energy of an isolated neutral oxygen atom that is used as a reference. By contrast, the atomization energy of metal, ΔE_M, decreases as a result of the deficient electron density around metal. Thus, there is probably no essential difference between the present catalytic problem of the hydrogen desorption and the electrocatalytic problem of the hydrogen overpotential.

10.9 3*d* Transition Metal Compounds

As another example, the present energy analysis is applied to the 3*d* transition metal compounds (e.g., borides, carbides, nitrides, fluorides, sulfides). They exhibit a variety of chemical and physical properties such as superconductivity, ferroelectricity, ferromagnetism, proton conductivity, and hydrogen storage as well. For example, there is a superferromagnetic $Nd_2Fe_{14}B$ [32]. Not only metal elements but also nonmetal elements are key factors for

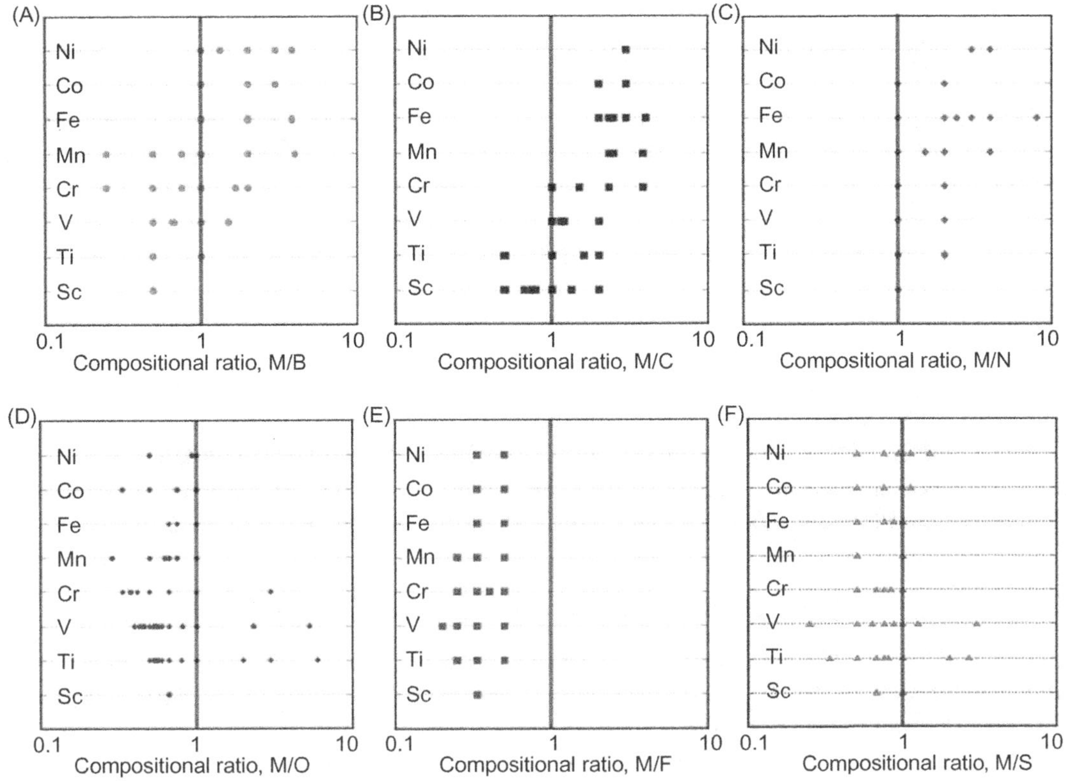

FIGURE 10-17 Compositional range for 3*d* transition metal compounds: (A) borides, (B) carbides, (C) nitrides, (D) oxides, (E) fluorides, and (F) sulfides.

the emergence of new functions in these compounds. For example, there are nonmetal elements, O and F, etc., in the Fe-based superconductor, $LaFeAsO_{1-x}F_x$ [33].

As shown in Fig. 10-17, there are a variety of 3*d* transition metal compounds. Each compound exists stably in its own compositional range [34]. In the figure, the compositional ratio of M/X is expressed, where, M's are 3*d* transition metals, and X's are B, C, N, O, F, and S. For example, borides are stable in both the metal rich and deficient compositional ranges. Carbides and nitrides are stable mainly in the metal rich compositional range, whereas oxides, fluorides, and sulfides are stable mainly in the metal deficient compositional range.

However, to the author's limited knowledge, the nature of the chemical bond between atoms has been investigated individually in each specific type of compound. To broaden our horizons it is needed to find out a more general trend of the chemical bond in a variety of compounds using the atomization energy concept. The knowledge or information of the characters or roles of whole nonmetal elements (B, C, N, O, F, S, etc.) in the compound formation will provide us a new insight for materials design from the viewpoint of nonmetal elements. To this end, a whole feature of the $3d$ transition metal compounds is investigated following the atomization energy concept.

10.9.1 Atomization Energy

For the binary compound expressed as M_mX_n, the atomization energy is defined as,

$$\Delta E_{\mathrm{M}} = \left(E_{\mathrm{M}}^{\mathrm{atom}} - E_{\mathrm{M}}^{\mathrm{M}_m\mathrm{X}_n}\right)\left(\frac{m}{n}\right), \tag{10.12}$$

$$\Delta E_{\mathrm{X}} = E_{\mathrm{X}}^{\mathrm{atom}} - E_{\mathrm{X}}^{\mathrm{M}m\mathrm{X}n}. \tag{10.13}$$

Here, a factor, m/n, is multiplied in Eq. (10.12), so ΔE_{M} is the energy of M atom to be counted per X atom. This normalization allows us to treat any compounds with different compositions in a similar way. Here, M atoms are Ti, Cr, Fe, and X atoms are B, C, N, O, F, S. Here, Ti, Cr, and Fe are selected as the representatives of the $3d$ transition metals, since they are located in the left, middle, and right side of the periodic table of elements.

10.9.2 Atomization Energy Diagram for all the Compounds

In Fig. 10-18, the calculated results are plotted on the ΔE_{M} vs ΔE_{X} diagram. Every compound is located on this diagram, despite the significant differences in the nature of chemical bonds among the compounds.

On this diagram, all the compounds except for CrN (no. 16) are located in the energy range where $\Delta E_{\mathrm{X}} > 0$, indicating that X atom is energetically more stable in the compound

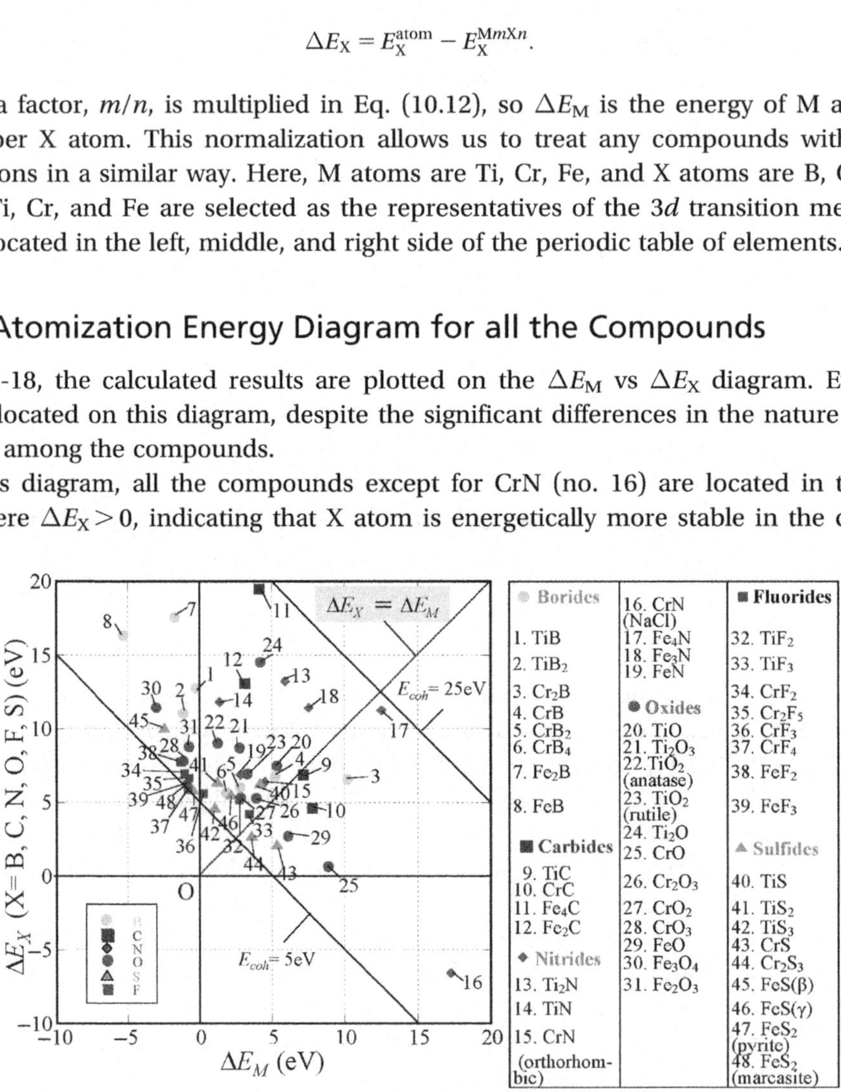

FIGURE 10-18 Atomization energy diagram for all the metal compounds, M_mX_n, where M = Ti, Cr, Fe, and X = B, C, N, O, F, S.

than in the isolated atom state. Also, ΔE_X is larger than ΔE_M in most compounds, so the nonmetal atom, X, contributes significantly to the compound formation.

In addition, the cohesive energy per X atom, E_{coh}, lies in the wide range of 5–25 eV. For example, E_{coh} is about 24 eV per X atom for Fe_4C (no. 11) and Fe_4N (no. 17), but the average cohesive energy per atom is $24/5 = 4.8$ eV/atom, which is comparable to the cohesive energy, 4.3 eV/atom for pure Fe [35]. The values of E_{coh} vary depending on the composition as well as the kind of compounds. For example, the M-rich compounds such as carbides and nitrides are located in a higher ΔE_X region where the cohesive energy is large. On the other hand, the X-rich compounds such as oxides, fluorides, and sulfides are located in a lower ΔE_X region where the cohesive energy is small. The borides existing in both the M-rich and X-rich compositional ranges are distributed over a boundary region between above two groups of compounds. These differences among the type of compounds arise from the definition of ΔE_M, since it is weighted by a factor, m/n, of M_mX_n in Fig. 10-18. The value of m/n is large for the M-rich compound and small for the X-rich compound.

Further explanation is presented of several compounds such as carbides, nitrides, and fluorides. Also, the oxides are adopted for comparison.

10.9.3 Iron-Based Carbides and Nitrides

10.9.3.1 Atomization Energy Diagram

For example, the atomization energy diagram for binary Fe-based carbides and nitrides is shown in Fig. 10-19 together with the result of binary Fe oxides [2]. They have larger cohesive energy than binary Fe oxides, owing to the Fe-rich composition in them as explained previously.

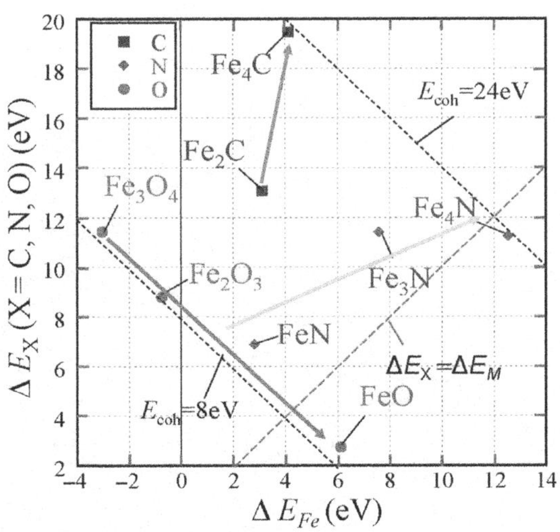

FIGURE 10-19 Atomization energy diagram for iron-based carbides, nitrides, and oxides.

In the figure, a vector is drawn in each compound system to show how the atomization energy changes with the M composition in the compound. As indicated by the vector in Fig. 10-19, both $\Delta E_{\rm Fe}$ and $\Delta E_{\rm C}$ (or $\Delta E_{\rm N}$) increase with increasing Fe composition in the Fe carbides (or nitrides). However, in the Fe oxides $\Delta E_{\rm Fe}$ increases and $\Delta E_{\rm O}$ decreases with increasing Fe composition. Thus, the increase of $\Delta E_{\rm Fe}$ is compensated for the decrease of $\Delta E_{\rm O}$. This implies that the Fe–O interaction is important in the Fe oxides [2]. However, the Fe carbides and nitrides are very different from the Fe oxides in the atomization energy change with the compositions. This is probably attributable to the difference in the chemical interactions arising from the compositional differences between them. So, the further analysis is continued to account for the increase in both $\Delta E_{\rm Fe}$ and $\Delta E_{\rm N}$ (or $\Delta E_{\rm C}$) with the Fe composition in the nitrides (or carbides).

10.9.3.2 Atomization Energy Diagram Normalized by Composition

Another atomization energy diagram is made by normalizing the composition, and the result is shown in Fig. 10-20A for binary Fe nitrides, Fe_mN. On this diagram, the atomization energies, $\Delta E_{\rm N}$ and $\Delta E_{\rm Fe}$, are normalized by the total number of atoms, p, in the chemical formula, Fe_mN, namely, $p = m + 1$, (e.g., $p = 5$ for Fe_4N), and defined as, $\Delta E'_{\rm N}$ and $\Delta E'_{\rm Fe}$, respectively. For example, in case of Fe_4N, $\Delta E'_N = (1/p)\cdot\Delta E_N = (1/5)\cdot\Delta E_N = (1/5)\cdot(E_{\rm N}^{\rm atom} - E_{\rm N}^{\rm Fe_4N})$ and $\Delta E'_{\rm Fe} = (1/p)\cdot\Delta E_{\rm Fe} = (1/5)\cdot\Delta E_{\rm Fe} = (4/5)\cdot(E_{\rm Fe}^{\rm atom} - E_{\rm Fe}^{\rm Fe_4N})$, following Eqs. (10.12) and (10.13). So, $\Delta E'_{\rm N}$ and $\Delta E'_{\rm Fe}$, are the compositionally averaged energies of N and Fe atoms, respectively. As shown in Fig. 10-20A, the positions of binary Fe nitrides are aligned along a dotted line. The point at the intersection of a dotted line and the horizontal axis at $\Delta E'_{\rm N} = 0$ falls on the cohesive energy, 4.3 eV, for pure Fe [35]. In this way, the Fe atom state is stabilized with increasing Fe composition in Fe_mN owing to the increase in the chemical interaction between Fe atoms.

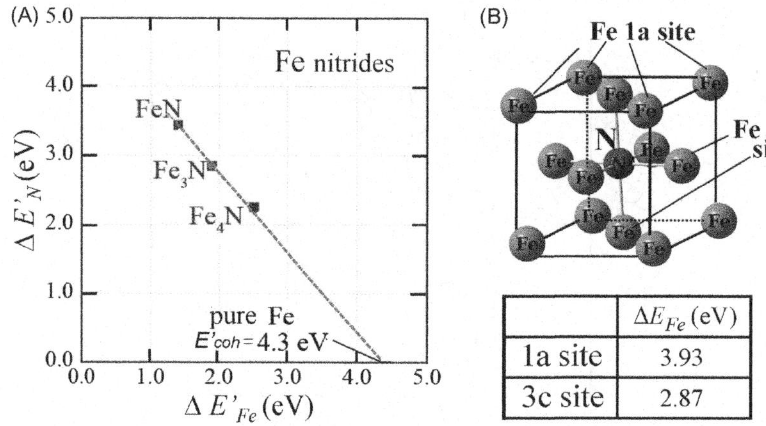

	ΔE_{Fe} (eV)
1a site	3.93
3c site	2.87

FIGURE 10-20 (A) Atomization energy, $\Delta E'_{\rm N}$ and $\Delta E'_{\rm Fe}$, normalized by compositions. For example, in case of Fe_4N, $\Delta E'_{\rm N} = 1/5\cdot\Delta E_{\rm N}$ and $\Delta E'_{\rm Fe} = 4/5\cdot\Delta E_{\rm Fe}$, (B) crystal structure of Fe_4N and the $\Delta E_{\rm Fe}$ change with 1a and 3c sites.

The crystal structure of Fe_4N is shown in Fig. 10-20B [36]. In Fe_4N, there are two Fe sites, namely, 1a site and 3c site. The 1a site is located near the Fe site, but the 3c site is located near the N site. The calculated atomization energy for Fe atom at the 1a site, ($E_{Fe}^{atom} - E_{Fe}^{Fe_4N} = 3.93$ eV), is higher than that for Fe atom at the 3c site, ($E_{Fe}^{atom} - E_{Fe}^{Fe_4N} = 2.87\,eV$). This indicates that the Fe atom state at the 1a site is stabilized by the strong interaction with the neighboring Fe atoms as explained above. The Fe atom state at the 3c site neighboring N atom is, however, destabilized to some extent, but instead the N atom state is stabilized by the energy transfer from the Fe to the N atom. This stabilization of N atoms progresses further as the Fe composition increases. As a result, both ΔE_N and ΔE_{Fe} increase with increasing Fe composition, and the sum of them, namely the cohesive energy, E_{coh}, becomes large as shown in Fig. 10-19.

This is also the case in the carbides. For example, as shown in Fig. 10-19, the cohesive energy is similar between Fe_4C and Fe_4N. But ΔE_C is much larger than ΔE_N, but instead ΔE_{Fe} is smaller in Fe_4C than in Fe_4N. This means that the energy transfer occurs more significantly from Fe to C in Fe_4C than from Fe to N in Fe_4N, indicating that the F—C interaction in Fe_4C is probably stronger than the F—N interaction in Fe_4N, in agreement with the previous calculation [37]. In addition, the slope of the vector shown in Fig. 10-18 is rather steeper in the Fe carbides than in the Fe nitrides. So, the increment in the ΔE_X with the Fe composition is larger in the Fe carbides than in the Fe nitrides. In other words, as the Fe composition increases, the Fe—C interaction tends to be more enhanced than the Fe—N interaction. Instead, judging from the ΔE_{Fe} values, the Fe—Fe interaction is supposed to be larger in Fe_4N than in Fe_4C.

Also, although the result is not shown here for the Ti and Cr carbides (or nitrides), both ΔE_M and ΔE_X increase with the M composition, so that they are very similar to the Fe carbides (or nitrides). As a whole, the atomization energy change with the composition is found to be very different between the M-rich carbides (or nitrides) and the M-deficient oxides (or fluorides).

10.9.4 Change in the Atomization Energy of Oxides with Density

A correlation of the atomization energy with the density is shown in Fig. 10-21 for several binary metal oxides [2]. There is a clear trend that ΔE_M increases and ΔE_O decreases with increasing density, as shown in (A) for Ti oxides, (B) for Cr oxides, and (C) for Fe oxides. Any crystal structure of oxides is made of MO_6 octahedron or MO_4 tetrahedron linked by sharing either vertex or edge or face, and their three-dimensional arrangement determines the density [38]. Such a long-range interaction operating between atoms also reflects the values of the atomization energy.

As shown in Fig. 10-21D, this is the case even in zirconia (ZrO_2), in which there are three polymorphs: monoclinic, tetragonal, and cubic phases. Phase transition occurs from monoclinic to tetragonal phase at about 1300K, and from tetragonal to cubic phase at about 2600K. As shown in Fig. 10-21D, the density increases monotonously with these phase transitions. This clearly indicates that the atomization energies vary significantly with the phase

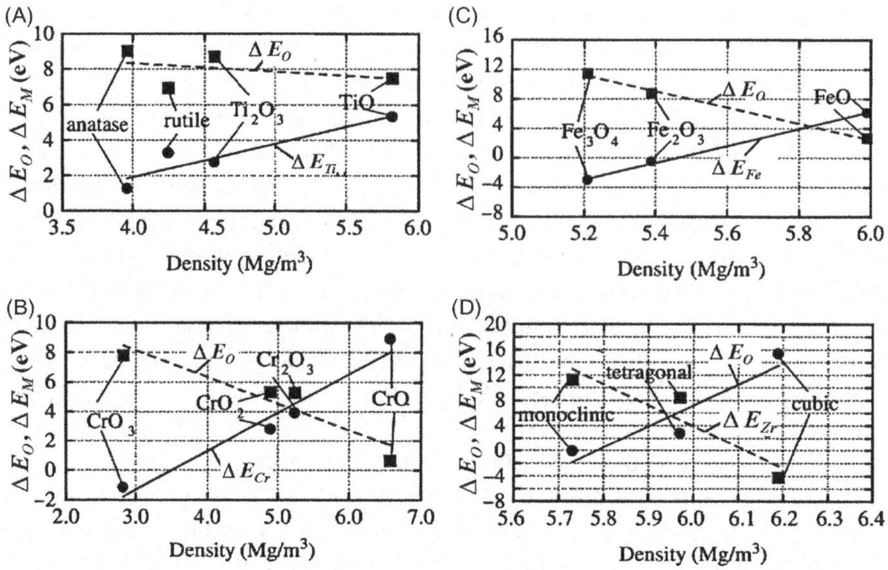

FIGURE 10-21 Correlation between atomization energy and density for (A) Ti oxides, (B) Cr oxides, (C) Fe oxides, and (D) zirconia (ZrO_2).

transition. The phase transition of the perovskite-type oxides is treated with the aid of the atomization energy [2].

10.9.5 Comparison Between Fluorides and Oxides

It is known that fluorine (F) resembles oxygen (O) rather than other halogen atoms such as Cl, Br, and I. In fact, both F and O atoms are similar in the atomic radius (F: 0.064 nm, O: 0.074 nm), 6-coordinated ionic radius (F: 0.133 nm, O: 0.140 nm), and electronegativity (F: 3.98, O: 3.44) [39]. Both fluorides and oxides exist stably in the M-deficient compositional range. Also, transition metal fluorides and oxides have similar crystal structures, for example, TiF_2 and FeF_2 has the rutile (TiO_2) type structure and FeF_3 has the ReO_3 type structure [39]. In addition, F is often substituted for O in the compound, for example, in the Fe-based superconductor, $LaFeAsO_{1-x}F_x$ [33]. So, the fluoride is compared with the oxide in view of the atomization energy.

10.9.5.1 Atomization Energy Diagram

The atomization energy diagram for the fluorides is shown in Fig. 10-22A. For comparison, the results of Cr oxides are also shown in Fig. 10-22B. The cohesive energy, E_{coh}, of the fluorides lies in the range of 5–8 eV, which is lower than the E_{coh} value of 6–11 eV for Cr oxides [2], so the chemical bond strength is weaker in the fluorides than in the oxides.

In each figure, a vector is drawn to show how the atomization energy changes with the M composition in each metal fluoride or Cr oxide. For example, in case of the Fe fluorides,

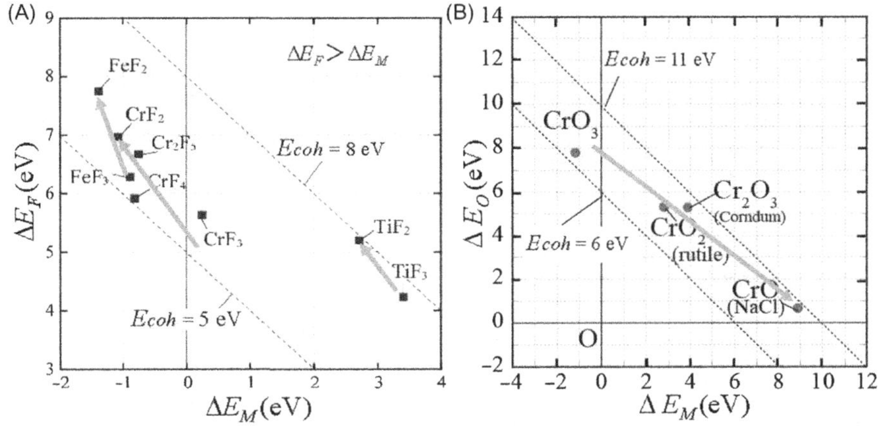

FIGURE 10-22 Atomization energy diagrams for (A) fluorides and (B) Cr oxides.

FeF_2 and FeF_3, ΔE_F increases and ΔE_M decreases with increasing M (=Fe) composition following a vector direction. This is also the case in the Cr and Ti fluorides. Thus, there is a clear trend that ΔE_F increases and ΔE_M decreases with increasing M composition in any fluorides. Such adjustment of the atomization energy between M and F atoms implies the importance of the F-M chemical bond in the fluoride formation. By contrast, as shown in Fig. 10-22B, ΔE_O decreases and ΔE_M increases with increasing M (=Cr) composition in the oxides. Thus, the oxides are completely reverse of the fluorides in respect to the atomization energy change with the M composition.

10.9.5.2 Correlation of Atomization Energy with Interatomic Distance and Density

Both the local structure and the average (crystal) structure reflect the atomization energies of compounds [2]. As shown in Fig. 10-23, the interatomic distance is chosen as a parameter to represent the local structure, since it is associated with the short-range interaction between atoms. On the other hand, the density is a parameter to represent the average crystal structure. It varies depending on the long-range atomic arrangement in crystal space. In this meaning, the density is associated with the long-range interaction between atoms. Fig. 10-23A and B shows the results of Cr fluorides, and (A') and (B') show the results of Cr oxides.

As is evident from Fig. 10-23A, ΔE_F increases and ΔE_{Cr} decreases with the Cr−F interatomic distance in the fluorides. On the other hand, as shown in Fig. 10-23A', ΔE_O decreases and ΔE_{Cr} increases with the Cr−O interatomic distance in the oxides. In this regard, the fluorides are reverse of the oxides. Also, as shown in Fig. 10-23B, ΔE_F increases and ΔE_{Cr} decreases with the density of the fluorides. But as shown in Fig. 10-23B', ΔE_O decreases and ΔE_{Cr} increases with the density of the oxides. Thus, the fluoride changes in a completely reverse way as does the oxide.

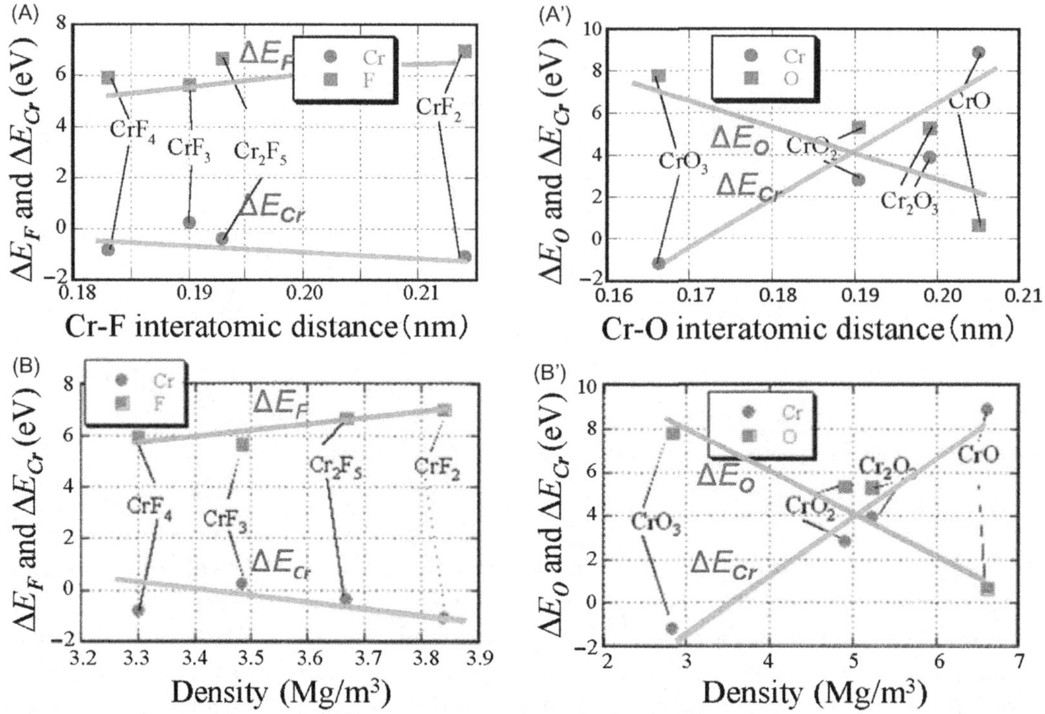

FIGURE 10-23 Correlation of atomization energy with (A) the Cr–F interatomic distance and (B) density of Cr fluorides, and with (A′) the Cr–O interatomic distance and (B′) density of Cr oxides.

10.9.5.3 Unique Atomization Energy Change in Fluorides

The reason for this difference is attributable to the stability difference between the F sublattice in fluorides and the O sublattice in oxides. As shown in Fig. 10-24, the atomization energies change with (A) the F–F interatomic distance in the Fe fluorides and (B) the O–O interatomic distance in the Fe oxides. There is a resemblance in the both plots (A) and (B). However, the F–F interatomic distance increases with increasing F composition in the fluorides. This means that a sort of the F–F repulsive interaction is operating in the F sublattice. This is interpreted as due to the repulsive Coulomb interaction operating between the negatively charged F ions, since F is the most electronegative element among all the elements. In fact, it is known that such a F sublattice is no longer formed in the multicharged metal fluorides containing higher F contents, but instead monomer or molecular compounds with the low melting temperatures (T_m) are predominantly formed [e.g., MoF_6 ($T_m = 290K$) and WF_6 ($T_m = 275K$)] [39]. On the other hand, in the case of the oxides, the O-O interatomic distance decreases with increasing O composition as shown in Fig. 10-24B, and the attendant increase in ΔE_O is observed. This is because the O–O bond in the O sublattice is partially covalent in character and a sort of the attractive O–O interaction is operating to stabilize the O sublattice in the oxides.

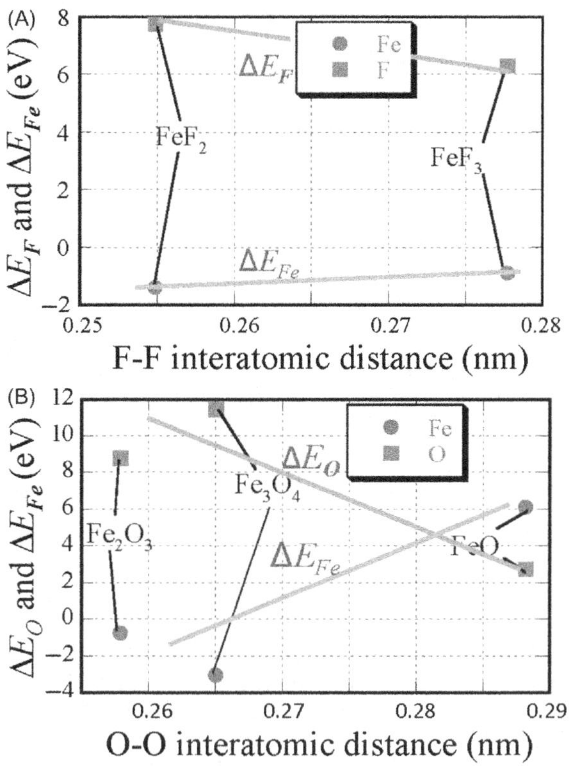

FIGURE 10-24 Correlation of atomization energy with (A) the F–F interatomic distance in Fe fluorides and (B) the O–O interatomic distance in Fe oxides.

Further analysis is carried out in the Ti chlorides, since Cl is the same halogen element as F. However, as shown in Fig. 10-25A and B, the result of chlorides is similar to that of oxides, but dissimilar to that of fluorides. Thus, the fluoride is very unique since it is contrary to the oxide or chloride in respect of the atomization energy change with the interatomic distance and the density.

10.9.5.4 Atomization Energy Change due to Charge Transfer From M to F Atom

As shown in Fig. 10-22A, ΔE_F is larger than ΔE_M in any fluorides regardless of the composition. This means that the energy transfers significantly from M to F atoms to strengthen the M–F interaction in the fluorides. In other words, charge transfer probably occurs in the direction from M to F following their electronegativity difference, since F is the most electronegative element among all the elements as described before. The resultant change in the electron density distributions causes ΔE_F to increase and ΔE_M to decrease. This is because the electron density increases near F as a result of electron transfer from M to F, and hence ΔE_F increases owing to the increase of the attractive Coulomb energy between the F nucleus

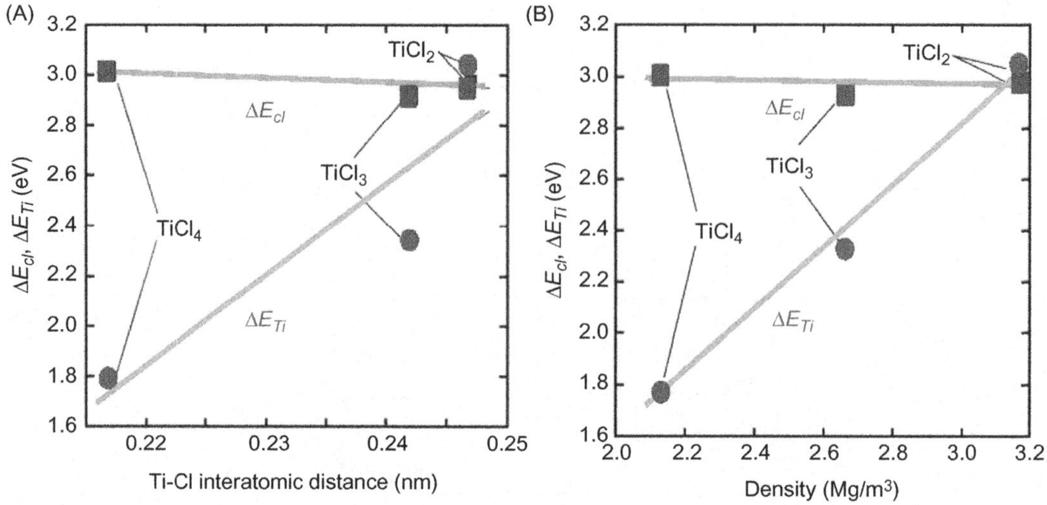

FIGURE 10-25 Correlation of atomization energy with (A) the Ti–Cl interatomic distance and (B) the density for Ti chlorides.

and the electrons. By contrast, the decrease in ΔE_M is due to the decrease of the attractive Coulomb energy between the M nucleus and the electrons owing to the deficient electron density around M.

10.10 Atomization Energy Approach to 3*d* Transition Metal Compounds

In Table 10-3, all the calculated results are summarized focusing on how ΔE_X and ΔE_M change with increasing M composition, interatomic distance, and density of the compounds. As previously explained, the interatomic distance and the density are employed as the parameter to show the short-range atomic interaction and the long-range atomic interaction, respectively. In the table, either up arrow (↑) or down arrow (↓) is put near ΔE_X and ΔE_M. An up arrow means the increase in ΔE_X (or ΔE_M) and a down arrow means the decrease in ΔE_X (or ΔE_M).

For example, for the borides (X = B), ΔE_X and ΔE_M increase with the M composition and the density. Also, ΔE_X and ΔE_M increase with increasing B–B interatomic distance, but decrease with increasing M–M interatomic distance. However, their correlation with the M–B interatomic distance is poor. In other words, B behaves nearly independent of M, and forms preferably the B–B pair or the B-layer in the structure, depending on the B composition. For example, the structure of CrB_2 is characteristic of the alternate stacking of Cr-layer and B-layer. The B–B interaction is strong in the B-rich borides like CrB_2, but the M–M interaction is strong in the M-rich borides.

For the carbides (X = C) and nitrides (X = N) formed in the M-rich compositional range, both ΔE_X and ΔE_M increase with the M composition. The M–M and M–C chemical bonds

Table 10-3 Trend of Atomization Energy Changes with the M Composition, Interatomic Distances, and Density of 3*d* Transition Metal Compounds

Type of Compounds	Composition (Changes in ΔE_X and ΔE_M With Increasing M Composition)	Interatomic Distances (Changes in ΔE_X and ΔE_M With Increasing Interatomic Distances)	Density (Changes in ΔE_X and ΔE_M With Increasing Density)
Borides	$\Delta E_X\uparrow$, $\Delta E_M\uparrow$	M–B: no correlation M–M: $\Delta E_X\downarrow$, $\Delta E_M\downarrow$ B–B: $\Delta E_X\uparrow$, $\Delta E_M\uparrow$	$\Delta E_X\uparrow$, $\Delta E_M\uparrow$
Carbides	$\Delta E_X\uparrow$, $\Delta E_M\uparrow$	M–C: $\Delta E_X\downarrow$, $\Delta E_M\downarrow$ M–M: $\Delta E_X\downarrow$, $\Delta E_M\downarrow$ C–C: $\Delta E_X\uparrow$, $\Delta E_M\uparrow$	$\Delta E_X\uparrow$, $\Delta E_M\uparrow$
Nitrides	$\Delta E_X\uparrow$, $\Delta E_M\uparrow$	M–N: no correlation M–M: $\Delta E_X\downarrow$, $\Delta E_M\downarrow$ N–N: $\Delta E_X\uparrow$, $\Delta E_M\uparrow$	No correlation
Sulfides	Ti: $\Delta E_X\uparrow$, $\Delta E_M\uparrow$ Cr: $\Delta E_X\downarrow$, $\Delta E_M\downarrow$	M–S: Ti : no correlation Cr: $\Delta E_X\downarrow$, $\Delta E_M\downarrow$	No correlation
Oxides	$\Delta E_X\downarrow$, $\Delta E_M\uparrow$	M–O: $\Delta E_X\downarrow$, $\Delta E_M\uparrow$ O–O: $\Delta E_X\downarrow$, $\Delta E_M\uparrow$	$\Delta E_X\downarrow$, $\Delta E_M\uparrow$
Fluorides	$\Delta E_X\uparrow$, $\Delta E_M\downarrow$	M–F: $\Delta E_X\uparrow$, $\Delta E_M\downarrow$ F–F: $\Delta E_X\downarrow$, $\Delta E_M\uparrow$	$\Delta E_X\uparrow$, $\Delta E_M\downarrow$

Up arrow, ↑, increase in $\Delta E_{X(\text{or M})}$; down arrow, ↓, decrease in $\Delta E_{X(\text{or M})}$.

are important in the carbides. There is a correlation of ΔE_X and ΔE_M with the density in the carbides, but a poor correlation in the nitrides.

On the other hand, for the oxides (X = O) formed in the O-rich compositional range, ΔE_X decreases and ΔE_M increases with the M composition and the density. Such a change in ΔE_X and ΔE_M is also seen in the M–O and O–O interatomic distances. For the fluorides (X = F) formed in the F-rich compositional range, ΔE_X increases and ΔE_M decreases with the M composition and the density. Such a change in ΔE_X and ΔE_M is also seen with the M–F interatomic distances. However, a reverse change in ΔE_X and ΔE_M is seen with the F–F interatomic distances. For the oxides and fluorides, the M–X chemical bond is predominant, and the energy is adjusted between M and X atoms as the composition changes. However, as to the compositional dependence of ΔE_X and ΔE_M, the oxide is completely reverse of the fluoride. As shown in Fig. 10-24, this is interpreted as due to the difference in the chemical stability between the O sublattice and the F sublattice.

For the sulfides (X = S) formed in the S-rich compositional range, any clear trends are not observed in the present analysis. But it is likely that ΔE_X decreases and ΔE_M increases as the M (=Cr) composition increases in the Cr sulfides, similar to the oxides. Also, both ΔE_X and ΔE_M tend to increase with the M composition in the Ti sulfides, similar to the borides. Thus, in view of the atomization energy, the sulfides seem to be the intermediate compounds between oxides and borides. As to the electronegativity value, S (2.58) is in between O (3.44) and B (2.04).

Thus, a whole feature of 3*d* transition metal compounds comes out through the atomization energy analysis. It is greatly expected to design new compounds with specific functions by taking into account a whole character of nonmetal elements and metal elements in every compound.

Finally, it is summarized that the atomization energy concept is useful not only for the fundamental understanding of materials, but also for materials design [40]. This atomization energy approach is applicable to any materials including alloys, ceramics, and semiconductors. This approach has been developed recently and it is still in progress.

References

[1] Y. Shinzato, H. Yukawa, M. Morinaga, T. Baba, H. Nakai, A unified approach to the analysis of the chemical bond in hydrides and hhydrocarbons, Acta Mater. 55 (2007) 6673–6680.

[2] Y. Shinzato, Y. Saito, M. Yoshino, H. Yukawa, M. Morinaga, T. Baba, et al., Energy expression of the chemical bond between atoms in metal oxides, J. Phys. Chem. Solids 72 (2011) 853–861.

[3] H. Nakai, Energy density analysis with Kohn–Sham orbitals, Chem. Phys. Lett. 363 (2002) 73–79.

[4] H. Nakai, Y. Kurabayashi, M. Katouda, T. Atsumi, Extension of energy density analysis to periodic boundary condition calculation: evaluation of locality in extended systems, Chem. Phys. Lett. 438 (2007) 132–138.

[5] Y. Shinzato, H. Yukawa, M. Morinaga, T. Baba, H. Nakai, New expression of the chemical bond in hydrides using atomization energies, Adv. Quant. Chem. 54 (2008) 145–160.

[6] M. Morinaga, H. Yukawa, H. Nakai, Energy expression of the chemical bond between atoms in hydrides and oxides and its application to materials design, in: T. Ishii, et al. (Eds.), The DV-Xα Molecular Orbital Calculation Method, Springer International Publishing Switzerland, 2015, pp. 183–213.

[7] H. Hirate, Y. Saito, I. Nakaya, H. Sawai, Y. Shinzato, H. Yukawa, et al., Quantitative approach to the understanding of catalytic effect of metal oxides on the desorption reaction of MgH_2, Int. J. Quant. Chem. 109 (2009) 2793–2800.

[8] H. Hirate, H. Sawai, H. Yukawa, M. Morinaga, Role of O–H bonding in catalytic activity of Nb_2O_5 during the course of dehydrogenation of MgH_2, Int. J. Quant. Chem. 111 (2011) 2251–2257.

[9] Y. Fukai, The Metal-Hydrogen System: Basic Bulk Properties, Springer-Verlag, New York, 1993.

[10] B. Baranowski, S.M. Filipek, 45 Years of nickel hydride-history and perspectives, J. Alloys Compd. 404-406 (2005) 2–6.

[11] H. Yukawa, Y. Takahashi, M. Morinaga, Alloying effects on the electronic structures of $LaNi_5$ containing hydrogen atoms, Intermetallics 4 (1996) S215–S224.

[12] Y. Takahashi, H. Yukawa, M. Morinaga, Alloying effects on the electronic structure of Mg_2Ni intermetallic hydride, J. Alloys Compd. 242 (1996) 98–107.

[13] H. Yukawa, M. Morinaga, The nature of the chemical bond in hydrogen storage compounds, Adv. Quant. Chem. 29 (1997) 83–108.

[14] T. Muramatsu, H. Yukawa, M. Morinaga, Alloying effects on the electronic structure of $ZrMn_2$ intermetallic hydride, J. Alloys Compd. 279 (1998) 192–200.

[15] H. Yukawa, Y. Takahashi, M. Morinaga, Electronic structures of hydrogen storage compound, TiFe, Comput. Mater. Sci. 14 (1999) 291–294.

[16] K. Nakatsuka, M. Yoshino, H. Yukawa, M. Morinaga, Roles of the hydride forming and non-forming elements in hydrogen storage alloys, J. Alloys Compd. 293 (1999) 222–226.

[17] H. Yukawa, M. Morinaga, Characteristics of electronic structures in hydrogen storage alloys, Adv. Quant. Chem. 37 (2000) 193–212.

[18] R.S. Mulliken, Electronic population analysis on LCAO–MO molecular wave functions. I, II, III, IV, J. Chem. Phys. 23 (1955) 1833–1840, 1841–1846, 2338–2342, 2343–2346.

[19] H. Yukawa, M. Takagi, A. Teshima, M. Morinaga, Alloying effects on the stability of vanadium hydrides, J. Alloys Compd. 330-332 (2002) 105–109.

[20] H. Yukawa, K. Nakatsuka, M. Morinaga, Design of hydrogen storage in view of chemical bond between atoms, Solar Energy Mater. Solar Cells 62 (2000) 75–80.

[21] W. Kohn, L.J. Sham, Self-consistent equations including exchange and correlation effects, Phys. Rev. 140 (1965) A1133–A1138.

[22] M.J. Frisch, G.W. Trucks, H.B. Schlegel, G.E. Scuseria, M.A. Robb, J.R. Cheeseman, et al., *Gaussian 03, Revision C.* 02, Gaussian. Inc., Wallingford CT, 2004.

[23] G.C. Pimentel, R.D. Spratley, Chemical Bonding Clarified Through Quantum Chemistry, Holden-day, Inc, San Francisco, 1969, pp. 193–314.

[24] M. Yoshino, K. Komiya, Y. Takahashi, Y. Shinzato, H. Yukawa, M. Morinaga, Nature of the chemical bond in complex hydrides, $NaAlH_4$, $LiAlH_4$, $LiBH_4$ and $LiNH_2$, J. Alloys Compd. 404-406 (2005) 185–190.

[25] K. Komiya, N. Morisaku, Y. Shinzato, S. Orimo, Y. Ohki, K. Tatsumi, et al., Synthesis and dehydrogenation of $M(AlH_4)_2$ (M = Mg, Ca), J. Alloys Compd. 446-447 (2007) 237–241.

[26] K. Komiya, N. Morisaku, R. Rong, Y. Takahashi, Y. Shinzato, H. Yukawa, et al., Synthesis and decomposition of perovskite-type hydrides, $MMgH_3$ (M = Na, K, Rb), J. Alloys Compd. 453 (2008) 157–160.

[27] H. Yukawa, N. Morisaku, Y. Liu, K. Komiya, R. Rong, Y. Shinzato, et al., Raman scattering and lattice stability of $NaAlH_4$ and Na_3AlH_6, J. Alloys Compd. 446-447 (2007) 242–247.

[28] S. Orimo, Y. Nakamori, J.R. Eliseo, A. Züttel, C.M. Jensen, Complex hydrides for hydrogen storage, Chem. Rev. 107 (2007) 4111–4132.

[29] Y. Nakamori, K. Miwa, A. Ninomiya, H. Li, N. Ohba, S. Towata, et al., Correlation between thermodynamical stabilities of metal borohydrides and cation electronegativites: first-principles calculations and experiments, Phys. Rev. B 74 (2006) 045126.

[30] G. Barkhordarian, T. Klassen, R. Bormann, Catalytic mechanism of transition-metal compounds on Mg hydrogen sorption reaction, J. Phys. Chem. B 110 (2006) 11020–11024.

[31] G. Barkhordarian, T. Klassen, R. Bormann, Fast hydrogen sorption kinetics of nanocrystalline Mg using Nb_2O_5 as catalyst, Scr. Meter. 49 (2003) 213–217.

[32] M. Sagawa, S. Fujimura, N. Togawa, H. Yamamoto, Y. Matsuura, New material for permanent magnets on a base of Nd and Fe, J. Appl. Phys. 55 (1984) 2083–2087.

[33] Y. Kamihara, T. Watanabe, M. Hirano, H. Hosono, Iron-based layered superconductor $La[O_{1-x}F_x]FeAs$ ($x = 0.05–0.12$) with Tc = 26 K, J. Am. Chem. Soc. 130 (2008) 3296–3297.

[34] H. Okamoto, Phase Diagrams for Binary Alloys, second ed., ASM International, Materials Park, OH, 2010.

[35] C. Kittel, Introduction to Solid State Physics, Fifth ed., John Wiley & Sons, Inc, New York, 1976.

[36] S. Matar, P. Mohn, G. Demazeau, B. Siberchicot, The calculated electronic and magnetic structures of Fe_4N and Mn_4N, J. Phys. 49 (1988) 1761–1768.

[37] M. Morinaga, N. Yukawa, H. Adachi, T. Mura, Electronic state of interstitial atoms (C, N, O) in FCC Fe, J. Phys. F: Met. Phys. 18 (1988) 923–934.

[38] W.D. Kingery, H.K. Bowen, D.R. Uhlman, Introduction to Ceramics, John Wiley & Sons, New York, 1976.

[39] The JSPS florine chemistry, No. 155 subcommittee (Ed.), Introduction to Florine Chemistry 2010, Sankyo Publishing Co. Ltd., Tokyo, 2010, pp. 1–40.

[40] M. Morinaga, Alloy design based on molecular orbital method, Mater. Trans. 57 (2016) 213–226.

11

Local Lattice Strains Around Alloying Elements in Metals

11.1 Local Lattice Strain Around Alloying Element in Magnesium

11.1.1 Alloying Elements and Local Lattice Strains in Mg

As explained in Chapter 6, Aluminum Alloys and Magnesium Alloys, magnesium is a nearly free electron metal with light weight. A variety of metals including simple metals, transition metals, and rare earth metals are alloyed into Mg metal to improve mechanical properties. For example, heat-resistant alloys [e.g., Mg–Zn–Nd–Zr alloy (EZ33A)] contain simple metal, rare earth metal, and transition metal as well [1–3]. High strength alloys (e.g., $Mg_{97}Zn_1Y_2$ alloy [4]) with the long-period stacking ordered (LPSO) structure contain simple metal (e.g., Zn) and rare earth metal (e.g., Y, Sm–Tm). In some cases, instead of simple metal (e.g., Zn), 3*d* transition metals such as Co, Ni, and Cu are alloyed into Mg to make the LPSO structure. Recently great attention has been directed toward the Mg alloy with this LPSO structure, since it is strengthened by kinks instead of ordinary dislocations [5]. It is said that the disclinations are connected to the kink formation in such Mg alloys [6,7].

As explained in Chapter 6, Aluminum Alloys and Magnesium Alloys, a new alloying parameter of the *s* orbital energy level, *Mk*, is determined from the molecular orbital calculations, and used for predicting the mechanical properties of Mg alloys [8]. In any solid solution alloy, local lattice strain is introduced into the crystal lattice due to the size mismatch between solute and solvent atoms. This causes the solid solution strengthening, for example, by the Cottrell pinning mechanism [9–11], as discussed in Chapter 6, Aluminum Alloys and Magnesium Alloys. Thus, information of the local lattice strain is important for alloy design, although full details still remain unknown.

According to previous investigations, such a local strain problem has been treated by the phenomenological approach [12], elastic approach using the Eshelby theory [13], and the first principle calculation as well [14]. However, no systematic calculations have been performed of the local lattice strain around solute atom in any metal despite the great importance in understanding alloy properties.

A series of calculations is firstly performed of the local lattice strains in hcp Mg metal using the pseudopotential method. Then, the energy density analysis (EDA) [15–18] is carried out to examine how the local lattice strain is introduced into the crystal lattice in a nearly free electron metal like Mg.

A Quantum Approach to Alloy Design. DOI: https://doi.org/10.1016/B978-0-12-814706-1.00011-X

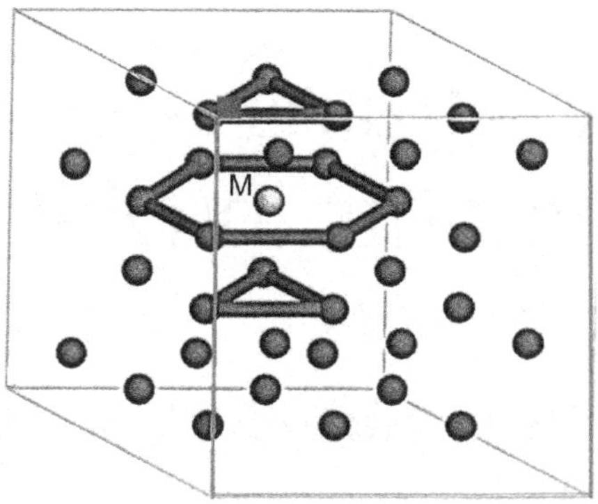

FIGURE 11-1 Supercell used for the calculation of hcp Mg with alloying element, M. The positions of the 12 first nearest neighbor Mg atoms from M atom are optimized, while keeping the positions of the other Mg atoms fixed.

11.1.2 Calculation Procedure

11.1.2.1 Calculation of Local Lattice Strain

As shown in Fig. 11-1, a supercell is made by the 3 × 3 × 2 stacking of the hcp Mg unit cell along the *a*- and *c*-crystal axes. It consists of 36 Mg atoms, and an alloying element, M, is substituted for one Mg atom in the supercell. The crystal structure of this supercell is optimized by the total energy minimization under the periodic boundary condition (PBC) by using the plane-wave pseudopotential method [19]. For this purpose, the first-principle calculations based the density functional theory (DFT) [20], are performed with a generalized gradient approximation (GGA) by Perdew et al. [21,22]. The implementation of DFT employed here combines a plane-wave basis set with the total energy pseudopotential method, as is embodied in the CASTEP code [19]. The present calculations used are based upon the ultrasoft pseudopotentials proposed by Vanderbilt [23]. The plane-wave cutoff energy is chosen to be 380 eV for the simple metals (e.g., Li Ca, Sn, etc.) and for the transition metals (e.g.,Ti, Mn, Fe, etc.), but 400–480 eV for the rare earth metals. For example, it is 400 eV for M = Gd, 440 eV for M = Nd, and 480 eV for M = Eu. In any case, fine convergent conditions are used; they are less than 5×10^{-6} eV/atom for the total energy, 0.1 eV/nm for the maximum force, and 5×10^{-3} nm for the maximum displacement. The sampling in the reciprocal space is done with the *k*-points grids proposed by Monkhorst and Pack [24].

In the calculation, the lattice parameter is first optimized for pure Mg. Then, by fixing the lattice parameter at constant, the positions of only 12 Mg atoms locating at the first nearest neighbor sites from M are relaxed and optimized by calculating the total energy, and the local strains of the 12 Mg atoms neighboring an isolated M atom are determined. These Mg atoms are illustrated in Fig. 11-1 by connecting bold lines between them.

11.1.2.2 Total Energy Calculation and Energy Density Analysis

The electronic structures of the optimized distorted crystal lattice are obtained by the DFT calculations [20] under the PBC by using the Gaussian03 program package [25], and the total energy is calculated. The EDA calculations [15,16] under the PBC are then performed by linking the original code for the EDA with Gaussian03 as explained in Chapter 10, Atomization Energy Approach to Alloys and Metal Compounds.

11.1.2.3 Atomization Energy

The respective atomic energy densities of M and Mg are obtained following the EDA. When the energy of the isolated neutral atom, $E_{\mathrm{M}}^{\mathrm{atom}}$ (or $E_{\mathrm{Mg}}^{\mathrm{atom}}$), is taken as a reference, the atomization energy, ΔE_{M} (or ΔE_{Mg}) is defined as,

$$\Delta E_{\mathrm{M}} = E_{\mathrm{M}}^{\mathrm{atom}} - E_{\mathrm{M}}^{\mathrm{MMg}}, \tag{11.1}$$

$$\Delta E_{\mathrm{Mg}} = E_{\mathrm{Mg}}^{\mathrm{atom}} - E_{\mathrm{Mg}}^{\mathrm{MMg}}. \tag{11.2}$$

Here, $E_{\mathrm{M}}^{\mathrm{MMg}}$ and $E_{\mathrm{Mg}}^{\mathrm{MMg}}$ are the atomic energy densities for M and Mg in alloyed Mg with M (denoted as MMg), respectively. The correlation of the atomization energies, ΔE_{M} (or ΔE_{Mg}) with the local strains around M is examined to understand how the local strains are introduced into the Mg lattice in a fundamental manner.

As explained in Chapter 10, Atomization Energy Approach to Alloys and Metal Compounds, the Mulliken population analysis [26] treats the electron density distribution over the crystal space in the alloy, whereas the EDA [15,16] treats the energy density distribution in the alloy. However, the information of the electron density is still involved in the atomization energy of each constituent atom in the alloy, as explained in Chapter 10, Atomization Energy Approach to Alloys and Metal Compounds.

11.1.3 Local Lattice Strain Around Alloying Element

First, the lattice parameter of pure Mg is optimized to be $a = 0.32137$ nm, $c = 0.51628$ nm, and $c/a = 1.607$. Then, local lattice strains are obtained by optimizing the positions of 12 Mg atoms at the first nearest neighbors from M at a center (see Fig. 11-1).

For example, the results are shown in Fig. 11-2A for M = Fe and (B) for M = Ca, where only the 12 Mg atoms around M are illustrated. The calculated local strains along the a- and c-axes of hcp Mg lattice are denoted as Δa_1 and Δc_1 for 6 Mg atoms sitting on a central hexagon, and as Δa_2 and Δc_2 for 3 Mg atoms sitting on the upper or lower triangle. The strain, Δa_1 (nm), along the a-axis for 6 Mg atoms on the hexagon is negative for Fe, but positive for Ca. The strain Δc_1 along the c-axis is null, $\Delta c_1 = 0$, from the symmetry around M. On the other hand, as indicated by the vectors along the c-axis in the figure, the Mg atoms on the upper triangle are displaced in the opposite direction of those on the lower triangle. Both the strains, Δa_2 (nm) and Δc_2 (nm), are negative for Fe, but positive for Ca. As a result of these strains, the whole Mg lattice contracts locally around Fe, but expands around Ca as shown in Fig. 11-2A–B. For any alloying element, M, the local strain mode is divided into either the Fe-type or the Ca-type without exception.

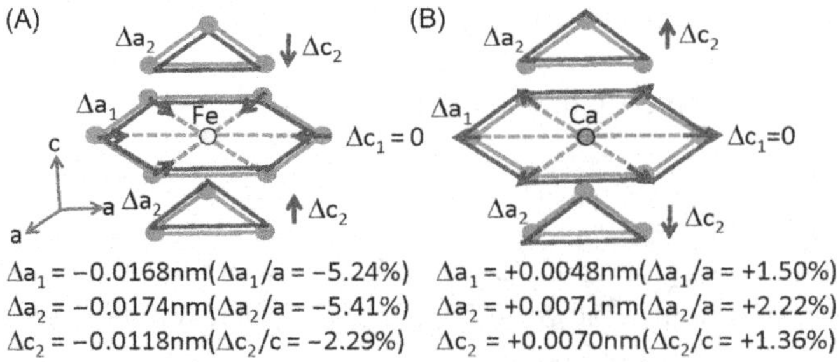

FIGURE 11-2 Calculated local lattice strains around alloying element, (A) Fe and (B) Ca, where *a* and *c* are lattice parameters.

All the calculated results are listed in Table 11-1 for a variety of alloying elements, M. The values of $\Delta a_1/a$, $\Delta a_2/a$, and $\Delta c_2/c$ are given in % units using the optimized lattice parameter, a and c. In addition, the M–Mg distances are given using the notation of d_1 or d_2. Here, d_1 is the distance between M and Mg atoms sitting on the six corners of the hexagon, and d_2 is the distance between M and Mg atoms sitting on the three corners of the upper or lower triangle. For pure Mg, their values are d_1=0.3214 nm and d_2=0.3179 nm. As listed in Table 11-1, the d_1 and d_2 distances change with the local strains around M, which will be discussed later.

For simplicity, the average strains are defined as, $\Delta a/a = [\Delta a_1/a + \Delta a_2/a]/2$ (%) and $\Delta c/c = 2\Delta c_2/c$ (%), where $\Delta c/c$ is set twice as large as $\Delta c_2/c$ by taking into account the effect of the opposite strain direction of the upper and lower triangles on the lattice parameter, *c* (see Fig. 11-2). The sum of these local lattice strains, $\Delta a/a + \Delta c/c$, are shown in Fig. 11-3 for various elements, M. The M's on the horizontal axis are arranged in regular order of the atomic number of M. It is apparent from this figure that the local lattice strains are negative for 3*d* transition elements (e.g., Mn, Fe), but positive for rare earth elements (e.g., Nd, Th). For example, the values of local lattice strains are relatively large for 3*d* transition elements (e.g., about −10% for Fe), whereas they are less than 5% for rare earth elements except for La. Thus, there is a tendency that the local Mg lattice contracts around 3*d* transition elements, but expands around rare earth elements [1].

11.1.4 Reliability of Calculated Local Strains

To show the reliability of the calculated strains, $\Delta a/a$ (i.e., $\Delta a_1/a$, $\Delta a_1/a$) and $\Delta c/c$, are compared with the measured lattice parameter change between Mg-1mol% M alloys and pure Mg. The results are shown in Fig. 11-4A–B. Here, only in the case of the Mg-1mol% Ca alloy, is the lattice parameter extrapolated from the data of the lower Ca alloys, since the solid solubility is as low as about 0.44 mol% for Ca [27]. As shown in Fig. 11-4, both the calculated values of $\Delta a/a$ in (A) and $\Delta c/c$ in (B) are about one order larger than the experimental values of the lattice parameter changes [28,29]. This is because the calculated strain is the value to be confined to the vicinities of M, whereas the lattice parameter is the value

Table 11-1 Calculated Local Strains of the First Nearest Neighbor Mg Atoms From Alloying Element M, and the M–Mg Bond Distances, d_1 and d_2, in a Distorted Lattice

M	$\Delta a_1/a$ (%)	$\Delta a_2/a$ (%)	$\Delta c_2/c$ (%)	d_1 (nm)	d_2 (nm)	M	$\Delta a_1/a$ (%)	$\Delta a_2/a$ (%)	$\Delta c_2/c$ (%)	d_1 (nm)	d_2 (nm)
Li	−0.53	−1.05	−0.59	0.3197	0.3143	La	2.06	2.62	1.62	0.3280	0.3275
Be	−2.30	−3.44	−1.60	0.3140	0.3074	Ce	1.40	1.79	1.49	0.3259	0.3263
Al	−0.76	−1.88	−1.11	0.3189	0.3112	Pr	1.46	1.61	1.01	0.3261	0.3239
Si	−1.26	−2.80	−1.34	0.3173	0.3092	Nd	1.58	1.67	1.00	0.3264	0.3239
Ca	1.50	2.22	1.36	0.3262	0.3260	Pm	1.38	1.59	1.20	0.3258	0.3244
Sc	−0.15	−0.35	−0.10	0.3209	0.3171	Sm	1.52	1.46	0.97	0.3263	0.3236
Ti	−1.76	−2.52	−1.28	0.3157	0.3098	Eu	1.36	1.42	1.23	0.3257	0.3246
V	−3.09	−3.69	−1.67	0.3114	0.3069	Gd	1.41	2.41	1.37	0.3259	0.3262
Cr	−4.01	−4.58	−1.98	0.3085	0.3047	Tb	0.97	2.61	1.32	0.3245	0.3263
Mn	−4.74	−5.11	−2.17	0.3061	0.3033	Dy	1.05	1.34	1.29	0.3247	0.3247
Fe	−5.24	−5.41	−2.29	0.3045	0.3025	Ho	0.99	1.53	1.23	0.3246	0.3247
Co	−4.90	−5.01	−2.10	0.3056	0.3037	Er	1.62	2.61	1.26	0.3266	0.3260
Ni	−4.11	−4.45	−1.88	0.3082	0.3052	Tm	1.08	1.25	0.66	0.3248	0.3220
Cu	−2.55	3.53	−1.56	0.3132	0.3076	Yb	1.14	1.47	1.02	0.3250	0.3238
Zn	−1.18	−2.35	−1.21	0.3176	0.3103	Lu	0.40	0.48	0.80	0.3227	0.3217
Ga	−0.67	−1.94	−0.31	0.3192	0.3111	Hg	0.51	−1.11	−0.64	0.3198	0.3140
Ge	−0.74	−2.17	−1.13	0.3190	0.3108	Tl	0.46	0.81	0.64	0.3228	0.3215
Sr	2.48	3.38	1.77	0.3293	0.3290	Pb	1.02	1.60	1.02	0.3246	0.3239
Y	0.96	1.43	1.04	0.3245	0.3238	Bi	1.03	0.98	1.06	0.3247	0.3234
Zr	−0.76	−0.87	−0.53	0.3189	0.3147	Ac	1.27	1.49	1.31	0.3255	0.3250
Ag	−1.63	−2.44	−1.24	0.3161	0.3101	Th	1.60	2.08	1.12	0.3265	0.3249
Cd	−0.29	−0.51	−0.36	0.3204	0.3159	Pa	0.06	0.41	0.67	0.3216	0.3212
In	0.21	0.05	0.08	0.3220	0.3183	U	0.58	0.66	1.06	0.3232	0.3231
Sn	0.33	0.14	0.16	0.3224	0.3187	Np	0.62	0.48	1.07	0.3234	0.3234
Sb	0.34	0.01	0.75	0.3225	0.3210	Pu	0.72	0.38	0.56	0.3237	0.3207
Ba	3.18	3.99	2.12	0.3316	0.3311						

to be averaged over the whole crystal lattice. Nevertheless, their variations with M are very similar as shown in Fig. 11-4A–B. Thus, there is good correspondence between the calculation and the experiment.

11.1.5 Axial Ratio *c/a* Change With Alloying Elements

In hcp Mg alloys, the axial ratio, c/a, is important, as this is a measure to show the anisotropy of the hcp crystal lattice. In an ideally isotropic case, the value of c/a is 1.633. For pure Mg, it is about 1.624 according to experiments [1], slightly smaller than the isotropic value. In case of the present optimization of the lattice parameter, it is about 1.607, being about 1% smaller than the experimental value. To improve the mechanical properties of Mg alloys, it is needed for the c/a value to approach 1.633 by alloying.

So, the change of the c/a value with M, $\Delta(c/a)$, is evaluated using a next equation,

$$\Delta(c/a) = [(c+\Delta c)/(a+\Delta a)]_{\mathrm{M}} - (c/a)_{\mathrm{Mg}}. \tag{11.3}$$

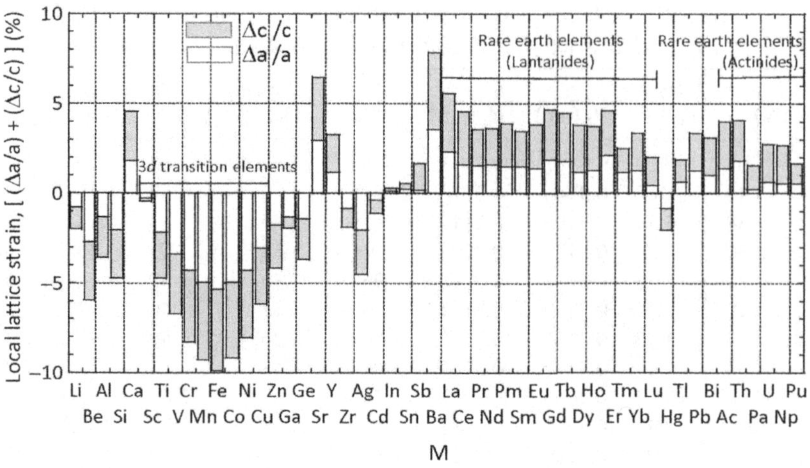

FIGURE 11-3 Local lattice strain, $(\Delta a/a) + (\Delta c/c)$, around a variety of alloying elements, M in hcp Mg.

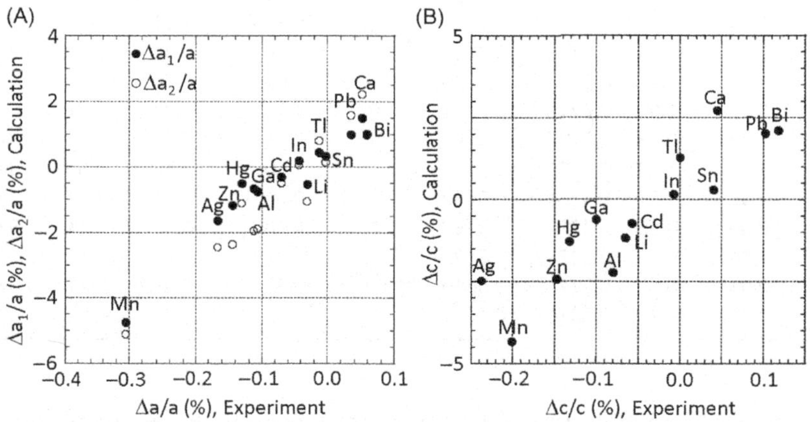

FIGURE 11-4 Comparison of calculated local lattice strains with the measured lattice parameter changes between Mg-1mol%M and pure Mg; (A) $\Delta a/a$ ($\Delta a_1/a$, $\Delta a_2/a$) and (B) $\Delta c/c$.

Here, $[(c + \Delta c)/(a + \Delta a)]_M$ is the c/a value for alloyed Mg with M, and $(c/a)_{Mg}$ is the c/a value for pure Mg. The result is shown in Fig. 11-5. Various alloying elements (e.g., rare earth elements) increase the c/a value. It is interesting that Y and Zn in $Mg_{97}Zn_1Y_2$ with the LPSO structure [4] have the nearly opposite $\Delta(c/a)$ value, as it is about $+0.014$ for Y and about -0.011 for Zn.

11.1.6 Energy Gained by Local Lattice Strain Around Alloying Element

The alloyed Mg with M becomes energetically stable by introducing the local strain into the crystal lattice around M. Such an energy gain is defined as the total energy difference before

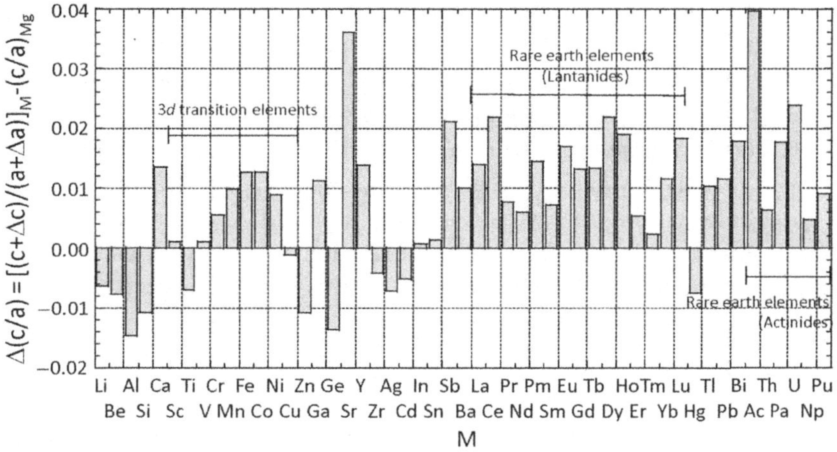

FIGURE 11-5 *c/a* ratio change with M, $\Delta\,(c/a) = [(c + \Delta c)/(a + \Delta a)]_M - (c/a)_{Mg}$.

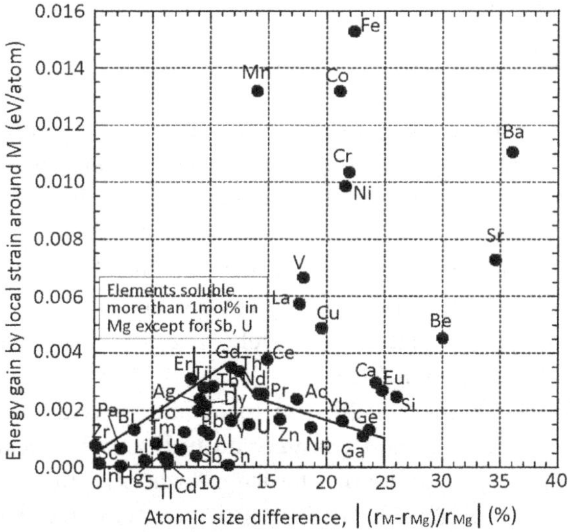

FIGURE 11-6 Energy gain by local strain vs. atomic size difference between M and Mg.

and after lattice relaxation. The calculated energy gains are already shown in Fig. 6-30 in Chapter 6, Aluminium Alloys and Magnesium Alloys. The energy gain varies significantly depending on M. It is large for those elements M that exhibit large negative or positive local strains. As shown in Fig.6-30, rare earth elements show the lower energy gain than 3*d* transition elements. The largest energy gain is about 0.015 eV/atom for M = Fe.

In Fig. 11-6 the energy gain is plotted against $|(r_M - r_{Mg})/r_{Mg}|$, where r_M and r_{Mg} are atomic radii of M and Mg, respectively. Therefore, $|(r_M - r_{Mg})/r_{Mg}|$ is a parameter to show the difference of atomic radii between M and Mg. Those elements that are soluble by more than 1 mol% in the Mg solid solution appear to be located inside a small region surrounded by solid lines in Fig. 11-6. Exception is, however, seen in Sb and U, both of which are scarcely soluble, but their positions are located in a soluble region near the origin.

In Fig. 11-6 there is a large separation in the locations between the transition metal group (e.g., Fe, Co, Cu) and the simple metal group (e.g., Ba, Sr, Ca). Thus, the correlation is not necessarily good between the energy gain and the atomic size difference parameter. So, the chemical bond between M and Mg atoms are further investigated using the atomization energy approach to understand how the local strains are introduced into the Mg lattice around M.

11.2 Atomization Energy Approach to Local Lattice Strain in Magnesium

11.2.1 Change in Atomization Energy Before and After Lattice Relaxation

Using the EDA explained in Chapter 10, Atomization Energy Approach to Alloys and Metal Compounds, the atomization energy is calculated to be $\Delta E_{Mg} = 0.800$ eV for pure Mg. This value corresponds to the cohesive energy for pure Mg, but it is about 0.7 eV lower than the cohesive energy, 1.51 eV [30]. The atomization energies of ΔE_M for M and ΔE_{Mg} for Mg are then calculated in the alloyed Mg with M before and after the lattice relaxation around M in the supercell.

For example, the calculated results are shown in Fig. 11-7A for M = Ca and (B) for M = Fe. In this figure, Mg atoms sitting on the central hexagon and on the upper or lower triangle are denoted as Mg1 and Mg2, respectively. The rest of the Mg atoms located in the region distant from M are denoted as Mg3.

As explained previously in Fig. 11-2B, the local lattice strain is large and positive for Ca, so that the local Mg lattice expands largely around Ca. As shown in the left figure of Fig. 11-7A, in the strain-free state before lattice relaxation, the atomization energies are $\Delta E_{Ca} = -1.879$ eV, $\Delta E_{Mg1} = 1.082$ eV, $\Delta E_{Mg2} = 1.078$ eV. On the other hand, in the strain state after lattice relaxation (see the right figure), they are $\Delta E_{Ca} = -1.698$ eV, $\Delta E_{Mg1} = 1.083$ eV, $\Delta E_{Mg2} = 1.074$ eV. Thus, the lattice relaxation changes the ΔE_{Ca} value to some extent, but scarcely changes the ΔE_{Mg1} and ΔE_{Mg2} values. Also, for the rest of Mg atoms, $\Delta E_{Mg3} = 0.815$ eV before lattice relaxation and $\Delta E_{Mg3} = 0.821$ eV after lattice relaxation, so that their change is small and the ΔE_{Mg3} value is comparable to the atomization energy for pure Mg, $\Delta E_{Mg} = 0.800$ eV.

On the other hand, as shown in Fig. 11-2A, the local lattice strain is large and negative for Fe, so that the local Mg lattice contracts largely around Fe. In the strain-free state before lattice relaxation (see the left figure of Fig. 11-7B), the atomization energies are $\Delta E_{Fe} = +3.146$ eV, $\Delta E_{Mg1} = 1.093$ eV, $\Delta E_{Mg2} = 1.060$ eV. In the strain state after lattice

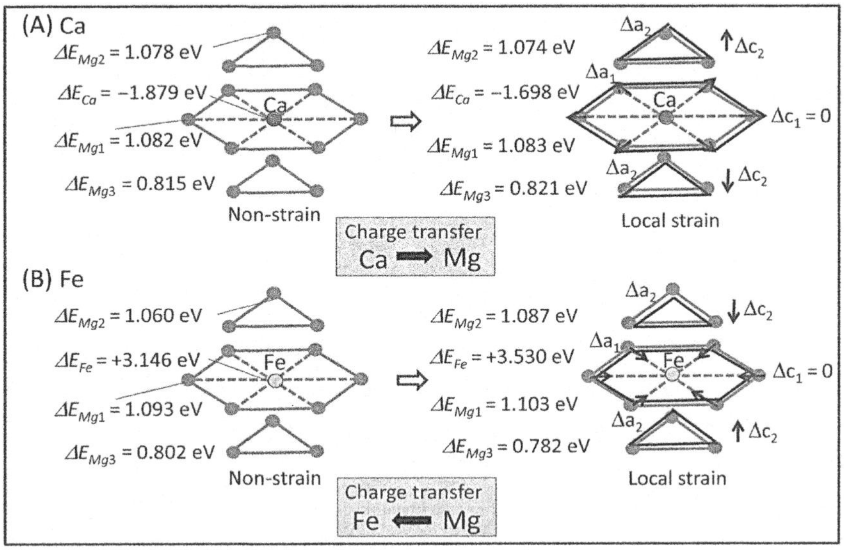

FIGURE 11-7 Atomization energy change before and after lattice relaxation around (A) Ca and (B) Fe in hcp Mg.

relaxation (see the right figure), they are $\Delta E_{\mathrm{Fe}} = +3.530$ eV, $\Delta E_{\mathrm{Mg1}} = 1.103$ eV, $\Delta E_{\mathrm{Mg2}} = 1.087$ eV. Thus, the lattice relaxation changes the ΔE_{Fe} value, but scarcely changes both the ΔE_{Mg1} and ΔE_{Mg2} values. Also, the atomization energy for the rest of the Mg atoms is $\Delta E_{\mathrm{Mg3}} = 0.802$ eV before lattice relaxation and $\Delta E_{\mathrm{Mg3}} = 0.782$ eV after lattice relaxation. So, this change is also very small.

11.2.2 Atomization Energy Change With Charge Transfer

As shown in Fig. 11-7A, in the case of Ca showing large and positive local lattice strain, charge transfers in the direction from Ca to Mg. This is because the electronegativity is $\phi = 1.00$ for Ca, which is smaller than $\phi = 1.31$ for Mg. As a result of the charge transfer, the electron density decreases near the Ca atom, which leads to the decrease in the Coulomb attractive energy between the Ca nucleus and the surrounding electrons, as compared with the energy of an isolated neutral Ca atom in gas state. So, ΔE_{Ca} becomes negative according to Eq. (11.1).

On the other hand, as shown in Fig. 11-7B, in the case of Fe showing large and negative local lattice strain, charge transfers in the direction from Mg to Fe. This is because the electronegativity is $\phi = 1.83$ for Fe, which is larger than $\phi = 1.31$ for Mg. As a result of the charge transfer, the electron density increases near the Fe atom, which leads to the increase in the Coulomb attractive energy between the Fe nucleus and the surrounding electrons, as compared with the energy of an isolated neutral Fe atom in gas state. So, ΔE_{Fe} becomes positive according to Eq. (11.1). Thus, charge transfers between Mg and M in Mg–M alloys and adjusts the size difference between them. In other words, the larger atom (e.g., Ca) than Mg

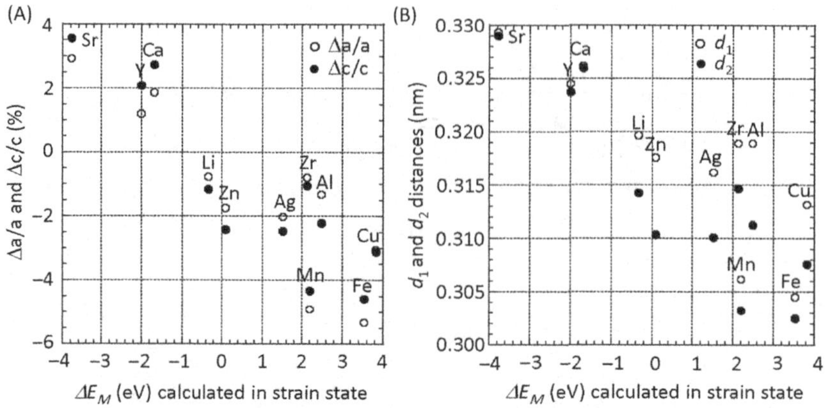

FIGURE 11-8 Correlation of atomization energy of M, ΔE, (A) with local lattice strains, $\Delta a/a$ and $\Delta c/c$ and (B) with the M–Mg bond distances, d_1 and d_2.

loses the electrons and decreases apparent atomic size, whereas the smaller atom (e.g., Fe) than Mg gains the electrons and increases apparent atomic size in the alloy.

To further confirm this interpretation, the local strains of $\Delta a/a$ and $\Delta c/c$ are plotted against the ΔE_M values in Fig. 11-8A. In the figure, local lattice strains, $\Delta a/a$ and $\Delta c/c$, vary monotonously with ΔE_M. There is a general trend that ΔE_M is negative for those M atoms (e.g., Ca, Sr) that expand the local Mg lattice (i.e., $\Delta a/a > 0$ and $\Delta c/c > 0$), whereas ΔE_M is positive for those M atoms (e.g., Fe, Mn) that contract the local Mg lattice (i.e., $\Delta a/a < 0$ and $\Delta c/c < 0$). The ΔE_M change with M is due simply to the charge transfer between M and Mg atoms. For example, in case electron transfers from M to Mg, the electron density decreases around M, and hence the Coulomb attractive energy between the electrons and M nucleus decreases, resulting in the negative ΔE_M. On the other hand, in case electron transfers from Mg to M, the situation is reverse, resulting in the positive ΔE_M. Thus, the main feature of the local lattice strain is well understood using a single parameter of the atomization energy, ΔE_M.

11.2.3 Atomization Energy of Mg Atom

From the above discussion, we may suppose that the atomization energies, ΔE_{Mg1} and ΔE_{Mg2}, of the first nearest neighbor Mg atoms change with M depending on the direction of the charge transfer. Nevertheless, ΔE_{Mg1} and ΔE_{Mg2} keep nearly unchanged regardless of the direction of the charge transfer between M and Mg. For example, ΔE_{Mg1} and ΔE_{Mg2} are 1.07–1.09 eV for M = Ca, and 1.06–1.10 eV for M = Fe. This indicates that the electron density change is not restricted to the neighborhood of the Mg1 and Mg2 atoms, but extended to the wide region distant from M. This is simply interpreted as due to the nearly free electron-like behavior of Mg metal, since a sea filled with the s and p electrons spreads over the Mg crystal lattice.

It is also noted that ΔE_{Mg3} is 0.78–0.82 eV for the Mg atoms locating in the region distant from M. This ΔE_{Mg3} value is definitely lower than the ΔE_{Mg1} and ΔE_{Mg2} value, 1.06–1.10 eV, for the first nearest neighbor Mg atoms from M. Therefore, the electron density is supposed to be higher around the first nearest neighbor Mg atoms than around the Mg atoms located in the region distant from M. These results are interpreted as due to the electron screening that modifies the electron density distribution around M so as to moderate an influence of the M impurity doped into a nearly free electron metal like Mg.

In the present calculation, the local displacements of Mg atoms are confined to the first nearest neighbor Mg atoms from M at a center because of the computational limitation in the calculation. However, as to the charge distribution around an impurity, it is known that the charge density, $\Delta\rho$, changes with the distance r from an impurity at the center, following an expression, $\Delta\rho = \rho_0 \cos(2k_F r + \eta)/r^3$ [31]. Here ρ_0 is a constant charge density, k_F is the Fermi wave-vector, and η is the phase shift. This is the so-called Friedel oscillation [31]. This charge oscillation arises from the sharp cut-off in the electron wavelength at k_F, which leads to an anomaly in the electron screening at $k = 2k_F$. As a result, haloes of charge density appear around an impurity. In case of the electron spin, the spin polarization of s electrons oscillates in a similar way around a magnetic impurity due to the R.K.K.Y. interaction [31]. Even in the present case, such an oscillation of the local lattice strain may appear around a solute atom owing to the electron screening anomaly in a nearly free electron metal like Mg.

11.2.4 Change in the M–Mg Bond Distance With Alloying Elements

The M–Mg bond distances, d_1 and d_2, are shown in Fig. 11-9 for a variety of elements, M, since both d_1 and d_2 distances are directly related to the chemical bond between M and Mg atoms. As is illustrated in an inset in Fig. 11-9, d_1 and d_2 are the interatomic distances between M and its first nearest neighbor Mg atoms. The d_1 and d_2 values are also listed in Table 11-1. For pure Mg, d_1 is longer than d_2. However, d_1 is shorter than d_2 for M = Ce, Gd, Tb, Dy, Ho, Np. Thus, the bond distances change largely depending on M. For example, in case of Y, d_1 is close to d_2, so that the Y addition reduces the anisotropy of the hcp Mg lattice to some extent [32].

Also, as shown in Fig. 11-8B, there is a correlation between the M–Mg bond distances, d_1, d_2, and the atomization energy, ΔE_{M}. There is a tendency that the M–Mg bond distances, d_1 and d_2, become short as the ΔE_{M} increases. This is very similar to the result shown in Fig. 11-8A. Thus, both the local lattice strains around alloying elements and the M–Mg bond distances are understood consistently using a single parameter of the atomization energy, ΔE_{M}.

As explained previously in Fig. 11-6, there is a large separation in the plots between the transition metal group (e.g., Fe, Mn, Cu) and the simple metal group (e.g., Sr, Ca). However, such a separation is no longer observed in Fig. 11-8A–B. Thus, the atomization energy, ΔE_{M}, is a suitable parameter to express the local lattice strains around M in Mg alloys.

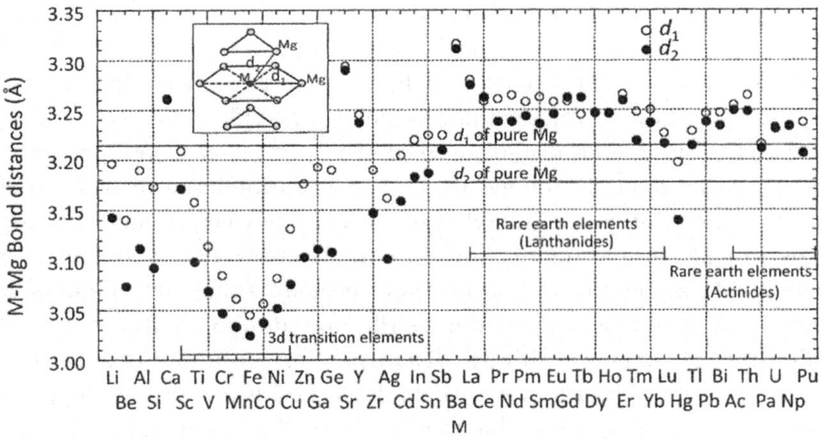

FIGURE 11-9 Change in the M–Mg bond distances, d_1 and d_2, with alloying elements, M.

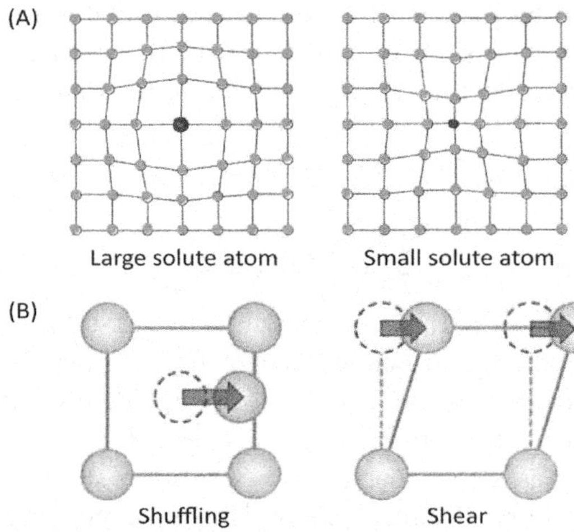

FIGURE 11-10 (A) Local lattice strain introduced into crystal lattice around solute atom due to the size difference between solute and solvent atoms. (B) Atomic displacements in martensitic transformation. *Courtesy of Prof. A. Kamegawa.*

11.3 Local Lattice Strain and Martensitic Transformation

As illustrated in Fig. 11-10A, local strains are introduced into the crystal lattice neighboring solute atom due to the size mismatch between solute and solvent atoms in alloy. Such a local lattice strain around the solute atom probably influences the martensitic transformation since it is a displacement-type transformation [33]. For example, as shown in Fig. 11-10B, local atomic displacements called shear and shuffling take place in the course of the martensitic transformation. Therefore, the locally displaced solvent atoms existing in the alloy before transformation could have a great influence on the martensitic transformation. In fact, there is an experiment in which the size of solute atom is associated with the martensitic transformation start temperature (*Ms*) in binary Ti alloys [34]. However, it is still unknown whether the transformation is either promoted or suppressed in the presence of such local lattice strains in the alloy.

Needless to say, the martensitic transformation is very important in iron alloys, since any steels are strengthened by this transformation. This is also the case in titanium alloys, since the martensitic transformation is related closely to the emergence of the shape memory and superelasticity in titanium alloys, as explained in Chapter 5, Titanium Alloys.

In this section, local lattice strains are first calculated around alloying elements, M, in both Ti and Fe metals using the pseudopotential method [19]. Then, it will be examined how the *Ms* temperature correlates with local lattice strains in binary Ti–M and Fe–M alloys.

11.4 Calculation Procedure for Titanium and Iron

11.4.1 Titanium

The supercell used for hcp Ti is the same one as used for hcp Mg as shown in Fig. 11-1 [2]. It consists of 36 Ti atoms, and an alloying element, M, is substituted for one Ti atom in the supercell. The calculation procedure used is also same as in the hcp Mg. Namely, the lattice parameter is first optimized to $a = 0.29461$ nm, $c = 0.46389$ nm ($c/a = 1.575$) for hcp pure Ti. Then, while keeping this lattice parameter constant, the positions of 12 Ti atoms locating in the first nearest neighbor sites from M are optimized, and the local strains are obtained of these 12 Ti atoms around M.

In case of bcc Ti, as shown in Fig. 11-11A, a supercell is constructed by stacking 3 × 3 × 3 unit cells of bcc Ti along three cubic crystal axes. It consists of 54 Ti atoms, and an alloying

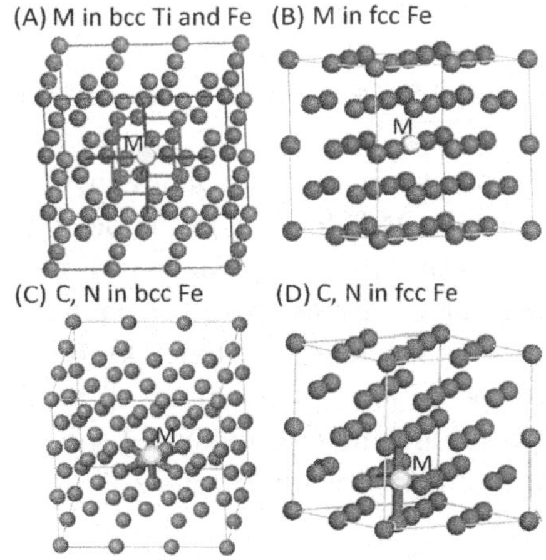

FIGURE 11-11 Supercells used for the calculations of substitutional alloying elements, M, in (A) bcc Ti and bcc Fe, (B) fcc Fe, and of interstitial C and N in (C) bcc Fe and (D) fcc Fe.

element, M, is substituted for one Ti atom in the supercell. The lattice parameter is first optimized to be $a = 0.32576$ nm for bcc pure Ti. Then, while keeping this lattice parameter constant, the positions of eight $Ti^{(1)}$ atoms locating in the first nearest neighbor sites and of six $Ti^{(2)}$ atoms locating in the second nearest neighbor sites from M are optimized, and the local strains are determined of these 14 Ti atoms. The positions of the second nearest neighbor $Ti^{(2)}$ atoms are optimized here, because the $M-Ti^{(2)}$ interatomic distance is only about 15% longer than the $M-Ti^{(1)}$ interatomic distance in bcc Ti.

In case of hcp Ti, the plane-wave cutoff energy is chosen to be 350 eV for the simple metals (e.g., Al, Si, Sn, etc.) and for the transition metals (e.g., V, Fe, Zr, etc.), but 400–480 eV for the rare earth metals. For example, it is 400 eV for M = Ce, 440 eV for M = Nd, and 480 eV for M = Eu. In case of bcc Ti, the plane-wave cutoff energy is chosen to be 380 eV for every metal, since any rare earth metals are not calculated.

11.4.2 Iron

In the case of bcc Fe, the supercell shown in Fig. 11-11A is employed [2]. It is the same supercell as used in bcc Ti. The optimized lattice parameter for bcc pure Fe is $a = 0.27469$ nm for bcc pure Fe. The local displacements of the first nearest neighbor $Fe^{(1)}$ atoms and the second nearest neighbor $Fe^{(2)}$ atoms around M in bcc Fe are obtained in the same way as in bcc Ti. On the other hand, in case of fcc Fe, as shown in Fig. 11-11B, a supercell is constructed by stacking $2 \times 2 \times 2$ unit cells of fcc Fe along three crystal axes. It consists of 32 Fe atoms, and an alloying element, M, is substituted for one Fe atom in the supercell. The plane-wave cutoff energy is chosen to be 380 eV for every element M in both fcc Fe and bcc Fe.

The optimized lattice parameter is $a = 0.34351$ nm for pure fcc Fe. Then, while keeping this lattice parameter constant, the positions of 12 $Fe^{(1)}$ atoms locating in the first nearest neighbor sites from M are optimized, and the local strains are determined of these 12 Fe atoms. Here, the lattice relaxation is confined only to the first nearest neighbor $Fe^{(1)}$ atoms from M, since the $M-Fe^{(2)}$ interatomic distance is about 41% longer than the $M-Fe^{(1)}$ interatomic distance in fcc Fe.

The supercell used for the calculation of interstitial alloying elements, M (=C, N) in bcc Fe is shown in Fig. 11-11C, in which M is located in the octahedral interstitial site in bcc Fe. The local strains are determined by optimizing the positions of two Fe atoms in the first nearest neighbor sites and of four Fe atoms in the second nearest neighbor sites from M (=C, N), while keeping the lattice parameter constant. On the other hand, in case of interstitial alloying elements, M (=C, N) in fcc Fe, the supercell used is shown in Fig. 11-11D, in which M (=C, N) is located in the octahedral interstitial site in fcc Fe. The local strains are then determined by optimizing only the positions of six Fe atoms in the first nearest neighbor sites from M.

Every calculation for Ti and Fe is performed by employing the fine convergent conditions as explained previously in the calculation for hcp Mg.

11.5 Local Lattice Strain Around Alloying Element in Titanium

11.5.1 Local Lattice Strain in hcp Ti

For example, the calculated results of hcp Ti are shown in Fig. 11-12A for M = V and (B) for M = Zr, where only the 12 Ti atoms around M are illustrated. Here, the same notations are used as in hcp Mg.

In the alloyed hcp Ti with M, the $Ti^{(1)}$ atoms in the first nearest neighbor sites from M are displaced in the direction indicated by the vector in Fig. 11-12. The hexagon either contracts (i.e., $\Delta a_1 < 0$) or expands (i.e., $\Delta a_1 > 0$) depending on M. Also, it is seen that the strain direction along the *c*-axis is opposite between the Ti atoms on the upper and lower triangles.

In case the hexagon contracts (i.e., $\Delta a_1 < 0$), the upper triangle and lower triangles also contact (i.e., $\Delta a_2 < 0$) and they approach each other along the *c*-axis (i.e., $\Delta c_2 < 0$). On the other hand, in case the hexagon expands (i.e., $\Delta a_1 > 0$), the upper triangle and lower triangles also expand (i.e., $\Delta a_2 > 0$) and they move away from each other along the *c*-axis (i.e., $\Delta c_2 > 0$). As shown in Fig. 11-12, V belongs to the former group (i.e., $\Delta a_1 < 0$, $\Delta a_2 < 0$, $\Delta c_2 < 0$), and Zr belongs to the latter group (i.e., $\Delta a_1 > 0$, $\Delta a_2 > 0$, $\Delta c_2 > 0$). Besides these, there is a very rare case, ($\Delta a_1 < 0$, $\Delta a_2 < 0$, $\Delta c_2 > 0$), in Au and Pd. Every strain is divided by the optimized lattice parameter of a and c, and the values of $\Delta a_1/a$, $\Delta a_2/a$ and $\Delta c_2/c$ are shown in % units in Fig. 11-12. These calculated local strains around alloying elements, M, in hcp Ti are summarized in Table 11-2 together with the M–Ti bond distances, d_1 and d_2. As explained previously, d_1 is the distance between M and Ti atoms sitting on the six corners of the hexagon, and d_2 is the distance between M and Ti atoms sitting on the three corners of the upper or lower triangle.

For simplicity, the average strains are defined as, $\Delta a/a = [\Delta a_1/a + \Delta a_2/a]/2$ (%) and $\Delta c/c = 2\Delta c_2/c$ (%). The sum of these local lattice strains, $\Delta a/a + \Delta c/c$, are shown in Fig. 11-13 for a variety of alloying elements, M. Local lattice strains change largely depending on M. For example, local lattice strains are negative for 3*d* transition elements (e.g., Mn, Fe), but positive for rare earth elements (e.g., Ce, Nd). Thus, there is a tendency that the local

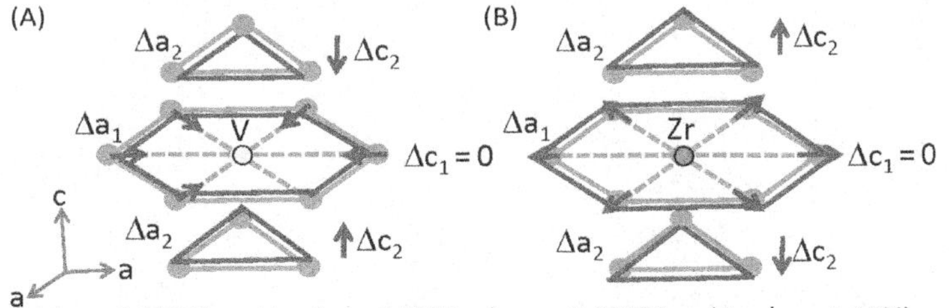

$\Delta a_1 = -0.00183$nm($\Delta a_1/a = -0.62\%$) $\Delta a_1 = +0.00352$nm($\Delta a_1/a = +1.19\%$)
$\Delta a_2 = -0.00395$nm($\Delta a_2/a = -1.34\%$) $\Delta a_2 = +0.00569$nm($\Delta a_2/a = +1.93\%$)
$\Delta c_2 = -0.00194$nm ($\Delta c_2/c = -0.42\%$) $\Delta c_2 = +0.00348$nm($\Delta c_2/c = +0.75\%$)

FIGURE 11-12 Calculated local lattice strains around alloying element, (A) V and (B) Zr in hcp Ti.

Table 11-2 Calculated Local Strains of the First Nearest Neighbor Ti Atoms From Alloying Element M, and the M–Ti Bond Distances, d_1 and d_2, in a Distorted hcp Ti Lattice

M	$\Delta a_1/a$ (%)	$\Delta a_2/a$ (%)	$\Delta c_2/c$ (%)	d_1 (nm)	d_2 (nm)	M	$\Delta a_1/a$ (%)	$\Delta a_2/a$ (%)	$\Delta c_2/c$ (%)	d_1 (nm)	d_2 (nm)
Al	−0.95	−1.16	−0.13	0.2918	0.2860	Ag	0.20	0.23	0.45	0.2952	0.2896
Si	−1.97	−3.19	−0.78	0.2888	0.2815	Sn	0.11	0.64	0.71	0.2949	0.2910
Sc	1.08	2.11	0.61	0.2978	0.2920	La	3.43	4.78	1.94	0.3047	0.2996
V	−0.62	−1.34	−0.42	0.2928	0.2847	Ce	1.49	3.83	1.45	0.2990	0.2967
Cr	−0.89	−2.31	−0.80	0.2920	0.2823	Pr	1.25	3.88	1.16	0.2983	0.2960
Mn	−0.72	−2.96	−1.17	0.2925	0.2803	Nd	1.15	4.00	1.09	0.2980	0.2958
Fe	−0.75	−3.51	−1.42	0.2924	0.2788	Pm	1.51	3.70	1.01	0.2991	0.2951
Co	−0.81	−3.06	−1.02	0.2922	0.2807	Sm	0.96	3.37	0.95	0.2975	0.2946
Ni	−0.91	−2.40	−0.66	0.2919	0.2828	Eu	0.92	3.19	0.88	0.2973	0.2942
Cu	−0.94	−1.58	−0.37	0.2919	0.2847	Hf	1.12	1.81	0.67	0.2979	0.2920
Y	2.34	3.90	1.27	0.3015	0.2963	Ta	0.37	0.42	0.37	0.2957	0.2894
Zr	1.19	1.93	0.75	0.2981	0.2924	W	−0.52	−1.06	−0.16	0.2931	0.2860
Nb	0.16	0.18	0.30	0.2951	0.2897	Re	−0.82	−2.01	−0.46	0.2922	0.2839
Mo	−0.29	−0.97	−0.11	0.2938	0.2863	Os	−0.63	−2.45	−0.69	0.2928	0.2826
Tc	−0.41	−1.81	−0.47	0.2934	0.2841	Ir	−0.55	−2.36	−0.58	0.2930	0.2831
Ru	−0.31	−2.14	−0.61	0.2937	0.2832	Pt	−0.34	−1.43	−0.71	0.2936	0.2860
Rh	−0.18	−1.67	−0.21	0.2930	0.2852	Au	0.06	−0.17	−0.09	0.2948	0.2889
Pd	−0.11	−0.95	0.09	0.2943	0.2871						

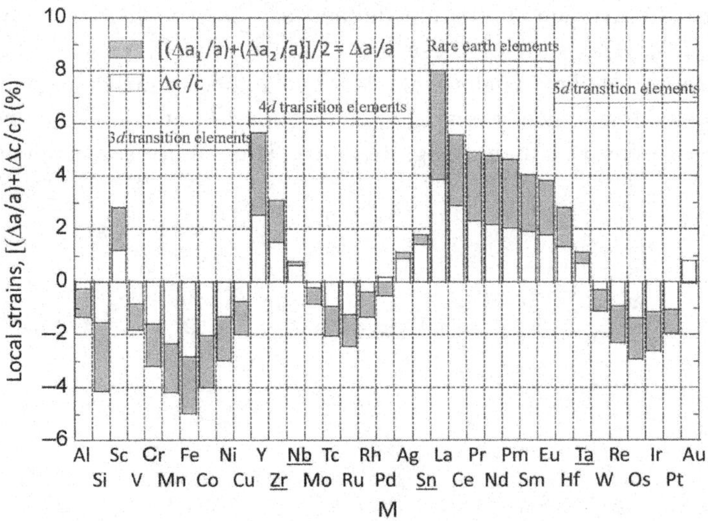

FIGURE 11-13 Local lattice strains, ($\Delta a/a + \Delta c/c$), around alloying elements M in hcp Ti.

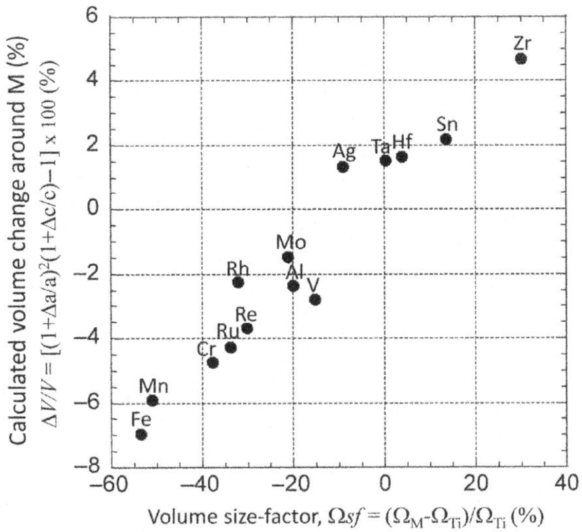

FIGURE 11-14 Comparison of calculated volume change around M with the volume size-factor, Ωsf.

Ti lattice contracts near 3*d* transition elements (i.e., $\Delta a/a < 0$, $\Delta c/c < 0$), but expands near rare earth elements (i.e., $\Delta a/a > 0$, $\Delta c/c > 0$).

11.5.2 Reliability of Calculated Local Strains

To compare the calculated local strains with experimental results, the parameter used is a volume size-factor, $\Omega sf = (\Omega_M - \Omega_{Ti})/\Omega_{Ti}$, where Ω_M and Ω_{Ti} are the respective atomic volumes of M and Ti. This Ωsf was proposed by King [12] on the basis of measured lattice parameters. For comparison, the volume change, $\Delta V/V = [(1 + \Delta a/a)^2(1 + \Delta c/c) - 1] \times 100$ (%), is estimated for various M by using the calculated local strains of $\Delta a/a$ and $\Delta c/c$. In Fig. 11-14, the calculated values of $\Delta V/V$ are plotted against the volume size-factor, Ωsf. There is relatively good correspondence between them.

11.5.3 Solid Solubility Change With M–Ti Interatomic Distance

It is known empirically that solute atoms are soluble in a terminal solid solution in case the size difference between solute and solvent atoms is less than about 10%, although the reason is unknown. This size effect may be related to the change in the M–Ti interatomic distances, d_1 and d_2, with M. As explained above, d_1 is the M–Ti interatomic distance where Ti atoms are locating on the six corners of the hexagon, and d_2 is the M–Ti interatomic distance where Ti atoms are located on the three corners of the upper or lower triangles (see an inset in Fig. 11-15). For hcp pure Ti, $d_1 = 0.2946$ nm, $d_2 = 0.2876$ nm, so that $d_1 > d_2$. These d_1 and d_2 values change largely with M, as shown in Fig. 11-15, In the figure, d_1 and d_2 are indicated by solid circle (●) and open circle (○), respectively.

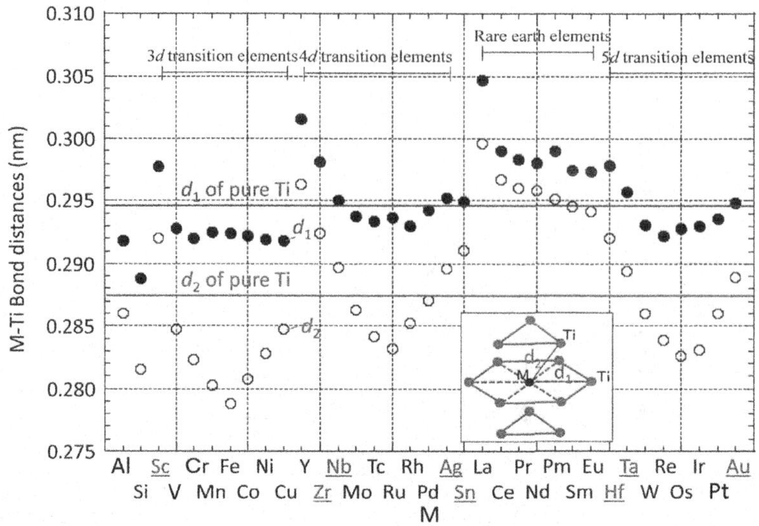

FIGURE 11-15 Change in the M–Ti bond distances, d_1 and d_2, with alloying elements M in hcp Ti.

Various alloying elements show positive local strains as shown in Fig. 11-13. Among them, some alloying elements (M = Sc, Zr, Nb, Ag, Sn, Hf, Ta, and Au) are more soluble than 2 mol% in the Ti solid solution [27]. These elements are indicated by drawing an underline on M in the horizontal axis of Fig. 11-15. It is apparent from this figure that for all these elements the d_2 distance (o) falls on the region intermediate between d_1 and d_2 for pure Ti. The levels of d_1 and d_2 for pure Ti are indicated by two horizontal lines in the figure. This result implies that there are certain allowable Ti–M bond distances for M to be soluble in the Ti solid solution. Such allowable Ti–M distances are attained by the proper local strains introduced into the lattice around M. This may be the essence of the atomic size effect. In fact, the atomic radii of all these elements range from 0.1405 nm for Sn to 0.1606 nm for Sc, being comparable to 0.1448 nm for Ti. Therefore, the size difference between solute and solvent atoms is less than about 11%. In this sense, the chemical bond strength between M and Ti atoms is important in understanding the solid solubility of M in the alloy.

However, one exception is seen at Eu, since Eu is a completely insoluble element, but the d_2 distance falls on the intermediate region. However, as shown in Fig. 11-15, the d_2 distance for Eu is very close to the d_1 distance for pure Ti, so that the Ti–Eu chemical bond is probably too weak to form a stable solid solution in the Ti–Eu alloy.

11.5.4 Correlation of Local Lattice Strain in hcp Ti With Martensitic Transformation Start Temperature (*Ms*)

The martensitic transformation of Ti alloys is expressed as [33],

$$\beta - \text{Ti(bcc)} \rightarrow \alpha'(\text{hcp}) \text{ or } \alpha''(\text{orthorhombic}).$$

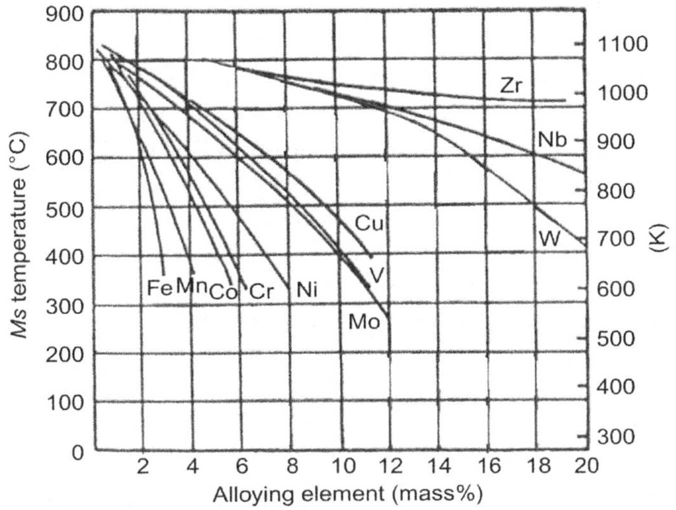

FIGURE 11-16 Compositional dependence of *Ms* temperature of binary Ti–M alloys [34].

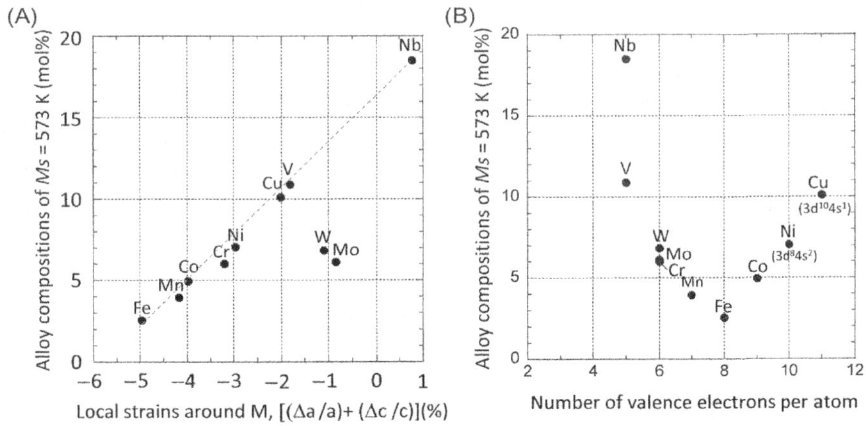

FIGURE 11-17 Correlation of (A) local lattice strains and (B) number of valence electrons per atom with alloy compositions of *Ms* = 573 K in binary Ti–M alloys.

As shown in Fig. 11-16, Sato et al. [34] reported the compositional dependence of the martensitic transformation start temperature (*Ms*) in various binary Ti–M alloys. In Fig. 11-17A, the measured alloy compositions where the *Ms* temperature is 573 K, are plotted against the calculated local lattice strains, $\Delta a/a + \Delta c/c$, shown in Fig. 11-13. In the figure, the *Ms* data of Nb is taken from Kim and Miyazaki [35].

It is very interesting that the alloy compositions of *Ms* = 573 K change linearly with the local lattice strains except for W and Mo. As the negative local lattice strains increase, the martensitic transformation takes place in the lower compositional alloys.

Sato et al. [34] reported that the *Ms* temperature is associated with the atomic size and the electrons-per-atom ratio, (*e/a*). So, in Fig. 11-17B, the alloy compositions of *Ms* = 573 K are plotted against *e/a*. The *Ms* temperatures are found to be nearly the same among Cr, Mo, and W, all belonging to the 6A group in the periodic table of elements. Any deviation is not observed in Mo and W, which is in contrast to the result shown in Fig. 11-17A. This implies the importance of the chemical bond between atoms in understanding the *Ms* temperature of binary Ti–M alloys. However, there is still a large difference in the alloy compositions of *Ms* = 573 K between V and Nb, despite that both of them are the 5A group elements in the periodic table. Thus, the question still remains, but the local lattice strain around M is probably one of the important factors in understanding the martensitic transformation of Ti–M alloys.

11.5.5 Strain Energy Analysis of Compositonal Dependence of *Ms* Temperature

To understand the compositonal dependence of *Ms* temperature in view of strain energy, the energy gained by the local strain around an isolated M atom (hereafter called strain energy) is shown in Fig. 11-18B for hcp Ti. For comparison, the calculated result for bcc Ti is shown in Fig. 11-18A, as explained later. Here the strain energy is defined as the total energy difference between the strain-free state and the strain state. The strain energy is smaller in hcp Ti than in bcc Ti. It is evident from Fig. 11-18A–B that the strain energy varies largely depending on M, and it is comparable to the thermal energy, $k_B T$.

The *Ms* temperature shown in Fig. 11-16 changes in the range of 800°C (1073 K) to 300°C (573 K). The strain energy is released from the alloy as the temperature increases. So, the alloy compositions showing the lowest *Ms* = 573 K are compared with the local strains around M in Fig. 11-17A.

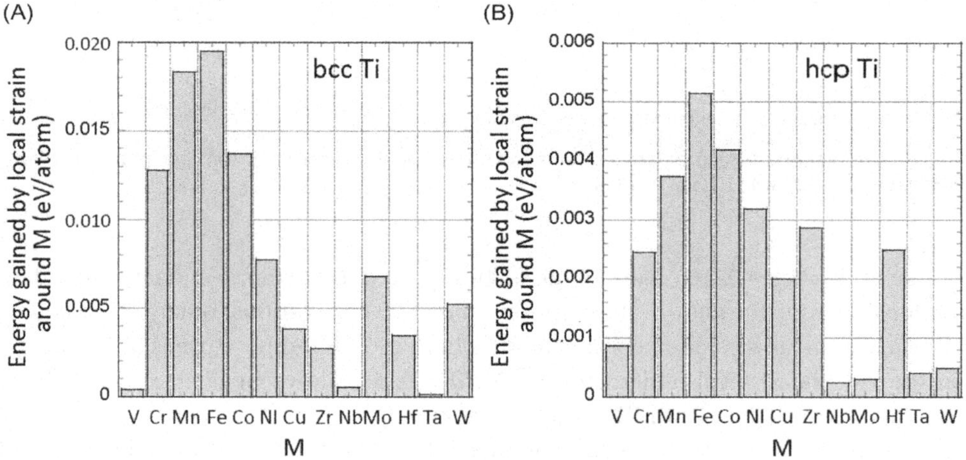

FIGURE 11-18 Comparison in the energy gained by local strain around M between (A) bcc Ti and (B) hcp Ti.

As is evident from Fig. 11-18A–B, the strain energy is large for M = Fe and small for M = Nb, irrespective of bcc Ti and hcp Ti. To compensate for the strain energy difference between them by alloying, the M composition of *Ms* = 573 K is inevitably low (about 2.5 mol %) for M = Fe because of the large strain energy per Fe atom, but high (about 18.5 mol%) for M = Nb because of the small strain energy per Nb atom. This is an explanation for the *Ms* temperature change with M, although it is still qualitative.

11.6 Local Lattice Strains in bcc Ti

11.6.1 Local Strain Change With Alloying Elements

Using the supercell shown in Fig. 11-11A, local lattice strains around M are calculated for bcc Ti. The results are, for example, shown in Fig. 11-19A for M = Fe and (B) for M = Zr. In case of (A) M = Fe, the first nearest neighbor $Ti^{(1)}$ atoms are displaced toward Fe along the <111> direction. The magnitude of the local strain is expressed as $\sqrt{3}\,\Delta a_3$, where Δa_3 is defined as the strain along the *a*-axis of bcc lattice. It is divided by the optimized lattice parameter, $a = 0.32576$ nm, and the value of $\sqrt{3}\,\Delta a_3/a$ is -6.56%. Also, the second nearest neighbor $Ti^{(2)}$ atoms move away from Fe along the <100> direction. Here, Δa_4 is defined as the strain of the second nearest neighbor $Ti^{(2)}$ atoms along the *a*-axis of the bcc lattice, and the value of $\Delta a_4/a$ is calculated to be $+2.86\%$ as shown in Fig. 11-19A. On the other hand, in the case of (B) M = Zr, both $Ti^{(1)}$ and $Ti^{(2)}$ atoms are displaced in opposite directions of (A) M = Fe. For convenience, all the calculated values are listed in Table 11-3.

The calculated local lattice strains are shown in Fig. 11-20A for alloying elements, M, in bcc Ti. For comparison, the results of hcp Ti are also shown in Fig. 11-20B by choosing the alloying elements, M, from Fig. 11-13. As a whole, the change in the local lattice strain with M appears to resemble between bcc Ti and hcp Ti. For example, the local lattice around Zr and Hf expands to some extent in both bcc Ti and hcp Ti.

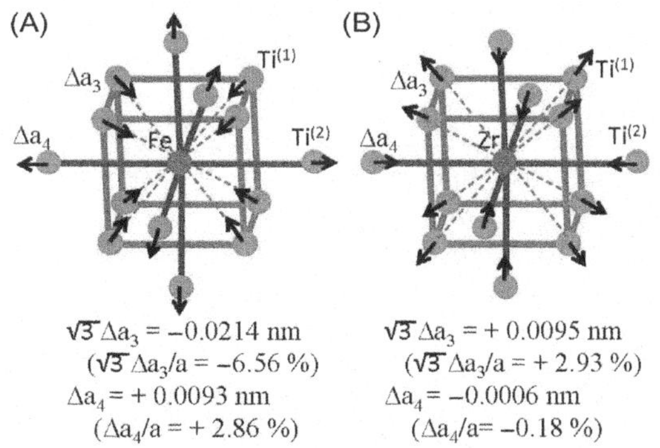

FIGURE 11-19 Calculated local lattice strains around alloying element, (A) Fe and (B) Zr in bcc Ti.

Table 11-3 Calculated Local Lattice Strains of the First and Second Nearest Neighbor Ti Atoms From Alloying Element M in bcc Ti

	bcc Ti	
M	$\sqrt{3}\Delta a_3/a$ (%)	$\Delta a_4/a$ (%)
V	−3.56	1.36
Cr	−5.84	2.55
Mn	−6.50	2.90
Fe	−6.56	2.86
Co	−5.51	2.85
Ni	−4.21	2.84
Cu	−2.85	2.98
Zr	2.93	−0.18
Nb	−1.11	1.19
Mo	−3.77	2.57
Hf	3.40	−1.08
Ta	−0.41	0.62
W	−3.42	2.04

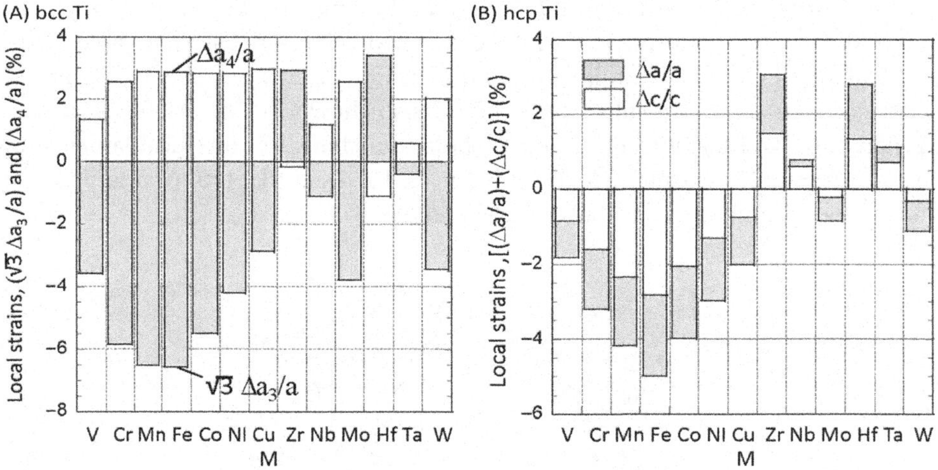

FIGURE 11-20 Comparison in the local lattice strains around M in between (A) bcc Ti and (B) hcp Ti.

According to Fig. 11-20A, Fe has the largest negative local lattice strain among the alloying elements in bcc Ti. Fe is soluble up to 22 mol% in bcc Ti [27]. So, the solid solution hardening will be enhanced by the addition of Fe into bcc Ti. This is confirmed experimentally [36].

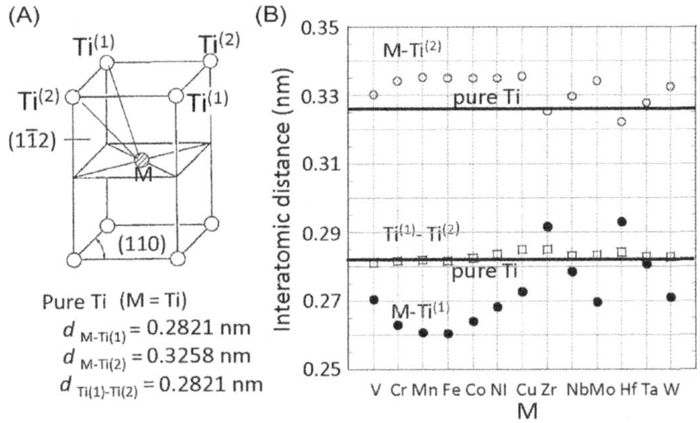

FIGURE 11-21 (A) Polyhedron containing M at a center and (B) change in the interatomic distances with M in bcc Ti.

11.6.2 Change in M–Ti and Ti–Ti Interatomic Distances With M

The M–Ti and Ti–Ti interatomic distances change with the local strains around M in bcc Ti. Shown in Fig. 11-21A is a polyhedron, in which M occupies a central position. As discussed later, a basal plane of this polyhedron is the (110) crystal plane, and a side plane is the (1$\bar{1}$2) crystal plane as indicated in the figure. The M–$Ti^{(1)}$, M–$Ti^{(2)}$, and $Ti^{(1)}$–$Ti^{(2)}$ interatomic distances in this polyhedron are presented in Fig. 11-21B, using symbols, •, ○, and □, respectively. As described before, $Ti^{(1)}$ and $Ti^{(2)}$ are the first nearest neighbor Ti atom and the second nearest neighbor Ti atom from M at the center, respectively. Also, in case of pure Ti, $Ti^{(2)}$ shown in Fig. 11-21A is located in the first nearest neighbor site from $Ti^{(1)}$, so that a relation of $d_{M-Ti(1)} = d_{Ti(1)-Ti(2)}$ is satisfied for pure Ti (i.e., M = Ti). The respective values of pure Ti are given together with the value of $d_{M-Ti(2)}$ below the polyhedron. The substitution of M for a central Ti introduces the local lattice strain around M, and the M–$Ti^{(1)}$, M–$Ti^{(2)}$, and $Ti^{(1)}$–$Ti^{(2)}$ interatomic distances change as shown in Fig. 11-21B.

The M–$Ti^{(1)}$ interatomic distance (•) is shorter than the distance in pure Ti except for Zr and Hf, whereas the M–$Ti^{(2)}$ interatomic distance (○) is longer except for Zr and Hf. The $Ti^{(1)}$–$Ti^{(2)}$ interatomic distance (□) is nearly constant, which is a great contrast to the large changes in both the M–$Ti^{(1)}$ and M–$Ti^{(2)}$ interatomic distances with M. Both Zr and Hf behave larger in size than Ti, judging from the longer M–$Ti^{(1)}$ interatomic distance, but instead they have the shorter the M–$Ti^{(2)}$ interatomic so as to keep the $Ti^{(1)}$–$Ti^{(2)}$ interatomic distance nearly same as in pure Ti. Thus, when local strains are introduced into the bcc lattice around M, the chemical bond between the neighboring Ti atoms is well balanced in a subtle way.

11.6.3 Correlation of Local Strains in bcc Ti With Martensitic Transformation Start Temperature (*Ms*)

As explained earlier, the martensitic transformation takes place in Ti alloys as follows: β-Ti(bcc)$\rightarrow\alpha'$(hcp) or α''(orthorhombic). The martensite phase changes the structure from α' to α'', as the alloy composition increases [33].

Here, the average local strain of Ti atoms around M in bcc Ti is defined as,

$$\text{Average local strain} = \left[\left(\sqrt{3}\Delta a_3/a\right) \times 8 + (\Delta a_4/a) \times 6\right] /14\,(\%). \tag{11.4}$$

This is an arithmetic average of the strains by weighting the coordination numbers around M. In Fig. 11-22, the alloy compositions where *Ms* is 573 K, are plotted against the average local strains in bcc Ti. As compared with Fig. 11-17A, data is scattered, and a linear relation is not seen so clearly in Fig. 11-22. No large deviation is seen even for Mo and W in bcc Ti. However, as the negative local strain increases, the martensitic transformation takes place in the lower compositional alloys. This trend is the same as in hcp Ti.

Fig. 11-23 is prepared to understand the reason for this trend. For comparison, two alloying elements, M_1 and M_2, are selected as shown in the figure. Let's assume that the

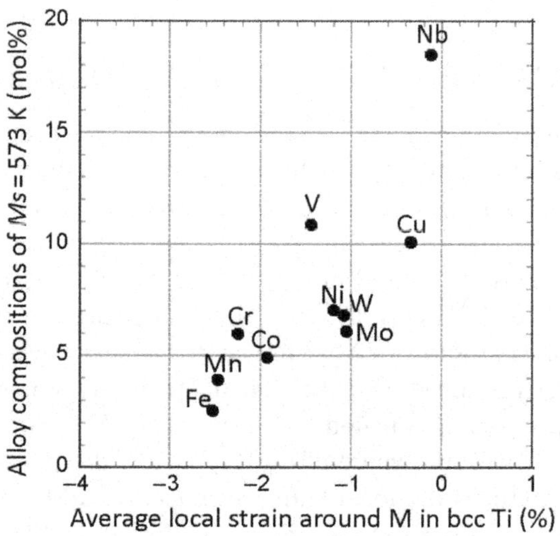

FIGURE 11-22 Correlation of average local strains around M with alloy compositions of *Ms* = 573 K in binary Ti–M alloys.

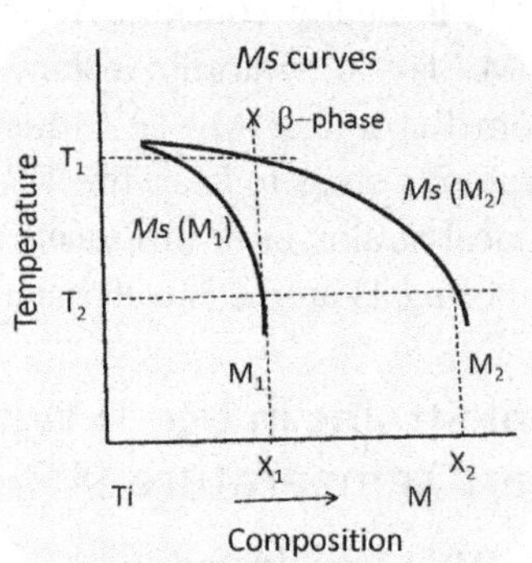

FIGURE 11-23 Correlation of the phase stability of bcc β phase with the *Ms* curves of Ti–M_1 and Ti–M_2 alloys.

Ms temperature is lower in the Ti–M_1 alloy than in the Ti–M_2 alloy. In this case, a high-temperature bcc β phase of the X_1 composition starts the martensitic transformation at the T_1 temperature in the Ti–M_2 alloy, but the bcc β phase still retains in the Ti–M_1 alloy down to the T_2 temperature. In other words, the bcc β phase is more stabilized in the Ti–M_1 alloy than in the Ti–M_2 alloy. So, the onset of the martensitic transformation is suppressed in the Ti–M_1 alloy rather than in the Ti–M_2 alloy. In Figs. 11-17A and 11-22, the alloy compositions of $Ms = 573$ K are shown in the vertical axis. Following the notations used in Fig. 11-23, they correspond to the X_1 composition in the Ti–M_1 alloy and the X_2 composition in the Ti–M_2 alloy of constant temperature, $T_2 = 573$ K. Therefore, for example, M_1 is Fe, and M_2 is Nb, according to Fig. 11-17A or Fig. 11-22. Thus, the onset of the martensitic transformation is suppressed in the alloy containing M with the negatively large local strains like Fe. This reason will be discussed in the next section following the transformation mechanism.

11.6.4 Correlation of Martensitic Transformation With Local Lattice Strains

The martensitic transformation mechanism of bcc Ti $\rightarrow \alpha'$(hcp) is proposed by Burgers, as illustrated in Figs. 11-24A–C [33]. First, in the process of (A) to (B), the bcc crystal shown in (A) undergoes the shear deformation along the $[\bar{1}11]$ direction on the $(1\bar{1}2)$ plane. As a result, a corner angle of the polyhedron drawn by thick solid lines in Fig. 11-24A–B changes from 70.5° to 60°. Here, 70.5° is an angle between the $[111]_{bcc}$ and $[\bar{1}11]_{bcc}$ directions. Then, in the process of (B) to (C), the shuffling occurs in every two $(110)_{bcc}$ planes (i.e., $(0001)_{hcp}$ basal plane) to form the hcp α' martensite phase. Therefore, the crystal orientation relationship is expressed as, $(110)_{bcc} \parallel (0001)_{hcp}$ and $[\bar{1}11]_{bcc} \parallel [11\bar{2}0]_{hcp}$ [33].

Let's consider next how the corner angle of the polyhedron prior to the shear deformation (hereafter called shear angle) changes by the local displacements of Ti atoms in the neighborhood of M. The shear angle is equal to 70.5° in pure Ti (see Fig. 11-24A). By the way, the

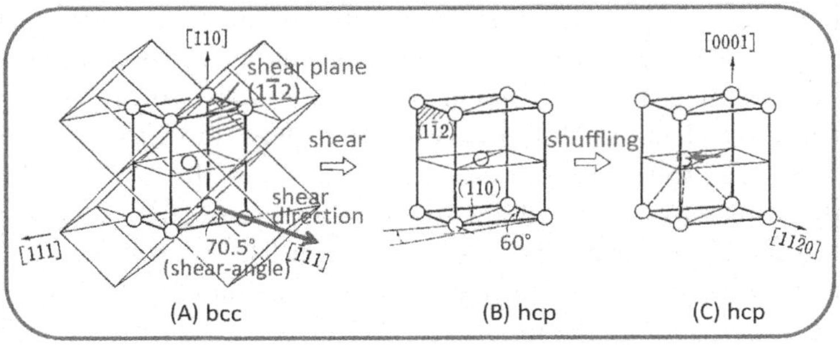

FIGURE 11-24 Martensitic transformation from bcc to hcp following the Burgers relation [33].

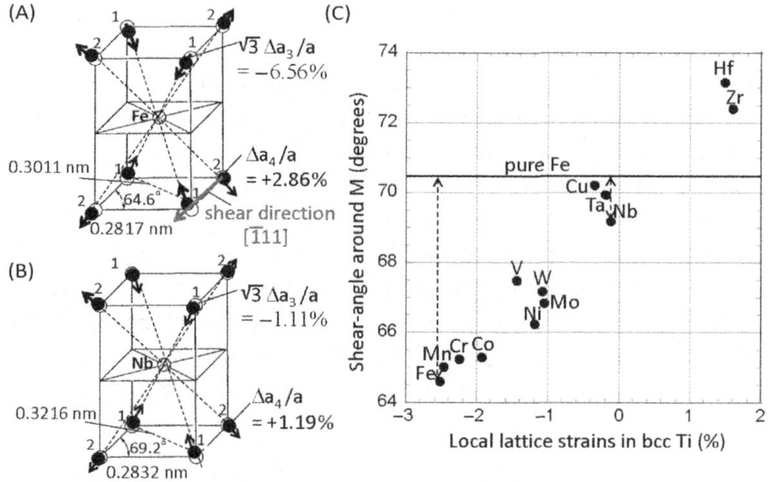

FIGURE 11-25 Calculated shear angles around alloying element M: (A) M = Fe, (B) M = Nb and (C) various M, which are plotted against local lattice strains in bcc Ti.

polyhedron shown in Fig. 11-24A is the same as the one previously shown in Fig. 11-21A. The shear angle is calculated by locating M at the center of this polyhedron. The results are, for example, shown in Fig. 11-25A for M = Fe and (B) for M = Nb. The Ti atoms to which one or two are numbered in the figure correspond to $Ti^{(1)}$ or $Ti^{(2)}$ in Fig. 11-21A, so that they are the first or second nearest neighbor Ti atoms from M. They are locally displaced in the direction indicated by the vectors in the figure.

The shear angle is calculated to be about 64.6° for (A) Fe and about 69.2° for (B) Nb. In either case it is smaller than the value of 70.5° for pure Ti. All the calculated results of the shear angles are summarized in Fig. 11-25C. As is apparent from this figure, shear angles are closer to 60° in those alloying elements (e.g., Fe) that have large and negative local strains. So, we may suppose that the martensitic transformation is progressed in such a Fe-containing alloy, since any large shear deformation is not necessarily required for the progress of the transformation. However, this is not the case. When we look over a whole crystal lattice of the Fe-containing alloy, the region of shear angle 70.5° is mixed together with the regions of smaller shear angles. In other words, various regions with the widely distributed shear angles coexist in the alloy. Furthermore, as illustrated in Fig. 11-25A, those Ti atoms numbering one or two in the neighborhood of Fe are displaced in the direction indicated by the vectors in the figure. Their displacement directions are very different from the shear direction, $[\bar{1}11]$, as shown in the figure (A). Thus, when there are large local distortions in the crystal lattice, it is rather difficult for a whole crystal to shear uniformly along the $[\bar{1}11]$ direction on the $(1\bar{1}2)$ plane. For this reason, large and negative local lattice strains work to suppress the onset of the martensitic transformation in Ti alloys.

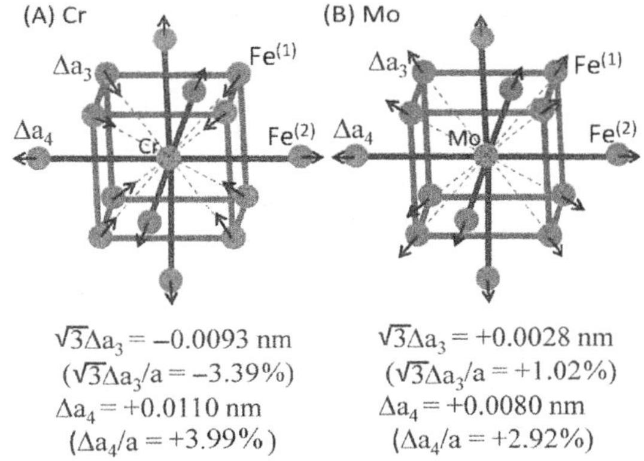

FIGURE 11-26 Calculated local lattice strains around alloying element, (A) Cr and (B) Mo in bcc Fe.

11.7 Local Lattice Strain Around Alloying Element in Iron

11.7.1 Local Lattice Strains in bcc Fe

Using the supercell shown in Fig. 11-11A, local lattice strains around substitutional alloying elements, M, are calculated in bcc Fe [2]. For example, the results are shown in Fig. 11-26A for M = Cr and (B) for M = Mo. In case of Cr, the first nearest neighbor $Fe^{(1)}$ atoms from M at a center are displaced toward Cr in the <111> direction. The magnitude of local strains is expressed as $\sqrt{3}\,\Delta a_3$ where Δa_3 is defined as the strain along the a-axis of bcc lattice. It is also divided by the optimized lattice parameter, $a = 0.27469$ nm, and the value of $\sqrt{3}\,\Delta a_3/a$ is about -3.39%. On the other hand, the second nearest neighbor $Fe^{(2)}$ atoms from M are displaced away from Cr in the <100> direction. The value of local strains expressed as $\Delta a_4/a$ is about $+3.99\%$. On the other hand, in the case of Mo, $Fe^{(1)}$ atoms are displaced away from M in the <111> direction. $Fe^{(2)}$ atoms are also displaced away from M in the <100> direction. Thus, local displacements vary depending on M in bcc Fe. All the calculated values are given in Table 11-4A.

Furthermore, using the supercell shown in Fig. 11-11C, local lattice strains are calculated around interstitial C and N in bcc Fe. The results are shown in Fig. 11-27A for C and (B) for N. In either result, two Fe atoms (the first nearest neighbor Fe atoms denoted as ① in the figure) sitting on the upper and lower corners of a central octahedron are displaced largely away from C or N in the <100> direction. The four Fe atoms (the second nearest neighbor Fe atoms denoted as ②) sitting on the other four corners of the octahedron are displaced slightly toward C or N in the <110> direction. The present result resembles other calculations [37].

All the results are shown in Fig. 11-28A for a variety of substitutional alloying elements, M. Local lattice strains change largely depending on M in bcc Fe. To show the validity of the present calculation, the average local strain calculated from Eq. (11.4) is plotted against the measured lattice parameter changes between Fe-1mol%M and pure Fe (%) [38] in Fig. 11-29A. There is a linear relationship between them.

Table 11-4 Local Lattice Strains Around Alloying Element M in bcc Fe and fcc Fe Obtained by (A) Non-Spin-Polarized Calculation and (B) Spin-Polarized Calculation

(A)				(B)		
	bcc Fe		fcc Fe		bcc Fe	
M	$\sqrt{3}\Delta a_3/a$ (%)	$\Delta a_4/a$ (%)	$\sqrt{2}\Delta a_7/a$ (%)	M	$\sqrt{3}\Delta a_3/a$ (%)	$\Delta a_4/a$ (%)
Al	−1.34	3.35	0.56	A1	0.67	0.24
Si	−4.11	4.02	0.78	Si	−0.74	0.61
Ti	−0.15	3.52	2.00	Ti	1.02	0.32
V	−2.37	3.93	1.05	V	−0.05	0.44
Cr	−3.39	3.99	0.39	Cr	0.01	−0.19
Mn	−2.03	2.97	0.27	Mn	0.07	−0.18
Co	−1.39	2.21	0.18	Co	0.49	0.26
Ni	−1.86	2.79	0.39	Ni	0.59	−0.28
Cu	−1.23	2.82	0.98	Cu	0.65	−0.04
Zr	7.76	−3.10	3.43	Zr	3.57	0.62
Nb	2.45	2.73	2.58	Nb	2.53	0.57
Mo	1.02	2.92	1.74	Mo	1.75	0.71
Hf	8.19	−2.55	3.86	Hf	4.28	1.05
Ta	7.82	−3.98	3.24	Ta	3.53	0.79
W	1.35	2.88	1.91	W	1.63	0.42

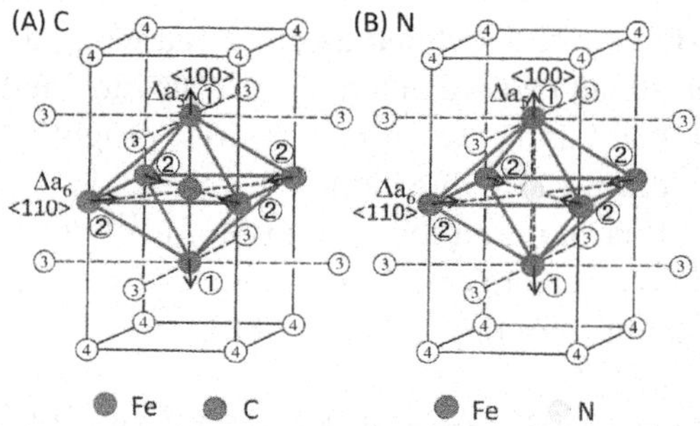

$\Delta a_5 = +0.0350$ nm ($\Delta a_5/a = +12.73\%$) $\sqrt{2}\Delta a_6 = -0.0044$ nm ($\sqrt{2}\Delta a_6/a = -1.62\%$)

$\Delta a_5 = +0.0334$ nm ($\Delta a_5/a = +12.16\%$) $\sqrt{2}\Delta a_6 = -0.0038$ nm ($\sqrt{2}\Delta a_6/a = -1.38\%$)

FIGURE 11-27 Calculated local lattice strains around (A) C and (B) N in bcc Fe.

As pure bcc Fe is ferromagnetic at lower temperature than about 1043 K, spin-polarized calculations are also performed for comparison. The calculated results of local strains around M are listed in Table 11-4B, and also shown in Fig. 11-28B. Their local strains are apparently different as compared with the non-spin-polarized results shown in Fig. 11-28B. The local strains are smaller in spin-polarized calculations than in non-spin-polarized calculations.

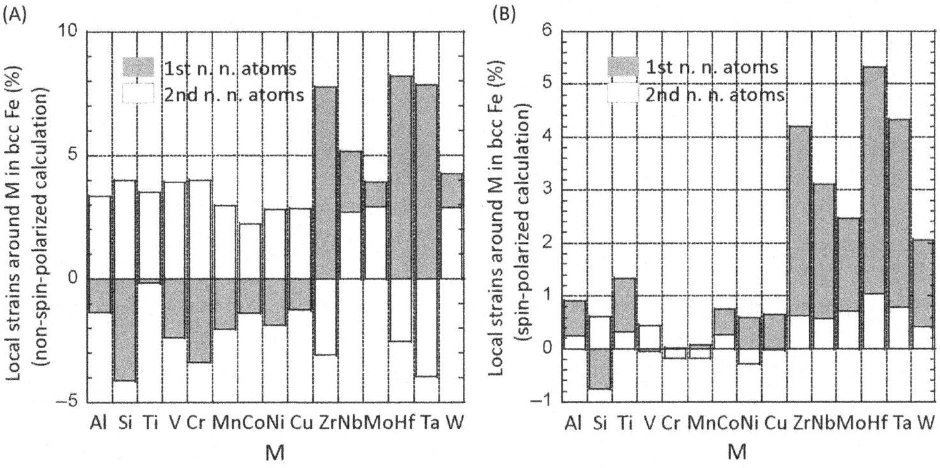

FIGURE 11-28 Comparison of local strains around M in bcc Fe, obtained by (A) non-spin-polarized and (B) spin-polarized calculations.

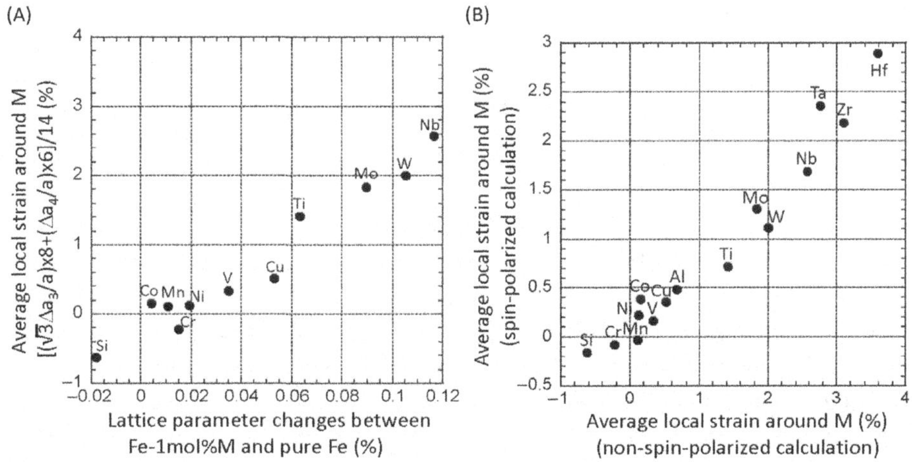

FIGURE 11-29 (A) Correlation of the average local strain around M in bcc Fe with lattice parameter changes between Fe-1mol%M and pure Fe (%), and (B) Comparison in the average local strain around M in bcc Fe between non-spin-polarized and spin-polarized calculations.

However, as shown in Fig. 11-29B, as to the average local strains calculated by using Eq. (11-4), there is an approximately linear relationship between them.

11.7.2 Local Lattice Strains in fcc Fe

Using the supercell shown in Fig. 11-11B, local lattice strains around substitutional alloying elements, M, are determined in fcc Fe. For example, the results are shown in Fig. 11-30A for

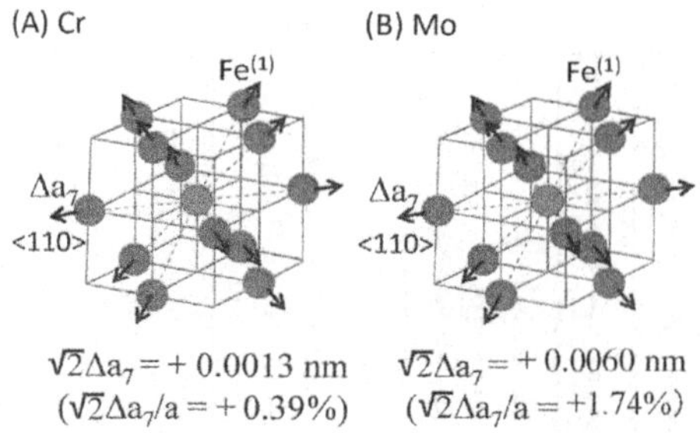

FIGURE 11-30 Calculated local lattice strains around alloying element, (A) Cr and (B) Mo in fcc Fe.

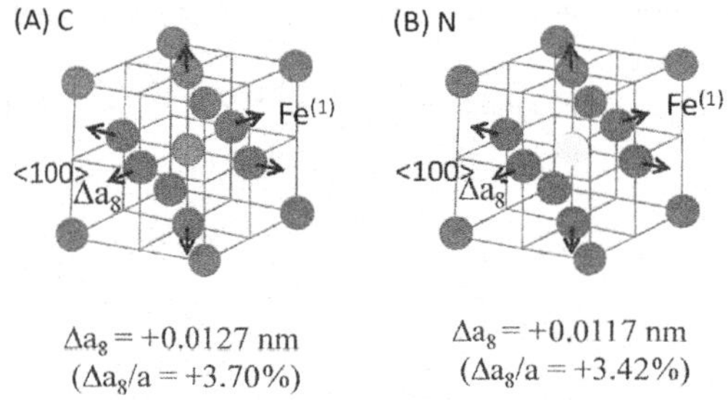

FIGURE 11-31 Calculated local lattice strains around (A) C and (B) N in fcc Fe.

M = Cr and (B) for M = Mo. In case of Cr, 12 first nearest neighbor $Fe^{(1)}$ atoms are displaced away from Cr in the <110> direction. The magnitude of local strains is expressed as $\sqrt{2}\Delta a_7$ where Δa_7 is defined as the strain along the a-axis of fcc lattice. It is divided by the optimized lattice parameter, $a = 0.34351$ nm, and the value of $\sqrt{2}\Delta a_7/a$ is about +0.39%. In case of Mo, 12 first nearest neighbor $Fe^{(1)}$ atoms are displaced away from Mo in the <110> direction. That is similar to the result of Cr, but the value of $\sqrt{2}\Delta a_7/a$ is about +1.74% for Mo, much larger than +0.39% for Cr. The calculated values in fcc Fe are given in Table 11-4A together with the results of bcc Fe.

Furthermore, using the supercell shown in Fig. 11-11D, local lattice strains are determined around interstitial C and N in fcc Fe. The results are shown in Fig. 11-31A for C and (B) for N. In either case, six first nearest neighbor $Fe^{(1)}$ atoms are displaced away from C or N in the <100> direction. The local strains, $\Delta a_8/a$, in fcc Fe is about 3%–4%, which is much smaller than the local strains in the <100> direction, 12%–13%, around C or N in bcc Fe as shown in Fig. 11-27. This difference probably causes the significantly larger solid solubility of C and N in fcc Fe (C:9.06 mol%, N:10.3 mol%) than in bcc Fe (C:0.096 mol%, N:0.4 mol%) [27].

All the calculated results for fcc Fe are summarized in Fig. 11-32. The local lattice strains change largely with M, but their values are always positive, irrespective of M.

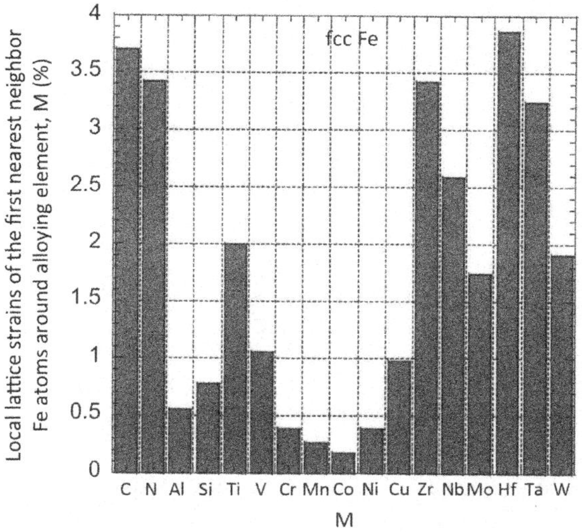

FIGURE 11-32 Local lattice strains of the first nearest neighbor atoms around alloying element M in fcc Fe.

In other words, the Fe atom behaves in fcc Fe as if it is the smallest element among various M's of interest in this study. Also, the local lattice strains are very small for the austenite forming elements such as Mn, Co, and Ni.

11.7.3 Strain Energy Change With Alloying Elements in bcc and fcc Fe

The energy gained by local lattice strains around the substitutional M (called strain energy) is compared between bcc Fe and fcc Fe in Fig. 11-33. The trend of the strain energy change with M resembles between them.

On the other hand, the strain energy of the interstitial element is about 0.128 eV/atom for C and about 0.139 eV/atom for N in bcc Fe, whereas it is about 0.033 eV/atom for C and about 0.029 eV/atom for N in fcc Fe. Thus, strain energies of C and N are much higher in bcc Fe than in fcc Fe.

11.7.4 Correlation of Local Strains With Martensitic Transformation

The martensitic transformation takes place in Fe alloys as follows, fcc (γ) $\rightarrow \alpha'$ (bcc or bct) [33]. The martensite phase, α', changes the structure from bcc to bct, as the alloy composition increases in Fe alloys.

There is a crystal orientation relation between the γ and α' phase. That is known as either the Kurdjumov–Sachs (K-S) relation (e.g., 1.4% C steel) or the Nishiyama (N) relation (e.g., Fe-30%Ni alloy) [33]. According to the classical theory, the martensitic transformation is interpreted as due to the shear deformation. In Fig. 11-34A, the crystal structures of the γ and α' phases are illustrated [33]. For the steel in which the K-S relation is satisfied, the shear progresses along the $[01\bar{1}]_\gamma$ direction on the $(111)_\gamma$ plane. For the alloy in which the

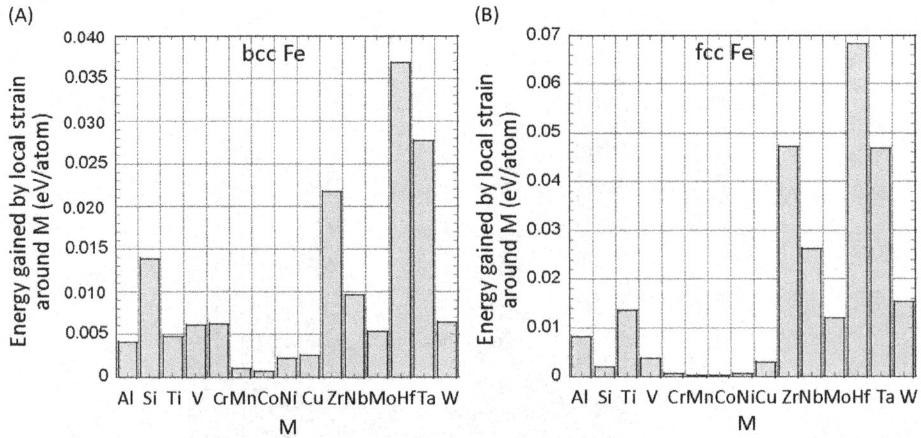

FIGURE 11-33 Comparison in the energy gained by local lattice strain around M between (A) bcc Fe and (B) fcc Fe.

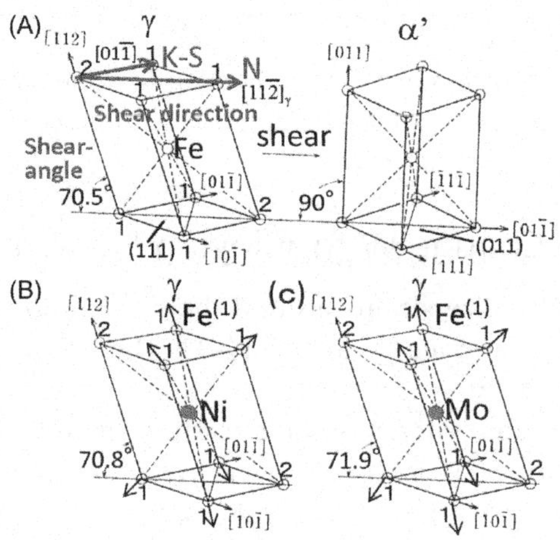

FIGURE 11-34 (A) Shear mechanism from fcc γ to α' martensite [33] and calculated shear angles around (B) Ni and (C) Mo in fcc γ Fe.

N-relation is satisfied, the shear progresses along the $[11\bar{2}]_\gamma$ direction on the $(111)_\gamma$ plane. As shown in Fig. 11-34A, the crystal orientation relationship is expressed as, $(111)_\gamma//(011)_{\alpha'}$ and $[01\bar{1}]_\gamma//[\bar{1}1\bar{1}]_{\alpha'}$ for the K-S relation and $(111)_\gamma//(011)_{\alpha'}$ and $[11\bar{2}]_\gamma//[01\bar{1}]_{\alpha'}$ for the N relation [33].

As a result of the shear deformation, a corner angle of the polyhedron changes from 70.5° in the γ phase to 90° in the α' (bcc or bct) phase, as shown in Fig. 11-34A. Here, 70.5° is the angle between $[112]_\gamma$ and $[11\bar{2}]_\gamma$ directions. This angle in the γ phase prior to the shear deformation is called the shear angle, which is the same term as used in Ti alloys shown in

Fig. 11-24. This shear angle changes by the local atomic displacements around M. So, the shear angles are calculated for various alloying elements, M. For example, the results are shown in Fig. 11-34B for M = Ni and (C) for M = Mo. The shear angle is 70.8° for Ni and 71.9° for Mo, both of which are not so different from 70.5° for pure Fe. For other alloying elements, the calculated shear angles lie in the range of 70.5–73.5°. Thus, the shear angle change is not so large for every M, so that it may not affect the onset of the martensitic transformation.

However, the first nearest neighbor $Fe^{(1)}$ atoms denoted as 1 in Fig. 11-34 are locally displaced in the direction indicated by vectors in the figure. It is apparent that the directions of local atomic displacements are very different from the shear direction indicated in (A). Thus, in case there are large local distortions in the crystal lattice, it seems difficult for a whole crystal to shear uniformly along either the $[01\bar{1}]_\gamma$ or the $[11\bar{2}]_\gamma$ direction on the $(111)_\gamma$ plane. However, further discussion may be needed since the present analysis is based on the classical theory, but not on the theory of invariant plane [33,39–42].

11.7.5 Correlation of Local Strains in Both bcc and fcc Fe with Martensitic Transformation Start Temperature (*Ms*)

Izumiyama et al. [43] reported the compositional dependence of the *Ms* temperature in high-purity binary Fe–M alloys. The measured alloy compositions where *Ms* is 873 K are plotted against the calculated local lattice strains in Fig. 11-35A for bcc Fe and (B) for fcc Fe. The *Ms* temperature of 873 K is relatively high. The local lattice strains are released from the alloy as the temperature increases, and hence the strain effect on the *Ms* temperature is supposed to become weak at high temperatures. But this *Ms* temperature of 873 K is adopted here to increase the number of data for M.

In case of bcc Fe, the average local strain of Fe atoms around substitutional M is calculated using Eq. (11.4). The average local strain around interstitial C or N in bcc Fe is defined as,

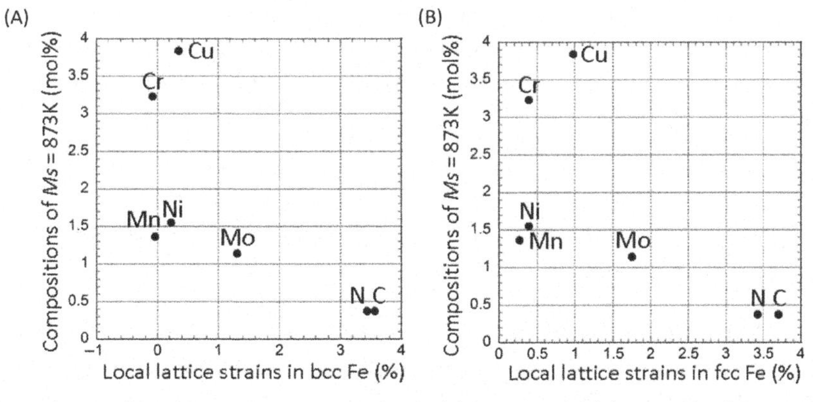

FIGURE 11-35 Correlation of local lattice strains around M in (A) bcc Fe and (B) fcc Fe with alloy compositions of *Ms* = 873 K in binary Fe–M alloys.

$$\text{Average local strain} = [(\Delta a_5/a) \times 2 + (\Delta a_6/a) \times 4]/6\ (\%). \tag{11.5}$$

This is an arithmetic average of the strains by weighting the coordination numbers around interstitial C or N. On the other hand, in the case of fcc Fe, it is expressed as, $\sqrt{2}\Delta a_7/a$ around substitutional M (see Fig. 11-30) and $\Delta a_8/a$ around interstitial C or N (see Fig. 11-31).

As shown in Fig. 11-35A–B, the measured alloy compositions where *Ms* is 873 K, decrease as the average local strains increase in both bcc Fe and fcc Fe. Namely, the large and *positive* local strains work to retard the onset of the martensitic transformation. As explained previously, in the case of Ti alloys the large and *negative* local strains work to retard the onset of the martensitic transformation as shown in Figs. 11-17A and 11-22. The positive strain in Fe alloys is reverse of the negative strain in Ti alloys, but in either alloy there is the same trend that the onset of the martensitic transformation is suppressed under the existence of large local lattice strains around M.

The reason why the *positive* strain in Fe alloys is reverse of the *negative* strain in Ti alloys is interpreted as follows. As mentioned previously, the bcc or bct α' martensite phase appears in the close-packed fcc γ phase in Fe alloys, so that the crystal lattice tends to expand by the martensitic transformation. For this reason, those alloying elements (e.g., C, N, Mo, etc.) that yield positive local lattice strains, cause significant changes in the *Ms* temperature of the Fe alloy. On the other hand, in case of Ti alloys, the close-packed hcp α' martensite phase appears in the bcc β–Ti phase, so that the crystal lattice tends to contract by the martensitic transformation. For this reason, those alloying elements (e.g., Fe, Mn, etc.) that yield negative local lattice strains cause significant changes in the *Ms* temperature of the Ti alloy. Thus, it is found that the onset of the martensitic transformation is suppressed by the local lattice strains around alloying elements in both Ti alloys and Fe alloys.

In the case of M = Co, the *Ms* temperature increases slightly with the Co content in binary Fe-Co alloys. Its *Ms* temperature is higher than *Ms* = 720°C (993 K) for pure Fe. As shown in Fig. 11-29A for bcc Fe and in Fig. 11-32 for fcc Fe, the local lattice strain is the smallest at Co among the alloying elements irrespective of bcc Fe or fcc Fe. Also, as explained previously, the primarily weak strain effect becomes even weaker at such a high temperature. This may be one of the reasons for the increase in the *Ms* temperature with the Co content in binary Fe-Co alloys.

11.8 Local Lattice Strains Around Alloying Elements in Other Metals

Using the pseudopotential method with the supercells shown in Figs. 11-1 and 11-11, a series of calculations is performed to obtain information on the local lattice strains around alloying elements M in Al, Ni, Cu, and Co metals. For convenience, only the calculated results are presented briefly here. Every notation to represent the local lattice strains is the same as used before in this chapter.

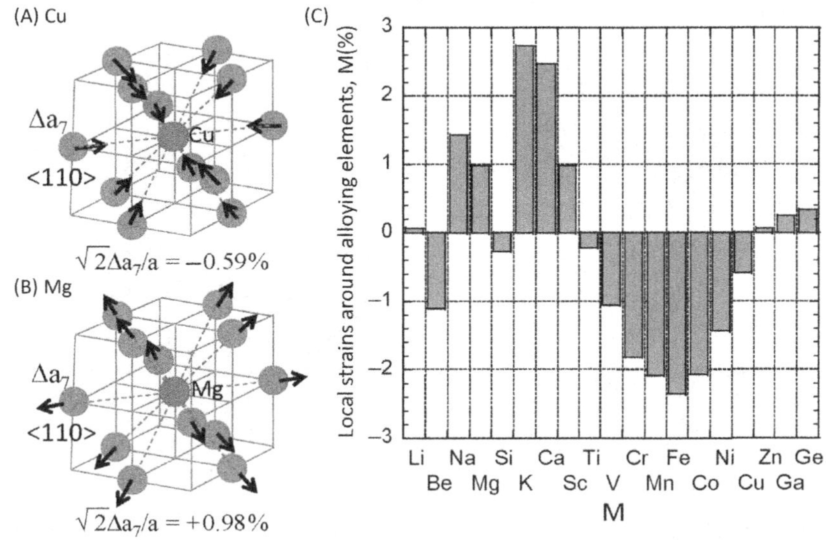

FIGURE 11-36 Local lattice strains around alloying elements, M, in fcc Al: (A) M = Cu, (B) M = Mg, and (C) various elements, M.

11.8.1 Aluminium

The optimized lattice parameter of fcc pure Al is $a = 0.40478$ nm. Using this lattice parameter, the calculated local lattice strain is expressed as $\sqrt{2}\,\Delta a_7/a\,(\%)$ as shown in Fig. 11-36. It is −0.59 % for (A) M = Cu and +0.98 % for (B) for M = Mg. All the results are given in (C).

11.8.2 Nickel

The optimized lattice parameter of fcc pure Ni is a = 0.35295 nm. Using this lattice parameter, the calculated local lattice strain is expressed as $\sqrt{2}\,\Delta a_7/a\,(\%)$ as shown in Fig. 11-37. It is −0.59 % for (A) M = Cr and +0.20 % for (B) for M = Al. All the results are given in (C).

11.8.3 Copper

The optimized lattice parameter of fcc pure Cu is a = 0.36258 nm. Using this lattice parameter, the calculated local lattice strain is expressed as $\sqrt{2}\,\Delta a_7/a\,(\%)$ as shown in Fig. 11-38. It is +0.28 % for (A) M = Zn and +2.25 % for (B) for M = Sn. All the results are given in (C).

11.8.4 Cobalt

There is an allotropic phase transition from hcp Co to fcc Co around 695 K. The optimized lattice parameter of fcc pure Co is a = 0.34855 nm. Using this lattice parameter, the calculated local lattice strain is expressed as $\sqrt{2}\,\Delta a_7/a\,(\%)$ as shown in Fig. 11-39. It is −0.14% for (A) M = Cr and +0.32 % for (B) for M = Ni. All the results are given in (C).

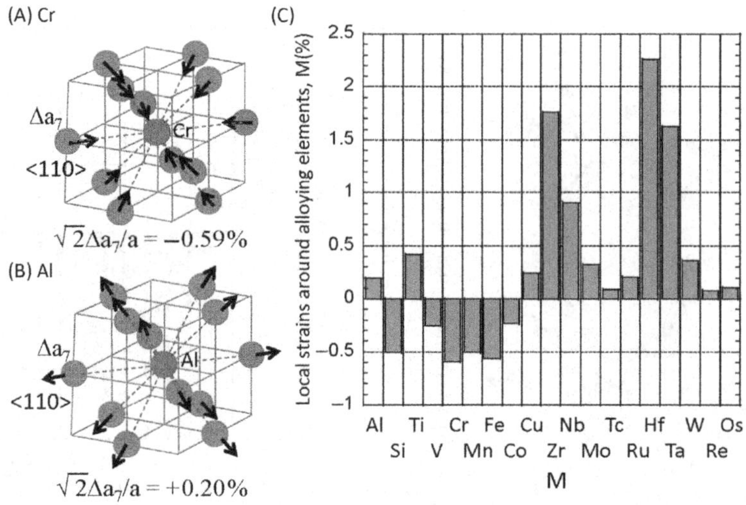

FIGURE 11-37 Local lattice strains around alloying elements, M, in fcc Ni: (A) M = Cr, (B) M = Al, and (C) various elements, M.

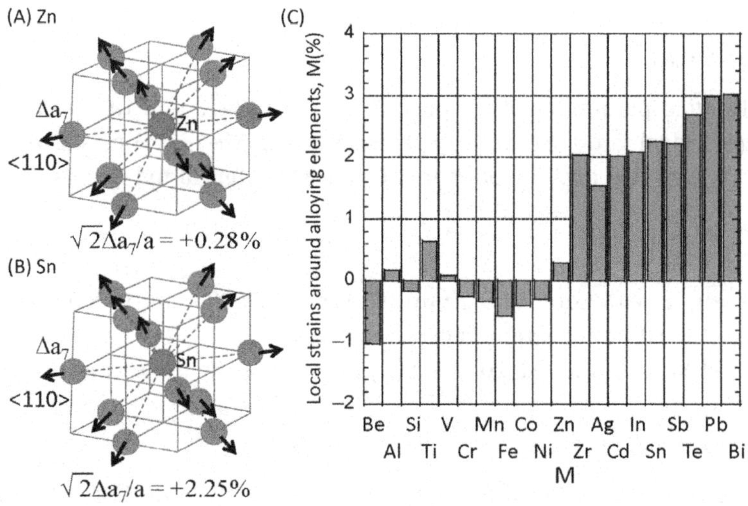

FIGURE 11-38 Local lattice strains around alloying elements, M, in fcc Cu: (A) M = Zn, (B) M = Sn, and (C) various elements, M.

The optimized lattice parameter of hcp pure Co is $a = 0.24757$ nm and $c = 0.39946$ nm. Using this lattice parameter, the calculated local lattice strains are expressed as $\Delta a/a$ and $\Delta c/c$ (%) as shown in Fig. 11-40. The values of $\Delta a/a$ and $\Delta c/c$ are -0.16% and -0.54% for (A) M = Cr, and +0.26 % and +0.52 % for (B) for M = Ni, respectively. All the results are given in (C).

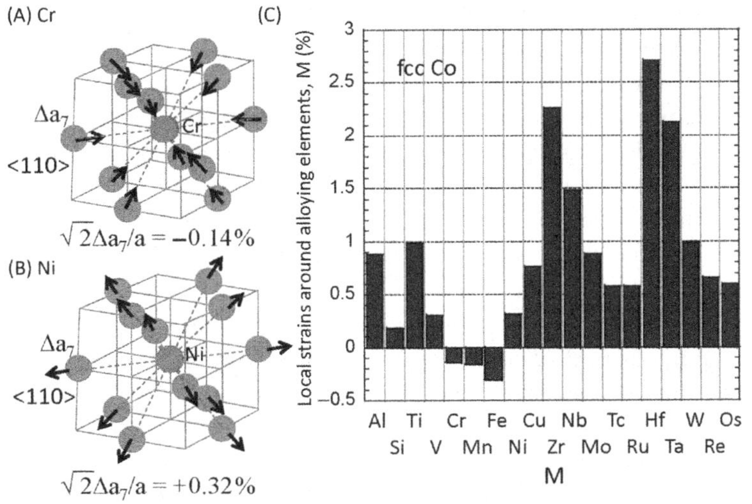

FIGURE 11-39 Local lattice strains around alloying elements, M, in fcc Co: (A) M = Cr, (B) M = Ni, and (C) various elements, M.

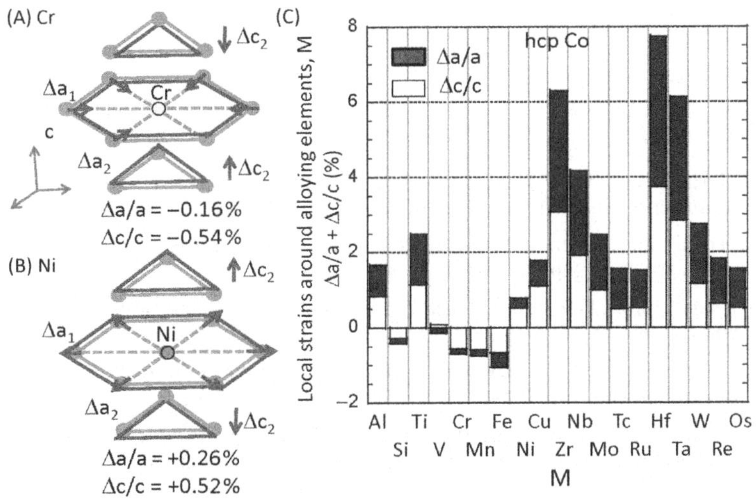

FIGURE 11-40 Local lattice strains around alloying elements, M, in hcp Co: (A) M = Cr, (B) M = Ni, and (C) various elements, M.

Finally, it is summarized that the local lattice strains around alloying elements, M, in various metals are calculated in a systematic way using the pseudopotential method. The calculated local lattice strains vary largely depending on M in any metal.

For example, the local lattice strains are negative for 3*d* transition elements, but positive for rare earth elements in hcp Mg. It is found from the atomization energy analysis that the charge transfer takes place between M and Mg atoms to adjust the atomic size between them. Also, the electron screening occurs around M so as to moderate the influence of M impurity doped into a nearly free electron metal like Mg.

Local lattice strains around alloying elements, M, are calculated in both Ti alloys and Fe alloys. Local lattice strains around M correlate with the martensitic transformation start temperature, *Ms*, in the both alloys. The existence of large local lattice strains around M retards the onset of the martensitic transformation in these alloys. This is because uniform shear deformation that leads to the martensitic transformation is probably suppressed in such a largely distorted crystal lattice in the alloy.

It is stressed that the local strain field around solute atom will interact with the other strain fields existing around dislocations, vacancies and other solute atoms in alloys. For instance, the solute–dislocation interaction causes the solid solution hardening (e.g., Cottrell pinning). The solute–vacancy interaction may influence the solute diffusivity. The solute–solute interaction may lead to the formation of the short-range order or cluster region in the alloy. For example, in the $Mg_{97}Zn_1Y_2$ alloy with the long-period stacking ordered (LPSO) structure [4], Zn and Y accompany the nearly opposite strain field around them as shown in Fig. 11-3. As a result, the Zn–Y elastic interaction may assist the formation of the $L1_2$-type Zn_6Y_8 cluster in the Mg alloy, despite that the $L1_2$-type phase does not appear in binary Zn–Y alloys [27]. Thus, besides the martensitic transformation, local lattice strains around alloying elements will influence alloy properties.

References

[1] M. Morinaga, Local lattice strains around alloying elements in magnesium and analysis of the chemical bond using atomization energy, Toyota Res. Rep. 69 (2016) 25–33.

[2] M. Morinaga, Local lattice strain around alloying element and martensitic transformation in titanium and iron alloys, Toyota Res. Rep. 70 (2017) 35–44.

[3] A.A. Nayeb-Hashemi, J.B. Clark (Eds.), Phase Diagrams of Binary Magnesium Alloys, ASM International, Metals Park, Ohio, 1988.

[4] Y. Kawamura, M. Yamasaki, Formation and mechanical properties of $Mg_{97}Zn_1RE_2$ alloys with long-period stacking ordered structure, Mater. Trans. 48 (2007) 2986–2992.

[5] E. Abe, Y. Kawamura, K. Hayashi, A. Inoue, Long-period ordered structure in a high-strength nanocrystalline Mg-1 at% Zn-2 at% Y alloy studied by atomic-resolution *Z*-contrast STEM, Acta Mater. 50 (2002) 3845–3857.

[6] E. Orowan, A type of plastic deformation new in metals, Nature 149 (1942) 643–644.

[7] W. Huang, T. Mura, Elastic fields and energies of a circular edge disclination and a straight screw disclination, J. Appl. Phys. 41 (1970) 5175.

[8] R. Ninomiya, H. Yukawa, M. Morinaga, K. Kubota, An electronic approach to the prediction of the mechanical properties of magnesium alloys, J. Alloys Compd. 215 (1994) 315–323.

[9] J.P. Hirth, J. Lothe, Theory of Dislocations, McGraw-Hill, Inc, New York, 1968.

[10] R.L. Fleischer, Substitutional solution hardening, Acta Metall. 11 (1963) 203–209.

[11] H. Borchers, P. Dropmann, Solid-solution hardening in aluminum single- crystals containing gallium, Z. Metallkd 63 (1972) 137–140.

[12] H.W. King, Quantitative size-factors for metallic solid solutions, J. Mater. Sci. 1 (1966) 79–90.

[13] J.D. Eshelby, Distortion of a crystal by point imperfections, J. Appl. Phys. 25 (1954) 255–261.

[14] T. Uesugi, K. Higashi, First-principles studies on lattice constants and local lattice distortions in solid solution aluminium alloys, Comput. Mater. Sci. 67 (2013) 1–10.

[15] H. Nakai, Energy density analysis with kohn-sham orbitals, Chem. Phys. Lett. 363 (2002) 73–79.

[16] H. Nakai, Y. Kurabayashi, M. Katouda, T. Atsumi, Extension of energy density analysis to periodic boundary condition calculation: Evaluation of locality in extended systems, Chem.Phys. Lett. 438 (2007) 132–138.

[17] Y. Shinzato, H. Yukawa, M. Morinaga, T. Baba, H. Nakai, A unified approach to the analysis of the chemical bond in hydrides and hydrocarbons, Acta Mater. 55 (2007) 6673–6680.

[18] Y. Shinzato, Y. Saito, M. Yoshino, H. Yukawa, M. Morinaga, T. Baba, et al., Energy expression of the chemical bond between atoms in metal oxides, J. Phys. Chem. Solids. 72 (2011) 853–861.

[19] V. Milman, B. Winkler, J.A. White, C.J. Pickard, M.C. Payne, E.V. Akhmatsukaya, et al., Electronic structure, properties, and phase stability of inorganic crystals: A pseudopotential plane-wave study, Int. J. Quant. Chem 77 (2000) 895–910.

[20] W. Kohn, L.J. Sham, Self-consistent equations including exchange and correlation effects, Phys. Rev 140 (1965) A1133–A1138.

[21] J.P. Perdew, K. Burke, Y. Wang, Generalized gradient approximation for the exchange-correlation hole of a many-electron system, Phys. Rev. B 54 (1996) 16533–16539.

[22] J.P. Perdew, J.A. Chevanry, S.H. Vosko, K.A. Jakson, M.R. Pederson, D.J. Singh, et al., Atoms, molecules, solids, and surfaces: Applications of the generalized gradient approximation for exchange and correlation, Phys. Rev. B 46 (1992) 6671–6687.

[23] D. Vanderbilt, Soft self-consistent pseudopotentials in a generalized eigenvalue formalism, Phys. Rev. B 41 (1990) 7892–7895.

[24] H.J. Monkhorst, J.D. Pack, Special points for brillouin-zone integrations, Phys. Rev. B 13 (1976) 5188–5192.

[25] M.J. Frisch, G.W. Trucks, H.B. Schlegel, G.E. Scuseria, M.A. Robb, J.R. Cheeseman, et al., Gaussian 03, revision C, 02, Gaussian. Inc., Wallingford CT, 2004.

[26] R.S. Mulliken, Electronic population analysis on LCAO–MO molecular wave functions. I, II, III, IV, J. Chem. Phys. 23 (1955) 1833–1840, 1841–1846, 2338–2342, 2343–2346.

[27] H. Okamoto, Desk Handbook, Phase Diagrams for Binary Alloys, 2nd Edition, ASM International, Materials Park, Ohio, 2010.

[28] R.S. Busk, Lattice parameters of magnesium alloys, J. Met. 2 (1950) 1460–1464.

[29] D. Hardie, R.N. Parkins, Lattice spacing relationships in magnesium solid solutions, Philos. Mag. 4 (1959) 815–825.

[30] C. Kittel, Introduction to Solid State Physics, 5th edition, John-Wiley, New York, 1976.

[31] J.M. Ziman, Principles of the Theory of Solids, 2nd Edition, Cambridge University Press, Cambridge, 1972.

[32] Y.M. Zhu, A.J. Morton, J.F. Nie, The 18R and 14H long-period stacking ordered structures in Mg-Y-Zn alloys, Acta Mater. 58 (2010) 2936–2947.

[33] Z. Nishiyama, Martensitic Transformation—Fundamentals, Maruzen, Tokyo, 1971.

[34] T. Sato, Y.C. Huang, S. Suzuki, Continuous-cooling-transformation diagrams of binary titanium alloys, Sumitomo Light Metal Techn. Rep. 3 (4) (1962) 314–335.

[35] H.Y. Kim, S. Miyazaki, Martensitic transformation and superelastic properties of Ti–Nb base alloys, Mater. Trans. 56 (2015) 625–634.

[36] K. Kondoh, J. Umeda, S. Kariya, M. Fukuo, Solid solution strengthened powder metallurgy Ti materials, Proceeding of the 4th Inter. Conference on Titanium Powder Metallurgy & Additive Manufacturing (PMTi 2017), 8–10 September 2017, Xi'an, P. R. China, p. 29.

[37] H. Ohtsuka, V.A. Dinh, T. Ohno, K. Tsuzaki, K. Tsuchiya, R. Sahara, et al., First-principles calculation of effects of carbon on tetragonality and magnetic moment in Fe-C system, ISIJ Int. 55 (2015) 2483–2491.

[38] W.B. Pearson, A Handbook of Lattice Spacings and Structures of Metals and Alloys, Pergamon Press, Oxford, 1958.

[39] D.S. Lieberman, M.S. Wechsler, T.A. Read, Cubic to orthorhombic diffusionless phase change-experimental and theoretical studies of AuCd, J. Appl. Phys. 26 (1955) 473–484.

[40] J.S. Bowles, J.K. Mackenzie, The crystallography of martensite transformations I, Acta Metall. 2 (1954) 129–137.

[41] J.K. Mackenzie, J.S. Bowles, The crystallography of martensite transformations II, Acta Metall. 2 (1954) 138–147.

[42] J.S. Bowles, J.K. Mackenzie, The crystallography of martensite transformations III. face-centred cubic to body-centred tetragonal transformations, Acta Metall. 2 (1954) 224–234.

[43] M. Izumiyama, M. Tsuchiya, Y. Imai, Effects of alloying element on supercooled A_3 transformation of iron, science reports of the research institutes, Tohoku Univ. A: Phys. Chem. Metall. 22 (2–3) (1970) 105–115.

12

Conclusions

12.1 Summary of Quantum Approach to Alloy Design

A quantum approach to alloy design is treated consistently in this book. This book covers various important problems in view of electron theory. The theory for alloy design works well in practice even for complex alloys with multiple components.

For example, for the design of transition metal–based alloys (e.g., Fe alloys, Ni alloys, and Ti alloys), two parameters, *Md* and *Bo*, are determined by the molecular orbital calculations of the electronic structures of alloys. The *Md* parameter is the *d*-orbital energy level of the alloying transition element, The *Bo* parameter is the bond order (overlap population) between M and mother metal X atoms in the alloy (see Chapter 2: Theory for Alloy Design). As these parameters are obtained in the same chemical environment as in the alloy, alloying effects of each element are involved inevitably in them. These parameters are in great contrast with the current parameters such as the electronegativity, the atomic radius, and the electrons-per-atom ratio (*e/a*), since all of them are assigned to pure metals without considering any alloying effects. Therefore, there is a great advantage to use these new parameters for alloy design instead of the current parameters. In fact, various alloys including Ni alloys, Fe alloys, and Ti alloys are designed following this approach (see Chapters 3: Nickel Alloys; 4: Iron Alloys; 5: Titanium Alloys).

Also, new PHACOMP has been developed to resolve various contradictions inherent to the PHACOMP (see Chapter 3: Nickel Alloys). Such a quantum approach is further extended to Al alloys and Mg alloys. Big data of their mechanical properties are put in order using a *Mk* parameter that is the *s* orbital energy level of alloying element, M. Using the *Mk* parameter the first trial is made to treat the mechanical properties in an energy scale (see Chapter 6: Aluminium Alloys and Magnesium Alloys).

A new concept is also developed for the hydrogen overpotential of alloys and compounds (see Chapter 7: Hydrogen Overpotential of Alloys and Intermetallic Compounds). The structure maps are proposed to predict crystal structures of intermetallic compounds (see Chapter 8: Crystal Structural Maps for Intermetallic Compounds). A universal relation in the electron density distribution in materials is discovered on the basis of a series of molecular orbital calculations (see Chapter 9: A Universal Relation in Electron Density Distribution in Materials).

In addition, an atomization energy method is newly developed to treat various materials in an energy scale (see Chapter 10: Atomization Energy Approach to Alloys and Metal Compounds). Following this method, every material is comparable with each other, even in the case where the nature of the chemical bond between atoms is very different among the materials. For example, metal hydrides, complex hydrides, and hydrocarbons for hydrogen

A Quantum Approach to Alloy Design. DOI: https://doi.org/10.1016/B978-0-12-814706-1.00012-1

storage are treated consistently with this method. The main characters of the chemical bond are also compared among various 3*d* transition metal compounds with the aid of this energy method. One of the applications of this method is the catalyst design. The catalytic activities of metal oxides (e.g., Nb_2O_5) are evaluated quantitatively on the dehydrogenation reaction of magnesium hydrides (MgH_2), $MgH_2 \rightarrow Mg + H_2$ (see Chapter 10: Atomization Energy Approach to Alloys and Metal Compounds).

In addition, local lattice strains around the alloying element in metals are calculated in a systematic manner (see Chapter 11: Local Lattice Strains Around Alloying Elements in Metals). Such local strains are introduced into the crystal lattice due to the size mismatch between solute and solvent atoms in the alloy. This is indeed a new topic that has never been treated in a quantitative way, despite the great importance in understanding alloy properties. For example, the martensitic transformation start temperature (*Ms*) is discussed in terms of the local lattice strains in both Ti and Fe alloys (see Chapter 11: Local Lattice Strains Around Alloying Elements in Metals).

It is added here that the energy gained by the local lattice strains around an alloying element (called strain energy) is comparable to the thermal energy, K_BT, where K_B is a Boltzmann constant and T is the absolute temperature (K) (see Chapter 6: Aluminium Alloys and Magnesium Alloys). Therefore, as stated previously, the martensitic transformation start temperatures (*Ms*) vary with the local lattice strains in Ti alloys. This is also the case in the other phase stability changes with temperatures. For example, according to New PHACOMP, the critical $\overline{Md}$ parameter to represent the $\gamma/\gamma + \sigma$ phase boundary in Ni alloys is expressed as, critical $\overline{Md} = 6.25 \times 10^{-5}\,T + 0.834$ (eV) (see Eq. (3-3) in Chapter 3: Nickel Alloys). The temperature coefficient, 6.25×10^{-5} (eV/K), is close to the value of k_B, 8.62×10^{-5} (eV/K). Thus, the energy variation on the $\gamma/\gamma + \sigma$ phase boundary with the temperature concerns the thermal energy, k_BT. This means that the phase stability depends largely on the temperature, although we already know this very well from a series of phase diagrams of alloys.

Also, as explained in Chapter 6, Aluminium Alloys and Magnesium Alloys, it is only natural that the tensile strength of the Al alloys decreases with increasing temperature. This is because local lattice strains that disturb dislocation motions are released from the alloy as the temperature increases.

Thus, the phase stability and alloy properties (e.g., tensile strength) change largely with temperatures. The thermal energy, k_BT, is about 0.02 eV/atom at 273 K (0°C) and about 0.11 eV/atom at 1273K (1000°C), much smaller than the cohesive energy (e.g., 3.39 eV/atom for Al, 4.85 eV/atom for Ti, and 4.28 eV/atom for Fe). To design superior alloys, we have to control the phase stability (or microstructure) and to optimize alloy properties in such a very small energy range of 0.02 to 0.11 eV/atom. This is a real feature of alloys.

The present molecular orbital approach presented in this book has been applied to various alloys as shown in Fig. 12-1. This is a tree of alloy design, and it is still growing up on the background of the molecular orbital theory. For example, there are Nb alloys [1], Mo alloys [1,2], and even perovskite-type oxides [3] in the tree.

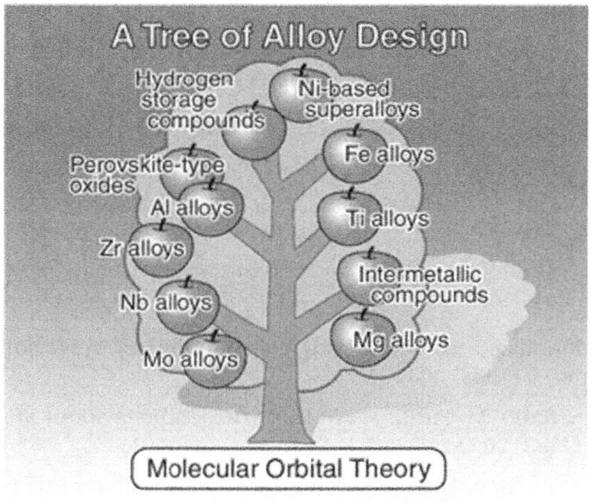

FIGURE 12-1 A tree of alloy design. It is growing on the ground of molecular orbital theory.

12.2 Impurity Problem

However, there are still some remaining problems in alloys that are not covered in this book. One of the important problems is the impurity problem in alloys. The minor impurities existing in alloys affect alloy properties significantly. For example, the presence of surfer (S) at about 0.005 mass% leads to the tremendous improvement of the steam oxidation resistance of high Cr ferritic steels at 923 K for 3.6 Ms without any detrimental effect on the mechanical properties [4]. This is a sort of warning to steel manufacturers that superpurification sometimes deteriorates the steel property. Another problem is the hydrogen embrittlement of high strength alloys. This problem still remains unsolved despite that enormous effort has been made for a long period of time. It is also one of the impurity problems, since the ppm order of hydrogen causes the embrittlement. It is rather difficult to detect impurities experimentally. However, a μSR experiment is now under way to investigate the hydrogen embrittlement of Fe alloys [5]. It will be needed to establish the "science and technology of impurity" in alloys to make materials as low cost as possible.

12.3 Other Calculation Methods

Finally, it is noted that there are several calculation methods that are not explained in this book. For example, the calculation of phase diagrams using the CALPHAD method [6,7] and the analysis of microstructural evolution using the phase field method [8–10] are now progressing steadily. The investigations of multiscale simulations [11] and materials informatics [12–14] are also in progress these days. In addition, there is a mathematical approach to the design of smart materials [15–17]. The collaboration between mathematics and materials science may help us to understand the relation between microscopic structures and macroscopic properties of materials. Thus, it is greatly expected that materials design could make remarkable progress by combining wide knowledge and information with the present quantum approach.

References

[1] S. Inoue, J. Saito, M. Morinaga, S. Kano, Alloying effect on the electronic structures of Nb and Mo, J. Phys.: Condens. Matter. 6 (1994) 5081–5096.

[2] J. Saito, M. Morinaga, S. Kano, M. Furui, K. Noda, Corrosion behavior of Mo-Re based alloys in liquid Li, J. Nucl. Mater. 264 (1999) 206–215.

[3] Y. Shinzato, Y. Saito, M. Yoshino, H. Yukawa, M. Morinaga, T. Baba, et al., energy expression of the chemical bond between atoms in metal oxides, J. Phys. Chem. Solids 72 (2011) 853–861.

[4] M. Morinaga, Y. Murata, R. Hashizume, Y. Sawaragi, Remarkable improvement in steam oxidation resistance due to the presence of sulfur in high Cr ferrtic steels, ISIJ Int. 41 (2001) 314–316.

[5] N. Nishida, M. Morinaga et al., to be published.

[6] Z.-K. Liu, First-principles calculations and CALPHAD modeling of thermodynamics, J. Phase Equilib. Diff. 30 (2009) 517–534.

[7] Y. Wang, S. Curtarolo, C. Jiang, R. Arroyave, T. Wang, G. Ceder, et al., Ab initio lattice stability in comparison with CALPHAD lattice stability, Comput. Coupl. Phase Diag. Thermochem. 28 (2004) 79–90.

[8] L.-Q. Chen, Phase-field models for microstructure evolution, Ann. Rev. Mater. Res. 32 (2002) 113–140.

[9] W.J. Boettinger, J.A. Warren, C. Beckermann, A. Karma, Phase-field simulation of solidification, Ann. Rev. Mater. Res. 32 (2002) 163–194.

[10] T. Miyazaki, T. Koyama, T. Kozakai, Computer simulations of the phase transformation in real alloy systems based on the phase field method, Mater. Sci. Eng. A 312 (2001) 38–49.

[11] W.A. Curtin, R.E. Miller, Atomistic/continuum coupling in computational materials science, Modell. Simul. Mater. Sci. Eng. 11 (2003) R33–R68.

[12] K. Rajan, Materials informatics, Mater. Today 8 (2005) 38–45.

[13] K. Rajan, Materials informatics: Materials "Gene" and big data, Ann. Rev. Mater. Res. 45 (2015) 153–169.

[14] S.R. Kalidindi, M. De Graef, Materials data science: Current status and future outlook, Ann. Rev. Mater. Res. 45 (2015) 171–193.

[15] M. Itoh, M. Kotani, H. Naito, T. Sunada, Y. Kawazoe, T. Adschiri, New metallic carbon crystal, Phys. Rev. Lett. 102 (2009) 055703.

[16] A. Hirata, L.J. Kang, T. Fujita, B. Klumov, K. Matsue, M. Kotani, et al., Geometric frustration of icosahedron in metallic glasses, Science 341 (2013) 376–379.

[17] M. Kotani, S. Ikeda, Materials inspired by mathematics, Sci. Technol. Adv. Mater. 17 (2016) 253–259.

Index

Note: Page numbers followed by "*f*" and "*t*" refer to figures and tables, respectively.

E

F

G

H

I

Made in the USA
Las Vegas, NV
08 April 2022

47112377R00162